Continuous Casting Technology
连铸生产技术

Yu Wansong

于万松　　主编

Beijing
Metallurgical Industry Press
2020

内容提要

本书共6个项目，主要包括连铸生产认知、连铸凝固过程认知、连铸设备操作、连铸生产工艺控制及操作、连铸生产事故处理、连铸坯缺陷控制等内容。

本书既可作为职业院校冶金相关专业的国际化教学用书，也可作为冶金企业员工的培训教材和有关专业人员的参考书。

图书在版编目(CIP)数据

连铸生产技术=Continuous Casting Technology：汉、英/于万松主编．—北京：冶金工业出版社，2020.8

国际化职业教育双语系列教材

ISBN 978-7-5024-8593-1

Ⅰ.①连…　Ⅱ.①于…　Ⅲ.①连续铸造—生产技术—双语教学—教材—汉、英　Ⅳ.①TG249.7

中国版本图书馆CIP数据核字(2020)第144417号

出版人　陈玉千
地　　址　北京市东城区嵩祝院北巷39号　邮编　100009　电话　(010)64027926
网　　址　www.cnmip.com.cn　电子信箱　yjcbs@cnmip.com.cn
责任编辑　俞跃春　杜婷婷　美术编辑　郑小利　版式设计　孙跃红　禹　蕊
责任校对　郑　娟　责任印制　李玉山
ISBN 978-7-5024-8593-1
冶金工业出版社出版发行；各地新华书店经销；三河市双峰印刷装订有限公司印刷
2020年8月第1版，2020年8月第1次印刷
787mm×1092mm　1/16；17.5印张；418千字；252页
58.00元

冶金工业出版社　投稿电话　(010)64027932　投稿信箱　tougao@cnmip.com.cn
冶金工业出版社营销中心　电话　(010)64044283　传真　(010)64027893
冶金工业出版社天猫旗舰店　yjgycbs.tmall.com

国际化职业教育双语系列教材

编 委 会

Foreword

With the proposal of the ' Belt and Road Initiative' , the Ministry of Education of China issued *Promoting Education Action for Building the Belt and Road Initiative* in 2016, proposing cooperation in education, including ' cooperation in human resources training' . At the Forum on China-Africa Cooperation (FOCAC) in 2018, President Xi proposed to focus on the implementation of the ' Eight Actions' , which put forward the plan to establish 10 Luban Workshops to provide skills training to African youth. Draw lessons from foreign advanced experience of vocational education mode, China's vocational education has continuously explored and formed the new mode of vocational education with Chinese characteristics. Tianjin, as a demonstration zone for reform and innovation of modern vocational education in China, has started the construction of ' Luban Workshop' along the ' Belt and Road Initiative' , to export high-quality vocational education achievements.

The compilation of these series of textbooks is in response to the times and it's also the beginning of Tianjin Polytechnic College to explore the internationalization of higher vocational education. It's a new model of vocational education internationalization by Tianjin, response to the ' Belt and Road Initiative' and the ' Going Out' of Chinese enterprises. Tianjin Polytechnic College and Uganda Technical College-Elgon reached a cooperation intention to establish the Luban Workshop to carry out vocational education cooperation on mechatronics technology and ferrous metallurgy technology major in 2019. The establishment of Luban Workshop is conducive to strengthen the cooperation between China and Uganda in vocational education, promote the export of high-quality higher vocational education resources, and serve Chinese enterprises in Uganda and Ugandan local enterprises. Exploring and standardizing the overseas operation of Chinese colleges, the expansion of international influences of China's higher vocational education is also one of the purposes.

The construction of ' Luban Workshop ' in Uganda is mainly based on the EPIP (Engineering, Practice, Innovation, Project) project, and is committed to cultivating high-quality talents with innovative spirit, creative ability and entrepreneurial spirit. To meet the learning needs of local teachers and students accurately, the compilation of these international vocational skills bilingual textbooks is based on the talent demand of Uganda and the specialty and characteristics of Tianjin Polytechnic.

These textbooks are supporting teaching material, referring to Chinese national professional standards and developing international professional teaching standards. The internationalization of the curriculums takes into account the technical skills and cognitive characteristics of local students, to promote students' communication and learning ability. At the same time, these textbooks focus on the enhancement of vocational ability, rely on professional standards, and integrate the teaching concept of equal emphasis on skills and quality. These textbooks also adopted project-based, modular, task-driven teaching model and followed the requirements of enterprise posts for employees.

In the process of writing the series of textbooks, Wang Xiaoxia, Li Rui, Wang Zhixue, Ge Huijie, Yu Wansong, Wang Lei, Li Xiujuan, Gong Na, Zhang Zhichao, Jia Yanlu, Chen Baoling and other chief teachers, professional teams, English teaching and research office have made great efforts, receiving strong support from leaders of Tianjin Polytechnic College. During the compilation, the series of textbooks referred to a large number of research findings of scholars in the field, and we would like to thank them for their contributions.

Finally, we sincerely hope that the series of textbooks can contribute to the internationalization of China's higher vocational education, especially to the development of higher vocational education in Africa.

Principal of Tianjin Polytechnic College Kong Weijun

May, 2020

序

随着“一带一路”倡议的提出，2016 年中华人民共和国教育部发布了《推进共建“一带一路”教育行动》，提出了包括“开展人才培养培训合作”在内的教育合作。2018 年习近平主席在中非合作论坛上提出，要重点实施“八大行动”，明确要求在非洲设立 10 个鲁班工坊，向非洲青年提供技能培训。中国职业教育在吸收和借鉴发达国家先进职教发展模式的基础上，不断探索和形成了中国特色职业教育办学模式。天津市作为中国现代职业教育改革创新示范区，开启了“鲁班工坊”建设工作，在“一带一路”沿线国家搭建“鲁班工坊”平台，致力于把优秀职业教育成果输出国门与世界分享。

本系列教材的编写，契合时代大背景，是天津工业职业学院探索高职教育国际化的开端。“鲁班工坊”是由天津率先探索和构建的一种职业教育国际化发展新模式，是响应国家“一带一路”倡议和中国企业“走出去”，创建职业教育国际合作交流的新窗口。2019 年天津工业职业学院与乌干达埃尔贡技术学院达成合作意向，共同建立“鲁班工坊”，就机电一体化技术专业、黑色冶金技术专业开展职业教育合作。此举旨在加强中乌职业教育交流与合作，推动中国优质高等职业教育资源“走出去”，服务在乌中资企业和乌干达当地企业，探索和规范我国职业院校“鲁班工坊”建设和境外办学，扩大中国高等职业教育的国际影响力。

中乌“鲁班工坊”的建设主要以工程实践创新项目（EPIP：Engineering，Practice，Innovation，Project）为载体，致力于培养具有创新精神、创造能力和创业精神的“三创”复合型高素质技能人才。国际化职业教育双语系列教材的编写，立足于乌干达人才需求和天津工业职业学院专业特色，是为了更好满足当地师生学习需求。

本系列教材采用中英双语相结合的方式，主要参照中国专业标准，开发国际化专业教学标准，课程内容国际化是在专业课程设置上，结合本地学生的技术能力水平与认知特点，合理设置双语教学环节，加强学生的学习与交流能力。同时，教材以提升职业能力为核心，以职业标准为依托，体现技能与质量并重的教学理念，主要采用项目化、模块化、任务驱动的教学模式，并结合企业岗位对员工的要求来撰写。

本系列教材在撰写过程中，王晓霞、李蕊、王治学、葛慧杰、于万松、王磊、李秀娟、宫娜、张志超、贾燕璐、陈宝玲等主编老师、专业团队、英语教研室付出了辛勤劳动，并得到了学院各级领导的大力支持，同时本系列教材借鉴和参考了业界有关学者的研究成果，在此一并致谢！

最后，衷心希望本系列教材能为我国高等职业教育国际化，尤其是高等职业教育走进非洲、支援非洲高等职业教育发展尽绵薄之力。

天津工业职业学院书记、院长　孔维军

2020年5月

Preface

Tianjin Polytechnic College and Uganda Technical College–Elgon reached a cooperation intention to establish the Luban workshop to carry out vocational education cooperation on mechatronics technology and ferrous metallurgy technology major in 2019. In order to strengthen the cooperation between China and Uganda in vocational education, the two colleges plan to compile a series of international vocational skills bilingual textbooks.

This book is one of the international vocational skills bilingual textbooks. This book starts from the continuous casting production process, the project comes from the real production process, combining simulation software and real operation to improve the skill level. The main content includes six parts: continuous casting production, continuous casting solidification process, operation of continuous casting equipment, continuous casting production process control and operation, accident treatment of continuous casting and defect control of continuous casting billet. Task preparation includes knowledge points and skill points needed to complete relevant tasks. Task implementation is a specific operation process. It reconstructs the contents needed to be mastered by each post and process of continuous casting, emphasizes the application and transfer of knowledge, combines the current advanced production technology and equipment, pays attention to the integration of traditional process technology and efficient continuous casting technology, and trains students' engineering practice innovation Ability.

This book is completed by the teaching team of ferrous metallurgy technology of Tianjin Polytechnic College. Yu Wansong is responsible for project 2, task 3 of project 3 and project 4~6, Li Bilin is responsible for project 1, Bai Junli is responsible for task 1~2 of project 3, Zhang Zhichao is responsible for task 4~7 of project 3. This book is unified by Yu Wansong. Lin Lei puts forward many valuable opinions on this book. During the compilation, it referred to a large number of metallurgical references, monographs and journals, and thank you very much!

Due to the limited level of editors, and I hope that readers will provide valuable comments and suggestions.

The editor

June, 2020

前言

2019年天津工业职业学院与乌干达埃尔贡技术学院达成合作意向，共同建立“鲁班工坊”，就机电一体化技术专业、黑色冶金技术专业开展职业教育合作，双方计划编撰国际化职业教育双语系列教材。

本书是国际化职业教育双语系列教材之一。本书从连铸生产过程出发，项目来源于真实生产过程，结合仿真软件和真实操作提高技能水平。本书共6个项目，主要内容包括连铸生产、连铸凝固过程、连铸设备操作、连铸生产工艺控制及操作、连铸生产事故处理、连铸坯缺陷控制等。任务准备包括完成相关任务所需要的知识点和技能点，任务实施则是具体的操作过程，将连铸各岗位和工序所需要掌握的内容进行重构，强调知识的运用和迁移，结合当前先进的生产技术和设备，注重传统工艺技术和高效连铸技术的融合，培养学生的工程实践创新能力。

本书由天津工业职业学院黑色冶金技术专业教学团队完成。于万松负责编写项目2、项目3的任务3、项目4~项目6，李碧琳负责编写项目1，白俊丽负责编写项目3的任务1、任务2，张志超负责编写项目3的任务4至任务7，全书由于万松统稿。林磊对本书提出很多宝贵意见，编写期间参考了冶金方面的资料、专著和期刊，在此一并表示衷心的感谢！

由于编者水平所限，书中不妥之处，敬请读者批评指正。

编　者

2020年6月

Contents

Project 1 Cognition of Continuous Casting Production ··· 1

Task 1.1 Cognition of Continuous Casting Process and Equipment ··· 1

1.1.1 Cognition of Continuous Casting Production ··· 1

1.1.2 Cognition of Continuous Casting Process ··· 2

1.1.3 Cognition of Continuous Casting Related Equipment ··· 3

1.1.4 Classification of Continuous Casters ··· 7

1.1.5 Comparison between Continuous Casting and Die Casting ··· 7

Task 1.2 Cognition of Continuous Casting Work Position and Technical Indexes ··· 10

1.2.1 Cognition of Continuous Casting Operation Post ··· 11

1.2.2 Economic and Technical Indexes of Continuous Casting Production ··· 13

Project 2 Cognition of Continuous Casting Solidification Process ··· 17

Task 2.1 Solidification Process Control of Molten Steel ··· 17

2.1.1 Solidification of Molten Steel ··· 17

2.1.2 Characteristics of Molten Steel Crystallization ··· 23

2.1.3 Grain Size Control after Crystallization ··· 25

2.1.4 Segregation ··· 26

2.1.5 Formation and Discharge of Gas ··· 28

2.1.6 Formation and Discharge of Inclusions ··· 28

2.1.7 Solidification Shrinkage ··· 29

2.1.8 Stress Control during Solidification and Cooling ··· 29

Task 2.2 Solidification Process Control of Billet ··· 31

2.2.1 Cognition of Continuous Casting Production Process ··· 31

2.2.2 Low Power Structure Characteristics of Billet ··· 35

2.2.3 Cognition of Billet Solidification Structure ··· 36

2.2.4 Control of Secondary Cooling on Slab Quality ··· 38

2.2.5 'Small Ingot' Structure Control ········ 39
2.2.6 Control of Billet Structure ········ 40

Project 3 Operation of Continuous Casting Equipment ········ 45

Task 3.1 Operation of Ladle and Ladle Turret ········ 45
3.1.1 Cognition of Ladle Equipment ········ 45
3.1.2 Cognition of Ladle Turret ········ 50
3.1.3 Installation of Ladle Long Nozzle ········ 51
3.1.4 Judge Whether the Ladle is Out of Service ········ 51
3.1.5 Ladle Cleaning and Maintain ········ 52
3.1.6 Use and Maintenance of Ladle Sliding Nozzle ········ 53
3.1.7 Operation of Ladle Baking ········ 54
3.1.8 Operation of Ladle Turret ········ 55
Task 3.2 Operation of Tundish and Tundish Car ········ 57
3.2.1 Cognition of Tundish ········ 57
3.2.2 Tundish Nozzle and Nozzle Control Mechanism ········ 61
3.2.3 Cognition of Tundish Car ········ 65
3.2.4 Installation of Tundish Stopper Rod ········ 67
3.2.5 Installation of Tundish Submerged Entry Nozzle ········ 68
3.2.6 Tundish Immersion Nozzle Position Adjustment ········ 69
3.2.7 Tundish Baking ········ 70
3.2.8 Inspection of Tundish Car ········ 72
Task 3.3 Operation of Mold and Mold Oscillation Device ········ 73
3.3.1 Cognition of Mold ········ 74
3.3.2 Cognition of Mold Lubrication Device ········ 82
3.3.3 Cognition of Mold Oscillation Device ········ 82
3.3.4 Inspection of Mold ········ 89
3.3.5 Maintenance of Mold ········ 91
3.3.6 Adjustment of Mold Width and Taper ········ 91
3.3.7 Inspection of Mold Oscillation Device ········ 92
3.3.8 Maintenance of Oscillation Device ········ 93
3.3.9 Inspection Method of Mold Oscillation Device ········ 94
Task 3.4 Operation of Secondary Cooling System ········ 95
3.4.1 Cognition of Secondary Cooling Area ········ 95
3.4.2 The Structure of Secondary Cooling Device ········ 96

3.4.3 Cognition of Secondary Cooling System ······ 100
3.4.4 Cooling Mode and Equipment ······ 102
3.4.5 Inspection of Secondary Cooling Device ······ 105
3.4.6 Inspection of Secondary Cooling Nozzle ······ 106
Task 3.5 Operation of Billet Withdrawing and Straightening Device ······ 108
3.5.1 Function of Tension Leveller ······ 109
3.5.2 Structure of the Tension Leveller ······ 109
3.5.3 Straightening Method of Continuous Casting Billet ······ 110
3.5.4 Inspection of Tension Leveller ······ 112
Task 3.6 Operation of Dummy Bar Device ······ 113
3.6.1 Cognition of Dummy Bar Device ······ 113
3.6.2 Structure and Classification of Dummy Bar ······ 114
3.6.3 Inspection of Dummy Bar Device ······ 117
3.6.4 Dummy Bar Feeding Operation ······ 119
3.6.5 Plug Dummy Bar Head Operation ······ 120
Task 3.7 Operation of Cutting Device ······ 122
3.7.1 Cognition of Billet Cutting Operation ······ 122
3.7.2 Flame Cutting Device ······ 123
3.7.3 Mechanical Cutting Device ······ 126
3.7.4 Inspection of Flame Cutting Device ······ 128
3.7.5 Use of Flame Cutting Device ······ 129
3.7.6 Use of Mechanical Cutting Device ······ 129

Project 4 Continuous Casting Process Control and Operation ······ 131

Task 4.1 Molten Steel Preparation ······ 131
4.1.1 Supply of Molten Steel ······ 132
4.1.2 Preparation of Molten Steel Temperature ······ 132
4.1.3 Composition Control of Molten Steel ······ 135
4.1.4 Control of Molten Steel Purity ······ 137
4.1.5 Control of Molten Steel Temperature and Fluidity ······ 139
4.1.6 Measures to Reduce the Temperature Drop of Molten Steel ······ 140
Task 4.2 Start Casting Operation ······ 141
4.2.1 Inspection and Preparation before Casting ······ 142
4.2.2 Key Points of Starting Casting Operation ······ 147
4.2.3 Operation of Starting Casting ······ 150

Task 4. 3 Normal Casting Operation …… 152
4. 3. 1 Control of Casting Speed …… 153
4. 3. 2 Liquid Level Control …… 153
4. 3. 3 Control of Cooling System …… 154
4. 3. 4 Operation of Dummy Bar Stripping …… 156
4. 3. 5 Cutting Operation …… 156
4. 3. 6 Sequence Casting Operation …… 157
4. 3. 7 Control Method of Casting Speed …… 158
4. 3. 8 Control of Cooling Water in Mold …… 159
4. 3. 9 Control of Cooling Water in Secondary Cooling Section …… 159
4. 3. 10 Operation of Dummy Bar Stripping …… 160
4. 3. 11 Operation of Replacing Ladle …… 161
4. 3. 12 Operation of Quick Replacement of Tundish …… 161
4. 3. 13 Operation of Sequence Casting of Different steel Grades …… 162
4. 3. 14 Sampling of Finished Products …… 163
Task 4. 4 Stop Casting Operation …… 165
4. 4. 1 Cognition of Stop Casting …… 165
4. 4. 2 Key Points of Stop Casting Operation …… 166
4. 4. 3 Cognition of Capping Operation …… 167
4. 4. 4 Operation Steps of Stop Casting …… 167
4. 4. 5 Operation Steps of Ladle Casting Finish …… 168
4. 4. 6 Operation Steps of Speed Reduction …… 168
4. 4. 7 Operation Steps of Capping …… 169
4. 4. 8 Operation Steps of Tail Billet Output …… 170
4. 4. 9 Cleaning and Inspection after Casting …… 170
Task 4. 5 Continuous Casting Process System Control …… 171
4. 5. 1 Casting Temperature Control …… 171
4. 5. 2 Control of Casting Speed …… 173
4. 5. 3 Cooling System Control …… 176
4. 5. 4 Control of Casting Speed …… 178
4. 5. 5 Control of Cooling Water in Mold …… 181
4. 5. 6 Control of Secondary Cooling Water in Different Stages …… 181
Task 4. 6 Continuous Casting Protective Casting …… 183
4. 6. 1 Protection Operation of Ladle to Tundish Injection Flow …… 183
4. 6. 2 Protection Operation of Tundish …… 184

4.6.3 Protection Operation of Casting Flow from Tundish to Mold ······ 184
4.6.4 Protection Operation in Mold ······ 184
4.6.5 Operation of Mold Flux Feeding ······ 187
4.6.6 Operation of Slag Dragging ······ 188

Project 5 Accident Treatment of Continuous Casting ······ 190

Task 5.1 Ladle Accident and Treatment ······ 190
5.1.1 Ladle Sliding Nozzle Blocked ······ 190
5.1.2 Steel Channeling of Ladle Sliding Nozzle ······ 191
5.1.3 Steel Flow Out of Control of Ladle Sliding Nozzle ······ 194
5.1.4 Ladle Stopper Nozzle Blocked ······ 195
5.1.5 Ladle Leakage Accident ······ 197
Task 5.2 Tundish Accident and Treatment ······ 202
5.2.1 Tundish Stopper Accident ······ 202
5.2.2 Submerged Nozzle Clogged ······ 204
Task 5.3 Mold Accident and Treatment ······ 208
5.3.1 Tundish Stopper Accident ······ 208
5.3.2 Mold Water Leakage Accident ······ 211
5.3.3 Mold Overflow Accident ······ 213
5.3.4 Other Continuous Casting Mold Accidents ······ 215
Task 5.4 Secondary Cooling System Accident and Treatment ······ 217
5.4.1 Treatment of Abrormal Situation of Secondary Cooling Water System ······ 217
5.4.2 Pull off Operation Accident ······ 218
Task 5.5 Breakout Accident and Treatment ······ 219
5.5.1 Cognition of Steel Leakage Accident ······ 219
5.5.2 Causes of Steel Leakage Accident ······ 220
5.5.3 Tool Preparation for Steel Leakage Accident Treatment ······ 221
5.5.4 Treatment of Steel Leakage in Small Section Billet ······ 222
5.5.5 Precautions ······ 224
Task 5.6 Treatment of Frozen Slab and Roller Blocking Billet Accident ······ 225
5.6.1 Cognition of Frozen Slab Accident and Roller Blocking Billet Accident ······ 225
5.6.2 Tool Preparation ······ 226
5.6.3 Treatment of Frozen Billet when Dummy Bar Still in the Tension Leveller ······ 226
5.6.4 Treatment of Roller Blocking Billet Accident ······ 228
5.6.5 Precautions ······ 228

Project 6 Defect Control of Continuous Casting Billet ························· 230

Task 6. 1 Quality Control of Continuous Casting Billet ························· 230

6. 1. 1 Cognition of Billet Quality ························· 230

6. 1. 2 Classification of Billet Defects ························· 231

Task 6. 2 Surface Defects of Continuous Casting Billet and Control ························· 233

6. 2. 1 Cognition of Billet Surface Defects ························· 233

6. 2. 2 Surface Cracks ························· 233

6. 2. 3 Oscillation Marks ························· 236

6. 2. 4 Gasholes and Bubbles ························· 237

6. 2. 5 Surface Slag Inclusion ························· 237

6. 2. 6 Other Surface Defects ························· 238

6. 2. 7 Prevention of Surface Cracks ························· 239

6. 2. 8 Prevention of Oscillation Marks ························· 240

Task 6. 3 Internal Defects of Continuous Casting Billet and Control ························· 241

6. 3. 1 Cognition of Internal Defects of Billet or Slab ························· 241

6. 3. 2 Internal Cracks ························· 242

6. 3. 3 Center Segregation ························· 244

6. 3. 4 Center Porosity ························· 244

6. 3. 5 Prevention of Internal Cracks ························· 244

6. 3. 6 Prevention of Center Segregation ························· 245

6. 3. 7 Prevention of Central Porosity ························· 245

Task 6. 4 Shape Defects of Continuous Casting Billet and Control ························· 246

6. 4. 1 Cognition of Billet Shape Defects ························· 247

6. 4. 2 Bulging ························· 247

6. 4. 3 Out of Square (Rhomboidity) ························· 248

6. 4. 4 Prevention of Bulging ························· 248

6. 4. 5 Prevention of Rhomboidity ························· 248

Task 6. 5 Control of Continuous Casting Billet Purity ························· 249

6. 5. 1 Cognition of Billet Purity ························· 250

6. 5. 2 Type of Inclusions ························· 250

References ························· 252

目　录

项目 1　连铸生产认知 …… 1

任务 1.1　连铸工艺和设备认知 …… 1

1.1.1　连铸生产认知 …… 2

1.1.2　连铸工艺认知 …… 2

1.1.3　连铸相关设备认知 …… 3

1.1.4　连铸机的分类 …… 7

1.1.5　连铸与模铸的比较 …… 7

任务 1.2　连铸岗位及技术指标认知 …… 10

1.2.1　连铸操作岗位认知 …… 11

1.2.2　连铸生产的经济技术指标 …… 13

项目 2　连铸凝固过程认知 …… 17

任务 2.1　钢液凝固过程控制 …… 17

2.1.1　钢液凝固认知 …… 18

2.1.2　钢液结晶的特点 …… 24

2.1.3　结晶后晶粒大小控制 …… 26

2.1.4　偏析控制 …… 26

2.1.5　气体的形成和排出 …… 28

2.1.6　夹杂物的形成和排出 …… 28

2.1.7　凝固收缩 …… 29

2.1.8　凝固、冷却过程中应力控制 …… 29

任务 2.2　连铸坯凝固过程控制 …… 31

2.2.1　连铸生产过程认知 …… 31

2.2.2　铸坯凝固的过程认知 …… 35

2.2.3　铸坯凝固结构认知 …… 36

2.2.4　二次冷却对铸坯质量的控制 …… 39

2.2.5　“小钢锭”结构控制 …… 40

2.2.6 铸坯结构的控制 …… 41

项目3 连铸设备操作 …… 45

任务3.1 钢包及钢包回转台操作 …… 45

3.1.1 钢包设备认知 …… 45

3.1.2 钢包回转台认知 …… 50

3.1.3 钢包长水口的安装 …… 51

3.1.4 钢包是否停用的判断 …… 52

3.1.5 钢包清理与维护 …… 53

3.1.6 钢包滑动水口的使用与维护 …… 53

3.1.7 钢包烘烤操作 …… 54

3.1.8 钢包回转台操作 …… 56

任务3.2 中间包及中间包车操作 …… 57

3.2.1 中间包认知 …… 57

3.2.2 中间包水口及水口控制机构 …… 61

3.2.3 中间包车认知 …… 65

3.2.4 中间包塞棒的安装 …… 67

3.2.5 中间包浸入式水口的安装 …… 68

3.2.6 中间包浸入式水口位置调整 …… 69

3.2.7 中间包烘烤 …… 70

3.2.8 中间包车检查 …… 72

任务3.3 结晶器及结晶器振动装置操作 …… 73

3.3.1 结晶器认知 …… 74

3.3.2 结晶器润滑装置认知 …… 82

3.3.3 结晶器振动装置认知 …… 83

3.3.4 结晶器的检查 …… 90

3.3.5 结晶器的维护 …… 91

3.3.6 结晶器宽度及锥度的调整 …… 92

3.3.7 结晶器振动装置的检查 …… 93

3.3.8 振动装置的维护 …… 93

3.3.9 结晶器振动装置检测方法 …… 94

任务3.4 二次冷却系统操作 …… 95

3.4.1 二次冷却区认知 …… 95

3.4.2 二次冷却装置结构 …… 96

3.4.3 二次冷却制度认知 …… 100

3.4.4 冷却方式及设备 …… 103
3.4.5 二冷装置的检查 …… 105
3.4.6 二冷喷嘴状态的检查 …… 107
任务 3.5 拉坯矫直装置操作 …… 108
3.5.1 拉矫装置的作用 …… 109
3.5.2 拉坯矫直机的结构形式 …… 109
3.5.3 连铸坯的矫直方式 …… 110
3.5.4 拉矫机检查 …… 112
任务 3.6 引锭装置操作 …… 113
3.6.1 引锭装置认知 …… 114
3.6.2 引锭杆的结构和分类 …… 114
3.6.3 引锭装置检查 …… 118
3.6.4 送引锭操作 …… 119
3.6.5 塞引锭头操作 …… 120
任务 3.7 切割装置操作 …… 122
3.7.1 铸坯切割操作认知 …… 122
3.7.2 火焰切割装置 …… 123
3.7.3 机械切割装置 …… 126
3.7.4 火焰切割装置的检查 …… 128
3.7.5 火焰切割装置的使用 …… 129
3.7.6 机械切割装置使用 …… 130
项目 4 连铸生产工艺控制及操作 …… 131
任务 4.1 钢液准备 …… 131
4.1.1 钢水的供应 …… 132
4.1.2 钢水的温度准备 …… 132
4.1.3 钢水的成分控制 …… 135
4.1.4 钢水纯净度控制 …… 137
4.1.5 钢液温度与流动性的控制 …… 140
4.1.6 减少钢水温度降低的措施 …… 140
任务 4.2 开浇操作 …… 141
4.2.1 浇铸前的检查与准备 …… 142
4.2.2 开浇操作要点 …… 148
4.2.3 开浇过程操作 …… 151

任务 4.3　正常浇铸操作 ······ 152
4.3.1　拉坯速度的控制 ······ 153
4.3.2　液面控制 ······ 154
4.3.3　冷却制度的控制 ······ 155
4.3.4　脱锭操作 ······ 156
4.3.5　切割操作 ······ 156
4.3.6　多炉连浇操作 ······ 157
4.3.7　拉速的控制方法 ······ 158
4.3.8　结晶器冷却水控制 ······ 159
4.3.9　二次冷却段冷却水控制 ······ 159
4.3.10　脱锭操作 ······ 160
4.3.11　更换钢包操作 ······ 161
4.3.12　快速更换中间包操作 ······ 162
4.3.13　异钢种连浇操作 ······ 163
4.3.14　成品取样操作 ······ 163
任务 4.4　停浇操作 ······ 165
4.4.1　停浇操作认知 ······ 166
4.4.2　停浇操作要点 ······ 166
4.4.3　封顶操作认知 ······ 167
4.4.4　停浇操作步骤 ······ 167
4.4.5　钢包浇完操作步骤 ······ 168
4.4.6　降速操作步骤 ······ 169
4.4.7　封顶操作步骤 ······ 169
4.4.8　尾坯输出操作步骤 ······ 170
4.4.9　浇铸结束后的清理和检查 ······ 170
任务 4.5　连铸工艺制度控制 ······ 171
4.5.1　浇铸温度控制 ······ 172
4.5.2　拉坯速度控制 ······ 173
4.5.3　冷却制度控制 ······ 176
4.5.4　连铸生产中拉速的控制 ······ 178
4.5.5　结晶器冷却水的控制 ······ 181
4.5.6　不同阶段的二次冷却水控制 ······ 182
任务 4.6　连铸保护浇铸 ······ 183
4.6.1　钢包到中间包铸流的保护操作 ······ 183
4.6.2　中间包的保护操作 ······ 184

4.6.3 中间包到结晶器铸流的保护操作 …… 184
4.6.4 结晶器的保护操作 …… 185
4.6.5 加结晶器保护渣操作 …… 188
4.6.6 捞渣操作 …… 188

项目5 连铸生产事故处理 …… 190

任务5.1 钢包事故及处理 …… 190
5.1.1 钢包滑动水口堵塞 …… 190
5.1.2 钢包滑动水口窜钢 …… 191
5.1.3 钢包滑动水口钢流失控 …… 194
5.1.4 钢包塞棒水口堵塞 …… 195
5.1.5 钢包穿漏事故 …… 197
任务5.2 中间包事故及处理 …… 202
5.2.1 中间包塞棒事故 …… 202
5.2.2 浸入式水口堵塞 …… 204
任务5.3 结晶器事故及处理 …… 208
5.3.1 结晶器断水事故 …… 208
5.3.2 结晶器漏水事故 …… 211
5.3.3 连铸结晶器溢钢 …… 213
5.3.4 其他连铸结晶器事故 …… 215
任务5.4 二冷事故及处理 …… 217
5.4.1 二冷水系统异常的处理 …… 218
5.4.2 拉脱操作事故 …… 218
任务5.5 连铸漏钢事故及处理 …… 219
5.5.1 漏钢事故认知 …… 220
5.5.2 漏钢事故产生的原因 …… 221
5.5.3 漏钢事故处理的工具准备 …… 222
5.5.4 小方坯发生漏钢时的抢救处理 …… 223
5.5.5 注意事项 …… 224
任务5.6 连铸冻坯和顶坯处理 …… 225
5.6.1 冻坯和顶坯事故认知 …… 225
5.6.2 工具准备 …… 226
5.6.3 引锭未出拉矫机的冻坯处理 …… 227
5.6.4 连铸顶坯处理 …… 228
5.6.5 注意事项 …… 229

项目 6　连铸坯缺陷控制……230
任务 6.1　连铸坯质量控制……230
6.1.1　铸坯质量认知……230
6.1.2　铸坯缺陷分类……231
任务 6.2　连铸坯表面缺陷及控制……233
6.2.1　铸坯表面缺陷认知……233
6.2.2　表面裂纹……234
6.2.3　振动痕迹……236
6.2.4　气孔和气泡……237
6.2.5　表面夹渣……237
6.2.6　其他表面缺陷……238
6.2.7　表面裂纹的预防……239
6.2.8　振动痕迹的预防……240
任务 6.3　连铸坯内部缺陷及控制……241
6.3.1　铸坯内部缺陷认知……241
6.3.2　内部裂纹……243
6.3.3　中心偏析……244
6.3.4　中心疏松……244
6.3.5　内部裂纹的预防……245
6.3.6　中心偏析的预防……245
6.3.7　中心疏松的预防……246
任务 6.4　连铸坯形状缺陷及控制……246
6.4.1　铸坯形状缺陷认知……247
6.4.2　鼓肚变形……247
6.4.3　脱方（菱变）……248
6.4.4　鼓肚的预防……248
6.4.5　脱方的预防……249
任务 6.5　连铸坯纯净度控制……249
6.5.1　铸坯纯净度认知……250
6.5.2　夹杂物的类型……250
参考文献……252

Project 1　Cognition of Continuous Casting Production
项目 1　连铸生产认知

Task 1.1　Cognition of Continuous Casting Process and Equipment
任务 1.1　连铸工艺和设备认知

Mission objectives:

任务目标:

(1) Know the type of continuous casting equipment and caster.

(1) 认识连铸的设备及连铸机的种类。

(2) Master the common continuous casting products and classification.

(2) 掌握连铸常见产品及分类。

(3) Be able to describe the production process and main equipment of continuous casting.

(3) 能够说出连铸的生产工艺流程和主要设备。

Task Preparation 任务准备

1.1.1　Cognition of Continuous Casting Production

There are two ways to make molten steel solidified and formed in the process of producing various kinds of steel products: ingot casting and continuous casting. The task of casting steel is to cast liquid steel with qualified composition into a certain shape (continuous casting billet or ingot) suitable for rolling and forging. Billet or ingot is the final forming process of steelmaking products, which is directly related to the yield and quality of steelmaking production. Therefore, steel casting as an important part of steel production process needs to be strictly controlled.

Ingot casting is the process of pouring molten steel into ingot mold made of cast iron and cooling solidification into ingot; continuous casting is the process of continuously pouring molten steel into a water-cooled mold to obtain the billet. Ingot casting has been basically replaced by continuous casting. Continuous casting has been developing rapidly because of its advantages.

1.1.1 连铸生产认知

在钢铁厂生产各类钢铁产品过程中，使钢水凝固成型有两种方法，即模铸法和连续铸钢法。铸钢的任务是将成分合格的钢液铸成适合于轧钢和锻压加工所需要的一定形状的钢块（连铸坯或钢锭）。铸坯或铸锭是炼钢产品最终成形的工序，直接关系到炼钢生产的产量和质量。因此，铸钢作为钢铁生产工序中重要一环，需要严格控制。

模铸是将钢液注入铸铁制作的钢锭模内，冷却凝固成钢锭的工艺过程；连铸是将钢液不断地注入水冷结晶器内，连续获得铸坯的工艺过程。模铸目前已基本被连铸所取代。连铸由于它所具有的一系列优越性，使得钢铁生产得到了迅猛的发展。

1.1.2 Cognition of Continuous Casting Process

The continuous casting process is shown in Figure 1-1. The molten steel from the steelmaking furnace is injected into the ladle, and then transported to the top of the continuous casting machine after secondary refining treatment. The molten steel is injected into the tundish through the ladle nozzle at the bottom of the ladle. The position of the tundish nozzle is preset to align with the mold below. After the tundish stopper (or sliding gate) is opened, the molten steel flows into the water-cooled mold blocked by the dummy bar head. In the mold, the molten steel gradually condenses into shell along its periphery. When there is a certain thickness of the shell at the outlet of the bottom of mold, the billet tension leveller and the mold oscillation device are started at the same time, so that the billet with liquid core can enter the arc-shaped guide section composed of several pinch rollers. The billet moves forward in the secondary cooling area and is forced to cool by the atomized water sprayed from the nozzle, and continues to solidify. After the dummy bar comes out of the straightener, it is separated from the billet. After the billets are straightened and completely solidified, they are cut into billets of fixed length by the cutting device, and finally transported to the designated place by the billet discharging device. With the continuous injection of molten steel, the billet continuously extends downward and is cut, forming the whole process of continuous casting.

1.1.2 连铸工艺认知

连铸工艺流程如图 1-1 所示，从炼钢炉出来的钢液注入钢包内，经炉外精炼处理后被运到连铸机上方，钢液通过钢包底部的水口再注入中间包内。中间包水口的位置被预先调好以对准下面的结晶器。打开中间包塞棒（或滑动水口）后，钢液流入下面由引锭杆头封堵的水冷结晶器内。在结晶器内，钢液沿其周边逐渐冷凝成坯壳。当结晶器下端出口处坯壳有一定厚度时，同时启动拉坯机和结晶器振动装置拉出铸坯，使带有液芯的铸坯进入由若干夹辊组成的弧形导向段。铸坯在二次冷却区域前进并受到喷嘴喷出雾化水的强制冷却，继续凝固。在引锭杆出拉坯矫直机后，将其与铸坯脱开。待铸坯被矫直且完全凝固后，由切割装置将其切成定尺铸坯，最后由出坯装置将定尺铸坯运到指定地点。随着钢液的不断注入，铸坯不断向下伸长，并被切割运走，形成了连续浇铸的全过程。

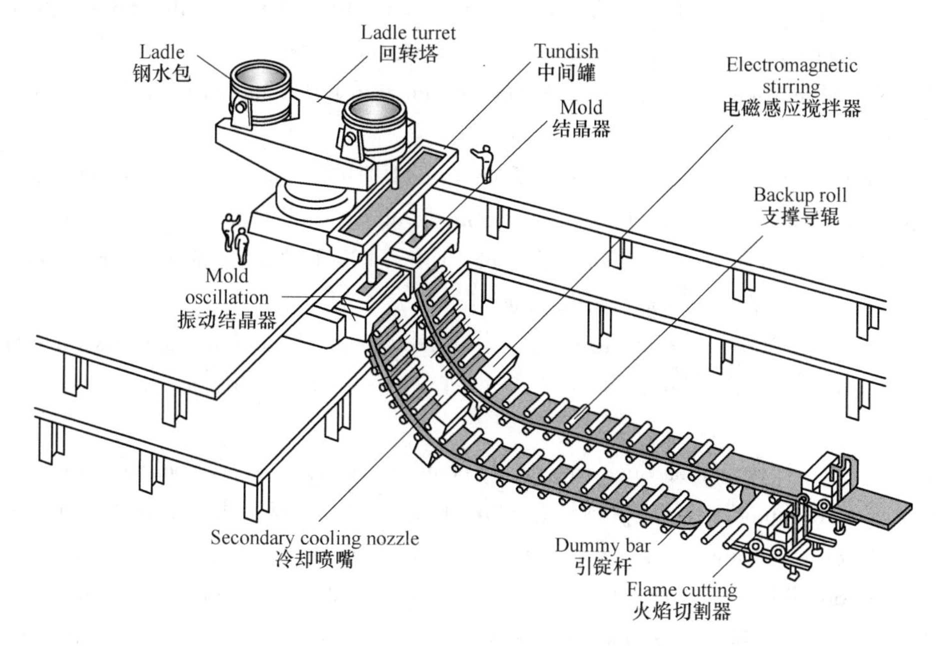

Figure 1-1　Process chart of continuous casting machine

图 1-1　连铸机工艺流程图

1.1.3　Cognition of Continuous Casting Related Equipment

1.1.3.1　Cognition of Basic Equipment of Continuous Casting Machine

The continuous caster is mainly composed of ladle, ladle carrier, tundish, tundish carrier, mold, mold oscillation device, secondary cooling device, straightening and withdrawal unit, dummy bar device, cutting device and billet delivery device.

1.1.3　连铸相关设备认知

1.1.3.1　连铸机主要设备认知

连铸机主要由钢包、钢包运载装置、中间包、中间包运载装置、结晶器、结晶器振动装置、二次冷却装置、拉坯矫直机、引锭装置、切割装置和铸坯运出装置等部分组成。

(1) Ladle is a device for holding and casting molten steel and is also a container for secondary refining.

(2) Ladle carrier is mainly ladle turret, its main function is to carry ladle and support ladle for casting operation. The ladle turret can be used to change the ladle quickly and realize multi ladles continuous casting.

(3) The tundish is also called middle tank, which is a transition device between the ladle and the mold to receive the molten steel, so that the molten steel has a reasonable flow and a prop-

er long stay time in the tundish, and the molten steel is divided by the tundish in the multi flow continuous casting machine; during the multi furnace continuous casting, the molten steel stored in the tundish plays a role of connection when replacing the ladle.

(4) Tundish carrier is mainly tundish car, which is used to support, transport and replace tundish.

(5) The mold is a special water-cooled steel mold. The molten steel is cooled in the mold, initially solidified and formed, and a certain thickness of the shell is formed, so as to ensure that when the shell is pulled out of the mold, the shell will not be leaked, deformed or cracked. So it is the key equipment of continuous casting machine.

(6) The oscillation device of the mold is to make the mold move up and down according to certain requirements, so as to prevent the initial shell from sticking to the mold and being cracked.

(7) The secondary cooling device is mainly composed of water spray cooling device and billet support device. Its function is to spray water directly to the billet or slab to make it completely solidified, and to prevent and limit the occurrence of bulging, deformation and leakage of steel by supporting and guiding the billet with liquid core by pinch roller and side guide roller.

(8) The function of the tension leveler is to pull out the billet smoothly during the casting process and straighten the arc billet. It also feeds the dummy bar unit into the mold before casting.

(9) The dummy bar device includes two parts: dummy bar head and dummy bar. Its function is to serve as the 'drop bottom' of the mold at the beginning of casting, block the bottom outlet of the mold, and make the molten steel solidify at the head of the dummy bar. Through the traction of the tension leveler, the casting billet is pulled out from the outlet of the mold along with the dummy bar. When the dummy bar is pulled out of the tension leveler, the dummy bar is removed and enters the normal state of billet casting.

(10) The purpose of the cutting device is to cut the billet into the required fixed length in the process of casting.

(11) The casting billet transportation equipment includes roller table, pusher, cooling bed, etc.

(1) 钢包是用于盛接钢液并进行浇铸的设备，也是钢液炉外精炼的容器。

(2) 钢包运载装置主要为钢包回转台，它的主要作用是运载钢包，并支撑钢包进行浇铸作业。采用钢包回转台还可快速更换钢包，实现多炉连铸。

(3) 中间包也称为中间罐，它是钢包和结晶器之间用来接受钢液的过渡装置，使钢液在中间包内有合理的流动和适当长的停留时间，在多流连铸机由中间包对钢液进行分流；在多炉连浇时，中间包中贮存的钢液在更换钢包时起到衔接的作用。

(4) 中间包运载装置主要是中间包车，它是用来支撑、运输、更换中间包的设备。

(5) 结晶器是一个特殊的水冷钢模，钢液在结晶器内冷却、初步凝固成形，并形成一定厚度的坯壳，以保证铸坯被拉出结晶器时，坯壳不被拉漏、不产生变形和裂纹等缺陷。因此它是连铸机的关键设备。

(6) 结晶器振动装置使结晶器能按一定的要求做上下往复运动，以防止初生坯壳与结晶器粘连而被拉裂。

（7）二次冷却装置主要由喷水冷却装置和铸坯支撑装置组成。它的作用是：向铸坯直接喷水，使其完全凝固；通过夹辊和侧导辊对带有液芯的铸坯起支撑和导向作用，防止并限制铸坯发生鼓肚、变形和漏钢事故。

（8）拉坯矫直机的作用是在浇铸过程顺利地将铸坯拉出，并对弧形铸坯进行矫直。在浇铸前，它还要将引锭装置送入结晶器内。

（9）引锭装置包括引锭头和引锭杆两部分，它的作用是在开浇时作为结晶器的“活底”，堵住结晶器的下口，并使钢液在引锭杆头部凝固；通过拉矫机的牵引，铸坯随引锭杆从结晶器下口拉出。当引锭杆拉出拉矫机后，将引锭杆脱去，进入正常拉坯状态。

（10）切割装置的作用是在铸坯行进过程中，将它切割成所需要的定尺长度。

（11）铸坯运出装置包括辊道、推钢机、冷床等。

1.1.3.2 Main Design Parameters of Continuous Casting Equipment

The main design parameters of the continuous casting equipment includes the size and shape of the section of the billet, the casting speed of the billet, the metallurgical length, the radius of curvature of caster, the number of the strands, the production capacity of the continuous casting machine, etc.

1.1.3.2 连铸设备的主要设计参数

连铸设备的主要设计参数有铸坯断面的尺寸和形状、拉坯速度（简称拉速）、冶金长度、基本圆弧半径、连铸机流数、连铸机生产能力等。

（1）The size and shape of the section of the billet. The size and shape of the billet section are the most basic design parameters of the caster, and other design parameters are selected according to it. The size and shape of the billet section are determined according to the shape and specification requirements of the next rolling process and the finished product, combined with the quality level that can be achieved in the current continuous casting production, as well as the capacity of steelmaking and smelting cycle of the steel-making production.

（1）铸坯断面的尺寸与形状。铸坯断面的尺寸和形状是连铸机最基本的设计参数、其他设计参数都是根据他来选定的。铸坯断面的尺寸和形状是按照下道轧钢工序与成品的形状、规格要求、结合当前连铸生产实际能达到的质量水平及炼钢生产的出钢量、冶炼周期来确定的。

（2）Casting speed. The casting speed is an important design parameter to determine the production capacity of the continuous casting machine. Appropriate casting speed can not only play the production capacity of the continuous casting machine, but also improve the surface quality of the continuous casting billet. There are many factors that affect the casting speed, such as the size and shape of the section of the casting billet, the steel grade, the casting temperature, the length of the caster, the resistance of the casting billet and the solidification thickness of the casting billet out of the mold outlet. The maximum casting speed must ensure that the casting billet has enough thickness after the casting billet out of the mold.

（2）拉坯速度。拉坯速度是决定连铸机生产能力的重要设计参数，合适的拉坯速度，既能发挥连铸机的生产能力，也能改善连铸坯的表面质量。影响拉坯速度的因素是多方面的，主要有铸坯断面的尺寸和形状、浇铸钢种、浇铸温度、机身长度、拉坯阻力以及铸坯出结晶器口的凝固厚度，最大拉坯速度必须保证铸坯出结晶器后铸坯有足够的厚度。

（3）Metallurgical length. Metallurgical length is the length of liquid phase calculated according to the maximum casting speed of continuous casting machine, which is related to the thickness, casting speed and cooling strength of the billet. Metallurgical length is related to the determination of the caster and the basic radius of curvature of continuous casting machine.

（3）冶金长度。冶金长度是按连铸机最大拉坯速度计算的铸坯液相长度，与铸坯厚度、拉坯速度以及铸坯的冷却强度有关。冶金长度关系到连铸机机身和基本圆弧半径的确定。

（4）The basic radius of curvature of continuous casting machine. The basic radius of curvature refers to the basic curvature radius of the external arc of the rollers of the caster, which affects the height of the caster, the maximum thickness of the casting billet, and the quality of the billet. The determination of the basic radius of curvature should meet the solidification requirements before straightening, the set surface temperature requirements, and the requirements of the allowable surface elongation during the internal arc straightening of the billet.

（4）基本圆弧半径。基本圆弧半径指连铸机辊列的外弧基本曲率半径，它影响连铸机机身高度、浇铸铸坯最大厚度以及铸坯质量。基本圆弧半径的确定应满足铸坯矫直前的凝固要求、设定的表面温度要求，并满足铸坯内弧矫直时允许的表面伸长率的要求。

（5）The number of the strands. The number of the continuous casting strands refers to the number of billets cast flow by one continuous caster at the same time. The determination of the number of strands should meet the requirements of casting capacity, casting cycle, steel-making production capacity, ladle capacity, etc. Generally, one machine with multiple strands is conducive to the production capacity of the equipment, but higher requirements are put forward for the equipment condition and operation level.

（5）连铸机流数。连铸机流数指一台连铸机同时浇铸的铸坯数量，连铸机流数的确定应满足连铸机浇钢能力、浇铸周期与炼钢生产能力、钢包容量等。一般一机多流有利于发挥设备的生产能力，但对设备状况、操作水平提出了更高要求。

（6）Production capacity of caster. The production capacity of continuous caster refers to the yield of billet of one caster or strand produced in unit time, generally expressed as hourly yield or annual yield. The production capacity of continuous caster mainly depends on such factors as the number of continuous caster strands, casting speed, section size and continuous casting operation rate. In order to improve the production capacity of continuous caster, multi furnace continuous casting should be organized. Therefore, the steel-making workshop must timely and quantitatively provide the molten steel with qualified composition and temperature in accordance with the production requirements of multi furnace continuous casting in continuous casting workshop, and achieve balanced production, stable rhythm, connection accuracy and quality assurance. In continuous casting and equipment, in order to realize multi-furnace continuous casting, ladle turret and large capacity tundish can be used, and equipment maintenance can be strengthened to avoid failure.

（6）连铸机生产能力。连铸机生产能力指一台或一流连铸机在单位时间内铸坯产量，一般以小时产量或年产量表示。连铸机生产能力主要取决于连铸机流数、拉坯速度、铸坯断面尺寸及连铸作业率等因素，为提高连铸机的生产能力，应组织多炉连浇，因此炼钢车间必须按照连铸车间多炉连浇的生产要求，准时、定量提供成分、温度合格的钢水，并做到生产均衡、节奏稳定、衔接准确、质量保证。在连铸及设备方面，为了实现多炉连浇，可采用钢包回转台、大容量中间包，并加强设备维护，避免故障产生。

1.1.4 Classification of Continuous Casters

Continuouscasters can be classified in a variety of ways:

(1) According to the configuration of the continuous caster structure, it can be divided into vertical continuous caster, vertical bending continuous caster, bow-type continuous caster (arc continuous caster), oval continuous caster, horizontal continuous caster, etc. Figure 1-2 is the schematic diagram of various types of caster.

(2) According to the shape and size of billet section, it can be divided into: Billet caster (if the section is not more than 150mm×150mm, it is called billet; if the section is more than 150mm×150mm, it is called bloom); Slab caster (the section of slab is rectangular, and its width thickness ratio is generally more than 3); Round billet caster (the section of the billet is round with a diameter of 60~400mm); Beam blank continuous caster (casting special section, such as profile, hollow tube, etc.); Continuous caster for both billet and slab (on one caster, it can cast both slab and billets); Thin slab caster (thin slab with thickness of 40~80mm), etc.

1.1.4 连铸机的分类

连铸机可以按多种方法进行分类：

（1）按连铸机结构的外形可分为立式连铸机、立弯式连铸机、弧形连铸机、椭圆形连铸机、水平连铸机等。图 1-2 为各种形式连铸机机型的示意图。

（2）按铸坯断面的形状和大小可分为：方坯连铸机（断面不大于 150mm×150mm 的小方坯、大于 150mm×150mm 的大方坯）；板坯连铸机（断面为长方形，其宽厚比一般在 3 以上）；圆坯连铸机（断面为圆形，直径 ϕ60~400mm）；异形坯连铸机（浇铸异形断面，如型材、空心管等）；方、板坯兼用连铸机（在一台铸机上，既能浇板坯，也能浇方坯），薄板坯连铸机（铸坯厚度为 40~80mm 的薄板坯料）等。

Task Implementation 任务实施

1.1.5 Comparison between Continuous Casting and Die Casting

1.1.5 连铸与模铸的比较

Compared with traditional ingot casting, continuous casting has the following advantages:
与传统的模铸相比，连铸有以下几方面的优势：

(1) Simplifies the production process and shortens the technology process. It can be seen

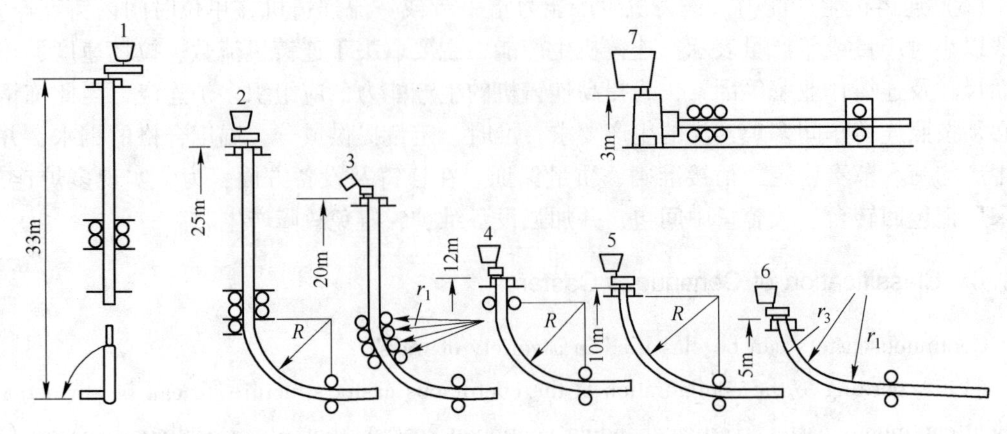

Figure 1-2 Schematic diagram of various caster models

1—Vertical continuous caster; 2—Vertical bending continuous caster;

3—Vertical machines (straight mold) with progressive bending and straightening;

4—Vertical machines (straight mold) with single point straightening;

5—Bow-type continuous caster; 6—Oval continuous caster; 7—Horizontal continuous caster

图 1-2 各种连铸机机型示意图

1—立式连铸机；2—立弯式连铸机；3—直结晶器多点弯曲连铸机；4—直结晶器弧形连铸机；

5—弧形连铸机；6—多半径弧形（椭圆形）连铸机；7—水平连铸机

from Figure 1-3 that the continuous casting process eliminates the processes of demolding, die-forming, ingot heating and blooming. The appearance of thin slab caster further simplifies the technology process. Compared with the traditional slab continuous casting (thickness of 150 ~ 300mm), thin slab continuous casting (thickness of 40 ~ 80mm) saves roughing mill process. The production cycle from molten steel to sheet metal is greatly shortened. Traditional slab continuous casting takes about 40 hours, while thin slab continuous casting only takes 1 ~ 2 hours.

(1) 简化了生产工序，缩短了工艺流程。从图 1-3 可以看出，连铸工艺省去了脱模、整

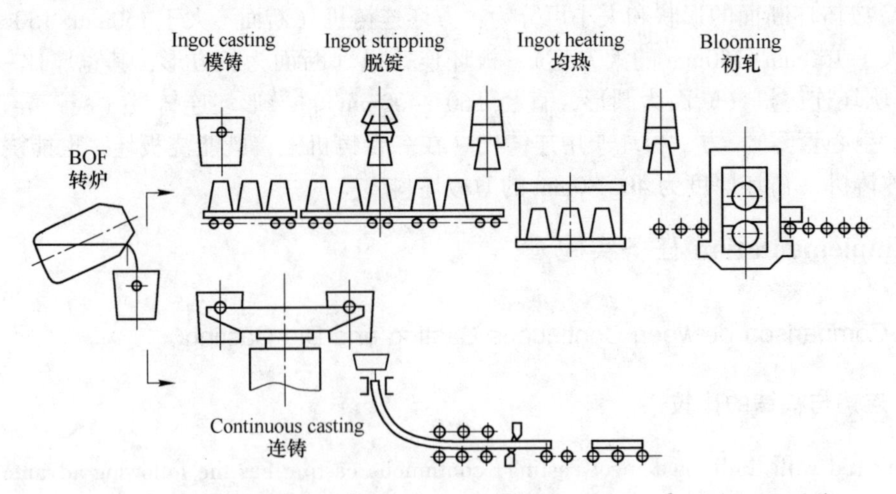

Figure 1-3 Comparison of production process of ingot casting and continuous casting

图 1-3 模铸与连铸生产流程比较

模、钢锭均热、初轧开坯等工序。薄板坯连铸机的出现，又进一步简化了工序流程。与传统板坯连铸（厚度为150~300mm）相比，薄板坯（厚度为40~80mm）连铸省去了粗轧机组，从钢水到薄板的生产周期大大缩短，传统板坯连铸约需40h，而薄板坯连铸仅为1~2h。

(2) Improved metal yield. By using the die casting process, the loss of cutting head and tail from molten steel to slab is 1% ~ 20%, while that of continuous casting is 1% ~ 2%, so the yield of metal can be increased by 10% ~ 14% (slab 10.5%, bloom 13%, billet 14%). If the steel plant with an annual capacity of 1 million tons of steel is calculated to increase by 10%, the production of 100 thousand tons of steel can be increased by adopting the continuous casting process. The economic benefits are considerable.

(2) 提高了金属收得率。采用模铸工艺，从钢水至铸坯的切头切尾损失达1%~20%，而连铸的切头切尾损失为1%~2%，故可提高金属收得率10%~14%（板坯10.5%、大方坯13%，小方坯14%）。如果以提高10%计算，年产100万吨钢的钢厂，采用连铸工艺，就可增产10万吨钢，带来的经济效益是相当可观的。

(3) Reduced energy consumption. The reheating process of soaking pit is eliminated by continuous casting, and the energy consumption can be reduced by 1/2~1/4. Usually, in the production of 1t billet, the continuous casting can save 400~1200MJ, which is equivalent to saving 10~30kg heavy oil fuel. If the continuous casting slab adopts hot delivery and direct rolling process, the energy consumption can be further reduced, and the processing cycle can be shortened.

(3) 降低了能源消耗。采用连铸省掉了均热炉的再加热工序，可使能量消耗减少1/2~1/4。通常生产1t铸坯，连铸比模铸一般可节能400~1200MJ，相当于节省10~30kg重油燃料。若连铸坯采用热送和直接轧制工艺，能耗还可进一步降低，并能缩短加工周期。

(4) Production process mechanization, high degree of automation. After adopting continuous casting, the labor environment has been fundamentally improved because of the improvement of equipment and operation level and the whole process computer control and management. Continuous casting operation automation and intelligence have become a reality.

(4) 生产过程机械化、自动化程度高。采用连铸后，设备和操作水平的提高以及采用全程计算机控制和管理，劳动环境得到了根本性的改善。连铸操作自动化和智能化已成为现实。

(5) Continuous casting steel grade expands, and the product quality is improving. At present, almost all kinds of steel can be produced by continuous casting. More than 500 kinds of continuous casting steel have been expanded to include super-clean steel (IF steel), high grade silicon steel, stainless steel, Z-direction steel, pipeline steel, heavy rail, hard wire, tool steel and alloy steel. Moreover, the quality of continuous casting slab is better than that of ingot.

(5) 连铸钢种扩大，产品质量日益提高。几乎所有的钢种都可用连铸生产，产品已扩大到包括超纯净度钢（IF钢）、高牌号硅钢、不锈钢、Z向钢、管线钢、重轨、硬线、工具钢以及合金钢等500多个。且连铸坯产品质量的各项性能指标大都优于模铸钢锭的轧材产品。

However, in the present case, continuous casting is not a complete substitute for the production of ingot casting, because the properties of some types of steel are not yet suitable for the production of continuous casting, or it is difficult to guarantee the quality of steel when using continuous casting (e. g. the rimming steel and high-speed steel with strong thermal sensitivity). For example, some small batch products, and some large-scale forging that must be forged parts (e. g. the main shaft of a 10000ton ship), and some large-scale rolled products, etc. Therefore, it is still necessary to retain part of the production mode of mold casting, and strive to improve the quality of steel.

但从目前的情况看，连铸尚不能完全代替模铸的生产，这是因为：有些钢种的特性还不能适应连铸的生产方式、或采用连铸时难以保证钢的质量（如沸腾钢以及热敏感性很强的高速钢等）；比如一些小批量产品，还有一些必须经锻造的大型锻造件（如万吨船只的主轴）；以及一些大规格的轧制产品等。所以仍需要保留部分模铸的生产方式，努力提高钢材的质量。

Exercises:

(1) What are the equipment and processes in the production process of continuous casting?

(2) What are the main products of continuous casting?

(3) Use the simulation software to complete the system login, select the production task, and observe the equipment layout of the continuous casting workshop.

思考与习题:

(1) 连铸的生产过程中的设备和流程是什么？

(2) 连铸生产的主要产品有哪些？

(3) 利用仿真软件完成系统的进入，选择生产任务，观察连铸车间设备布置情况。

Task 1.2 Cognition of Continuous Casting Work Position and Technical Indexes
任务1.2 连铸岗位及技术指标认知

Mission objectives:

任务目标:

(1) Understand the responsibilities and tasks of each post of continuous casting production.

(1) 理解连铸生产各岗位的职责和任务。

(2) Master the economic and technical indicators of continuous casting production.

(2) 掌握连铸生产的经济技术指标。

(3) Master the connection of each process of continuous casting production.

(3) 掌握连铸生产各工序的联系。

Task Preparation 任务准备

1.2.1 Cognition of Continuous Casting Operation Post

1.2.1 连铸操作岗位认知

(1) Director of continuous caster. The director is the person in charge of the steel casting operation, responsible for the production, organization and management of the unit, equipment maintenance and use, safety and civilized production, etc.; ensure the completion of the production tasks and technical and economic indicators issued by the workshop; guide the production strictly according to the requirements of the operation procedures in the production; carefully inspect and monitor the equipment to master the equipment of the unit.

(1) 连铸机长。机长是浇钢操作的负责人，负责机组的生产、组织管理、设备维护使用以及安全、文明生产等工作；保证完成车间下达的生产任务和各项技术经济指标；在生产中严格按操作规程要求指导生产；对设备要认真检查和监护，掌握本机组设备情况。

(2) Steel casting workers. The steel casting workers are the specific executor of the steel casting operation, which is composed of several workers. Each team member completes the following work: check the equipment condition of the unit before casting steel, prepare tools, instruments and raw materials; carry out various operations in strict accordance with the requirements of the operation procedures in the steel pouring production; cooperate with the caster supervisor in troubleshooting in case of accident; carefully check and maintain the equipment, and find out the equipment fault report to the caster supervisor immediately; conduct civilized production, keep the site clean and tidy; obey the supervisor's work arrangement and operation instructions.

(2) 浇钢工。浇钢工是浇钢操作的具体执行人员，由若干工人组成，各组员分工完成下列工作：浇钢前检查本机组的设备情况，准备好工器具和原材料；在浇钢生产中严格按照操作规程要求执行各项操作；事故状态下配合机长排除故障；认真检查和维护好设备，发现设备故障立即报告机长；做好文明生产，保持现场清洁、整齐；服从机长工作安排及操作指令。

(3) Dummy ingot operator. Dummy ingot operator is responsible for mastering the equipment status on duty, doing a good job of equipment inspection and preparation, contacting with relevant personnel in time in case of any problem to ensure normal production; controlling and monitoring various operation elements on the dummy ingot operation platform in the casting production; carrying out the operation of sending dummy ingot, casting off ingot and cutting head; contacting the captain in time in case of any abnormality in the production and taking corresponding measures.

(3) 引锭操作工。引锭操作工负责掌握当班设备状况，做好各项设备检查准备工作，发现问题及时与有关人员联系，确保正常生产；在浇铸生产中负责引锭操作台上各种操作元件的控制和监视；执行送引锭、开浇脱锭及切头等项操作；生产中出现异常时及时与机长联系并采取相应措施。

(4) Cutter. Cutter is responsible for mastering the integrity of the equipment on duty, mak-

ing preparations for the inspection of various equipment, contacting relevant personnel in time in case of any problems to ensure the normal production; controlling and monitoring various operating elements on the cutting operation platform in the pouring production; cutting the continuous casting billet in strict accordance with the specified size during the cutting operation; immediately contacting the supervisor and taking measures to solve in case of any abnormality in the production.

(4) 切割工。切割工负责掌握当班设备完好情况，做好各项设备检查的准备工作，发现问题及时与有关人员联系，确保正常生产；在浇铸生产中负责切割操作台上各种操作元件的控制和监视；切割操作时要严格按规定尺寸切割连铸坯；生产中出现异常时立即与机长联系并采取措施加以解决。

(5) Main control room operator. The operator in the main control room is responsible for the inspection, control and monitoring of various signals, indicator lights, instruments and buttons in the main control room; communicating various production instructions, process temperature and previous process time to all relevant posts; feeding back the equipment, production and accidents of the continuous caster; during the casting process, it is necessary to control the oscillation parameters of cooling water and mold according to the requirements of the operation procedures; record all kinds of production data and conditions truthfully, accurately and clearly.

(5) 主控室操作工。主控室操作工负责主控室内各种信号、指示灯、仪表及按钮的检查、控制和监视；向各有关岗位传达各项生产指令、过程温度及前道工序时间；反馈连铸机设备、生产和事故情况；浇铸过程中注意按操作规程要求控制好冷却水及结晶器的振动参数；真实、准确、清晰地记载各种生产数据及情况。

(6) Water distributor. The water distribution engineer shall check whether the water systems are normal and deal with the problems in a timely manner; adjust the water quantity according to the conditions of each steel grade to achieve reasonable water distribution; closely observe the condition of the slab and deal with and report the problems in a timely manner. Be responsible for filling in the post report and record. Be responsible for the sanitation of the area of the post. Strictly implement safety regulations.

(6) 配水工。配水工检查各处水系统是否正常，发现问题及时处理；按各钢种条件进行水量调节，达到合理配水；密切观察铸坯情况，发现问题及时处理并汇报。负责填写本岗位报表、记录。负责本岗位责任区域卫生。严格执行安全规程。

(7) Ladle operator. Operate casting equipment, check and replace overflow tank and emergency tank; prepare tools before casting steel, start casting of ladle, add heat preservation agent and control liquid level of tundish; use and replace ladle nozzle, and take temperature measurement and sampling for tundish. Be responsible for the sanitation of the responsible area of the post. Strictly implement safety regulations.

(7) 大包工。操作浇铸设备，对溢流罐及事故槽进行检查、更换；浇钢前工具准备，钢水包开浇、加保温剂及控制中间包液面；使用及更换钢包长水口，对中间包进行测温取样。负责本岗位责任区域卫生。严格执行安全规程。

(8) Drawing steel workers. Prepare all kinds of materials and tools for steel casting, fully prepare for production; check the speed control system, mold, tundish repair and baking; control

the casting speed according to the steel type requirements and temperature, and be responsible for pulling the molten steel out of the mold to be qualified billet; be responsible for tundish preparation and replacement, and mold replacement, and strictly implement the process discipline. Be responsible for the sanitation of the responsible area of the post. Strictly implement safety regulations.

(8) 拉钢工。准备各种浇钢用材料及工具，充分做好生产前的准备工作；对拉速控制系统、结晶器、中间包修砌及烘烤情况进行检查；按钢种要求及温度控制好拉钢速度，负责把钢水从结晶器中拉出合格钢坯；负责中间包准备与更换，结晶器的更换，严格执行工艺纪律。负责本岗位责任区域卫生。严格执行安全规程。

(9) Steel pushing worker. Check the operation of the operating systems such as roller table, cooling bed, steel moving machine, etc., contact relevant personnel in time in case of any problems; be responsible for the operation of roller table transporting billets, and transporting the cut billets to the cooling bed. Fill in the position record. Be responsible for the sanitation of the responsible area of the post. Strictly implement safety regulations.

(9) 推钢工。检查辊道、冷床、移钢机等操作系统运转情况，发现问题及时与有关人员联系；负责辊道运坯操作，将切割后的尾坯运送到冷床。负责填写本岗位记录。负责本岗位责任区域卫生。严格执行安全规程。

Task Implementation 任务实施

1.2.2 Economic and Technical Indexes of Continuous Casting Production

1.2.2 连铸生产的经济技术指标

1.2.2.1 Continuous Casting Billet Yield

The yield of continuous casting billet refers to the output of qualified billet within a specified period of time (generally in months, quarters and years). The calculation formula is:

Continuous casting billet yield = the total amount of production billet − the amount of waste after inspection − the amount of waste after rolling or returned by users.

The continuous casting billet must be produced in accordance with standards, or in accordance with the standards and technical agreements stipulated in the supply contract.

1.2.2.1 连铸坯产量

连铸坯产量是指在某一规定的时间内（一般以月、季、年为时间计算单位）合格铸坯的产量。计算公式为：

连铸坯产量 = 生产铸坯产量 − 检验废品量 − 轧后或用户退回废品量

连铸坯必须按照标准生产，或按供货合同规定标准、技术协议生产。

1.2.2.2 Continuous Casting Ratio

Continuous casting ratio refers to the percentage of qualified production of continuous casting billet in total steel production. It is one of the important marks of steelmaking process level and benefit, and also reflects the development of continuous casting production in enterprises or regions. The calculation formula is:

$$\text{continuous casting ratio} = \frac{\text{qualified continuous casting billet output}}{\text{total qualified steel output}} \times 100\%$$

1.2.2.2 连铸比

连铸比指的是连铸坯合格产量占总钢产量的百分比。它是炼钢生产工艺水平和效益的重要标志之一，也反映了企业或地区连铸生产的发展状况。计算公式为：

$$\text{连铸比} = \frac{\text{合格连铸坯产量}}{\text{总合格钢产量}} \times 100\%$$

1.2.2.3 Qualified Rate of Continuous Casting Billet

The qualified rate of continuous casting billet refers to the percentage of qualified continuous casting billet in the total inspection quantity of continuous casting billet, also known as quality index (generally in months and years as time statistical units). The calculation formula is:

$$\text{qualified rate} = \frac{\text{qualified billet after inspection}}{\text{total inspected billet}} \times 100\%$$

1.2.2.3 连铸坯合格率

连铸坯合格率指的是连铸合格坯占连铸坯总检验量的百分比，又称为质量指标（一般以月、年为时间统计单位）。计算公式为：

$$\text{合格率} = \frac{\text{合格铸坯产量}}{\text{铸坯总检验量}} \times 100\%$$

1.2.2.4 Yield of Continuous Casting Billet

The yield of continuous casting billets refers to the percentage of qualified production of continuous casting billets in the total amount of molten steel. It accurately reflects the consumption of continuous casting production and the recovery of molten steel. The calculation formula is:

$$\text{yield of continuous casting billet} = \frac{\text{qualified continuous casting billet output (t)}}{\text{total amount of molten steel in continuous casting (t)}} \times 100\%$$

The yield of continuous casting billets is related to the size of section, while the yield with small section is lower.

1.2.2.4 连铸坯收得率

连铸坯收得率是指合格连铸坯产量占连铸浇铸钢水总量的百分比。它比较精确地反映了连铸生产的消耗及钢液的收得情况。计算公式为：

$$\text{连铸坯收得率} = \frac{\text{合格连铸坯产量}}{\text{连铸浇铸钢液总量}} \times 100\%$$

铸坯收得率与断面大小有关，铸坯断面小则收得率低些。

1.2.2.5 Operation Rate Continuous Casting Machine

The operation rate of continuous caster refers to the percentage of the actual operation time of the caster in the total calendar time (generally calculated by month, quarter and year). It reflects the operation and production capacity of the continuous casting machine. The calculation formula is:

$$\text{operation rate of continuous caster} = \frac{\text{actual operation time of continuous caster (h)}}{\text{calendar time (h)}} \times 100\%$$

1.2.2.5 连铸机作业率

连铸机作业率是指铸机实际作业时间占总日历时间的百分比（一般可按月、季、年统计计算）。它反映了连铸机的作业及生产能力。计算公式为：

$$\text{连铸机作业率} = \frac{\text{连铸机实际作业时间 (h)}}{\text{日历时间 (h)}} \times 100\%$$

1.2.2.6 Design-level-reaching Rate of Continuous Caster

The design-level-reaching rate of continuous caster refers to the percentage of the actual output of continuous caster in the designed output of the continuous caster within a certain period of time (generally based on annual statistics). It reflects the equipment level of the caster. The calculation formula is:

$$\text{design-level-reaching rate} = \frac{\text{actual output of continuous caster (t)}}{\text{design output of continuous casting machine (t)}} \times 100\%$$

1.2.2.6 连铸机达产率

连铸机达产率是指在某一时间段内（一般以年统计），连铸机实际产量占该台连铸机设计产量的百分比。它反映了这台连铸机的设备发挥水平。计算公式为：

$$\text{连铸机达产率} = \frac{\text{连铸机实际产量 (t)}}{\text{连铸机设计产量 (t)}} \times 100\%$$

1.2.2.7 The Average Number of Sequence Continuous Casting

The average number of sequence continuous casting refers to the ratio of the number of ladles for casting molten steel to the number of times of continuous casting. It reflects the continuous operation ability of continuous caster. The calculation formula is:

$$\text{the average number of sequence continuous casting (ladle/time)} = \frac{\text{number of casting ladles}}{\text{number of casting starts of continuous caster}}$$

1.2.2.7 平均连浇炉数

平均连浇炉数是指浇铸钢液的炉数与连铸机开浇次数之比。它反映了连铸机连续作业的能力。计算公式为：

$$平均连浇炉数（炉/次）=\frac{浇铸钢液炉数}{连铸机开浇次数}$$

1.2.2.8 Average Sequence Casting Time

The average sequence casting time refers to the ratio of the actual operation time of the continuous casting machine to the number of casting starts of the continuous casting machine. It also reflects the continuous operation of the caster. The calculation formula is:

$$\text{average sequence casting time (h/time)} = \frac{\text{actual operation time of caster (h)}}{\text{number of casting starts of continuous caster}}$$

1.2.2.8 平均连浇时间

平均连浇时间是指连铸机实际作业时间与连铸机开浇次数之比。它同样反映了连铸机连续作业的情况。计算公式为：

$$平均连浇时间（h/次）=\frac{铸机实际作业时间（h）}{连铸机开浇次数}$$

1.2.2.9 Overflow Leakage Rate of Caster

The overflow leakage rate of caster refers to the percentage of the total strands of the caster in a certain period of time. The calculation formula is:

$$\text{overflow leakage rate of caster} = \frac{\text{total strand of overflow or leakage}}{\text{total casting strands} \times \text{number of strands owned by the caster}} \times 100\%$$

1.2.2.9 铸机溢漏率

铸机溢漏率是指在某一时间内连铸机发生溢漏钢的流数占该段时间内该铸机浇铸总流数的百分比。计算公式为：

$$铸机溢漏率=\frac{溢漏钢流数总和}{浇铸总炉数\times铸机拥有流数}\times100\%$$

Exercises:

(1) What are the posts in continuous casting? What are the main responsibilities of these posts?

(2) Describe the commonly used technical indicators of continuous casting production.

(3) Use simulation software to simulate the role of each post and check the equipment in charge.

思考与习题：

（1）连铸生产中有哪些岗位，其负责的主要工作有哪些？

（2）阐述常用的连铸生产技术指标。

（3）利用仿真软件模拟各岗位角色，检查所负责的设备。

Project 2　Cognition of Continuous Casting Solidification Process
项目 2　连铸凝固过程认知

Task 2.1　Solidification Process Control of Molten Steel
任务 2.1　钢液凝固过程控制

Mission objectives:

任务目标:

(1) Master the solidification process and conditions of metal.

(1) 掌握金属的凝固过程和凝固条件。

(2) Master the phenomenon of molten steel during solidification and the stress produced during cooling.

(2) 掌握钢液在凝固过程中的现象及冷却过程中产生的应力。

(3) According to the solidification principle of molten steel, measures can be provided for the development of continuous casting process and improvement of billet quality.

(3) 能够根据钢液的凝固规律，为制定连铸工艺和改善铸坯质量等提供措施。

Task Preparation 任务准备

2.1.1　Solidification of Molten Steel

The essence of molten steel casting is the process of molten steel change from liquid to solid. This is the crystallization process of steel, also known as steel solidification. The crystallization theory of molten steel mainly starts from the thermodynamic point of view, and studies crystal nucleation, growth, shape changes and the relationship between metal's structure and quality. The solidification theory of molten steel mainly starts from the perspective of heat transfer, and studies the size of the solidification zone, solidification time, the relationship between solidification characteristics and quality. Continuous casting process makes the casting to achieve the required size, shape, quality and structure by the process, quality requirements.

2.1.1 钢液凝固认知

钢液浇铸实质就是钢液由液态转变为固态，此为钢的结晶过程，也称为钢的凝固。钢液的结晶理论主要是从热力学的观点出发，研究晶体成核、长大、形状变化规律以及金属组织与质量的关系。钢液的凝固理论主要是从传热学的观点出发，研究凝固区的大小、凝固时间、凝固特性与质量的关系。连铸过程中按工艺、质量的要求，适当控制凝固过程，使铸坯达到规定尺寸、形状、质量和结构。

2.1.1.1 Thermodynamic Conditions of Molten Steel Crystallization

The crystallization process of the metal from liquid into solid, must meet certain thermal conditions to occur spontaneously, when the system can reduce the free energy. By the law of thermodynamics, the relationship between the liquid metal free energy G_L and the solid metal free energy G_S can be calculated. According to the principle of minimum free energy of the thermodynamics, it can proceed from the high free energy state to the lower state spontaneously. The free energy of liquid metal and solid metal decreases with the increasing temperature, and the temperature corresponding to the intersection of them is T_0, which is called the theoretical crystallization temperature.

2.1.1.1 钢液结晶的热力学条件

金属从液态转变为固态的结晶过程，必须满足一定的热力学条件才能自发进行，即系统的自由能降低。通过热力学定律，可以将液态金属自由能 G_L 和固态金属自由能 G_S 关系计算出来。根据热力学的最小自由能原理，过程能够自发地从自由能高的状态向较低的状态进行。液态金属和固态金属的自由能都随温度的升高而降低，两者相交时交点对应的温度为 T_0，称为理论结晶温度。

When the temperature is lower than T_0, $G_S < G_L$, the solid phase is stable, and the liquid metal can change into a solid state spontaneously. When $G_S > G_L$, the liquid phase is stable, and the solid metal will melt into a liquid state automatically. When $T = T_0$, $G_S = G_L$, both solid and liquid phases are in equilibrium and can coexist.

The results show that the liquid metal begins to crystallize when it cools to a certain temperature below the theoretical crystallization temperature T_0, which is called the actual crystallization temperature T_n. The difference between T_0 and T_n is called undercooling.

$$\Delta T = T_0 - T_n \tag{2-1}$$

当温度低于 T_0时，$G_S < G_L$，固相稳定，液态金属可以自发地转变成固态；反之，$G_S > G_L$时，液态稳定，固态金属将自动熔化为液态；而在 T_0时，$G_S = G_L$，两者处于平衡状态而共存。

研究表明，只有冷却到低于理论结晶温度 T_0以下的某一温度时，液态金属才开始结晶，这时的温度称为实际结晶温度 T_n。T_0与 T_n之差称为过冷度，即：

$$\Delta T = T_0 - T_n \tag{2-1}$$

A certain undercooling is not only the necessary condition for metal crystallization, but also the thermodynamic condition for crystallization. The greater the undercooling is, the greater the crystallization trend of the liquid phase is. In order to obtain the necessary undercooling for crystallization, the temperature of liquid metal must be reduced and the latent heat of crystallization must be released. Therefore, the crystallization of metals is an exothermic process.

The release and dissipation of latent heat is one of the important factors affecting the crystallization process. When the release of latent heat is equal or less than the heat radiated to the surrounding environment, the temperature will keep constant or fall continuously, and the crystallization can continue until it is completely solidified or reaches a new balance. When the release of latent heat is greater than the heat dissipated, the temperature will rise until the crystallization stops, and even the remelting phenomenon will occur in the local area.

具有一定过冷度是金属结晶的必要条件，也是结晶的热力学条件。过冷度越大，液相的结晶趋势越大。为得到结晶所必需的过冷度，必须使液态金属的温度降低，将结晶潜热散发出去。因此，金属的结晶过程是个放热过程。

结晶时潜热的释放和逸散，是影响结晶过程的重要因素之一。当潜热的释放等于或小于散发到周围环境中的热量时，温度将保持恒定或不断下降，结晶可以继续进行，直至完全凝固或达到新的平衡；当潜热的释放大于散发掉的热量时，温度会回升，直到结晶停止进行，甚至在局部区域还会发生重熔现象。

2. 1. 1. 2 Dynamic Conditions of Molten Steel Crystallization

The crystallization needs to be carried out under the condition of undercooling, and the dynamic condition of metal crystallization is nucleation and growth. Metal crystallization is mainly controlled by these two processes.

(1) Nucleation. Generally, there are two forms of nucleation: homogeneous nucleation (spontaneous nucleation) and heterogeneous nucleation (non spontaneous nucleation). Actually the mostly method of the nucleation is the heterogeneous nucleation in the metal crystallization.

Heterogeneous nucleation is also called non spontaneous nucleation. The solid particles and the wall with unsmooth surface in the liquid phase of the alloy can form the core and develop into the initial crystal core.

2. 1. 1. 2 钢液结晶的动力学条件

结晶需在过冷条件下进行，金属结晶的动力学条件是形核和长大。金属结晶主要由这两个过程进行控制。

(1) 晶核的形成。晶核的形成一般有两种形式：一种是均质形核（自发形核）；另一种是异质形核（非自发形核）。实际金属结晶时，大多数是以异质形核的方式进行的。

异质形核又称非均质形核，也称非自发形核。在合金液相中已存在的固相质点和表面不光滑的器壁均可作为形成核心的“依托”而发展成初始晶核。这一过程称为异质形核。

Molten steel is an alloy, and many solid particles with high melting point are suspended in the liquid. These heterogeneous impurities can easily develop into the natural cores, namely heterogeneous nucleation. The surface energy of nucleation caused by the nucleation attached to these particles increases very little, and the nucleation energy ΔG_k is also small, the critical size is reduced, so the undercooling in the crystallization is required less than before. It can greatly increase the nucleation rate, and makes more stable nucleus be formed easily. We can see the relationship between the rate of homogeneous and heterogeneous nucleation and the undercooling (ΔT) in the Figure 2-1. It is difficult to occur homogeneous nucleation if it has the condition of heterogeneous nucleation. The crystallization of pure metal can only be formed by homogeneous nucleation.

钢液是合金，其内部悬浮着许多高熔点的固态质点，或多或少都含有异相杂质，可以成为自然的核心，即异质形核。附着在这些质点上产生晶核所引起的晶核表面能增加极少，且形核功 ΔG_k 也较小，从而使临界尺寸减小，结晶过程中不需要太大的过冷度，大大增加了形核率，容易形成稳定的晶核。均质形核和异质形核速率与过冷度 ΔT 的关系如图 2-1 所示。只要存在非均质形核的条件，很难发生均质形核。纯金属的结晶只能靠均质形核。

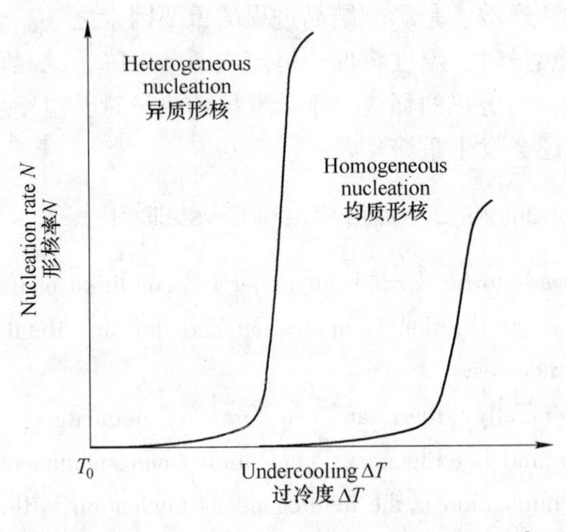

Figure 2-1 Relationship between homogeneous and heterogeneous nucleation and supercooling

图 2-1 均质形核和异质形核速率与过冷度的关系

(2) Growth of nucleus.

1) Crystal growth mode. The mode of crystal growth depends on two cases. One case is that there is a positive temperature gradient from the interface to the liquid phase (shown in Figure 2-2), the farther away the liquid phase from the solid - liquid interface, the higher the temperature is. The other case is that there is a negative temperature gradient from the interface to the liquid phase (shown in Figure 2-3), the farther away from the interface, the lower the temperature in the liquid phase is.

(2) 晶核的长大。

1) 晶体的生长方式。晶体长大的方式取决于结晶过程中固、液相界面附近的温度分布，即与液相中的过冷度有关。结晶前温度分布有两种情况：一种情况是由界面到液相具有正的温度梯度，如图 2-2 所示，固体结晶前沿的液相离固、液界面越远，温度越高；另一种情况是由界面向液相具有负温度梯度，如图 2-3 所示，离界面越远，液相中的温度越低。

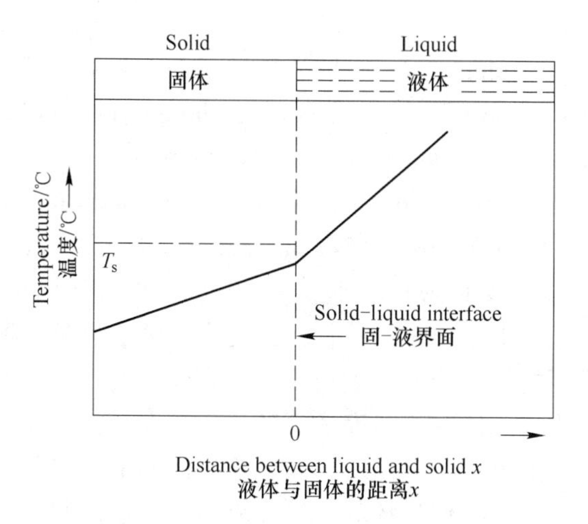

Figure 2-2 Distribution of crystallization front with positive temperature gradient

图 2-2 结晶前沿具有正温度梯度时的分布情况示意图

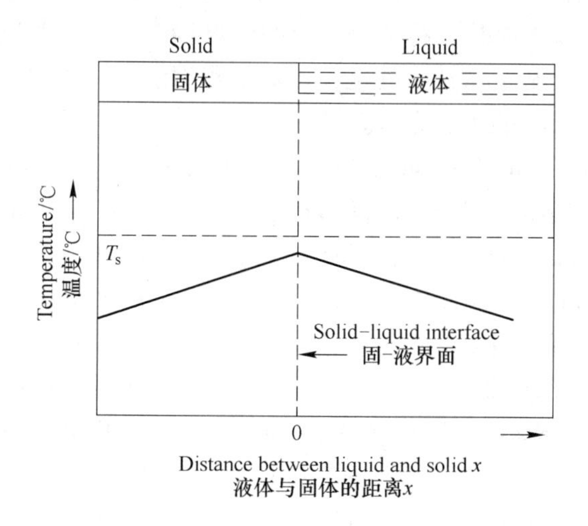

Figure 2-3 Distribution of crystallization front with negative temperature gradient

图 2-3 结晶前沿具有负温度梯度时的分布情况示意图

When the crystal grows under the condition of positive temperature gradient, the growth rate is completely controlled by the condition of heat dissipation. Due to the high temperature at the solid-liquid interface, it is impossible for the solid to extend into the liquid phase for growth. The surface of the solid seems to be smooth on the macro level, and the solid-liquid interface moves forward to the liquid phase in a plane shape. When the crystal grows up under the condition of negative temperature gradient, the occasional bulge on the solid surface can penetrate into the place with larger undercooling degree in the liquid. At this time, the plane shape of the solid-liquid interface becomes unstable, and the solid surface is no longer smooth, some elongated crystal columns will be formed and make the interface bulgy. If the undercooling near the interface is very large, the protruding part of the solid surface will soon extend into the supercooled liquid and grow up and form the dendrites, this kind of growth is called dendrite growth. If the undercooling is small, some occasionally protruding parts of the solid surface may extend into the supercooled area and grow, but they can't develop to the depth, so that the growing interface is between the plane and dendrite and form a kind of uneven cellular like structure called cellular tissue, which is called cellular growth. When other conditions are same, the alloy with fractional crystallization is more conducive to the growth of dendrites. For the real molten steel, due to the effect of impurity elements and alloy elements, the growth mode of dendrite is the most common.

晶体在正温度梯度条件下生长时，其长大速度完全由散热条件所控制。由于固、液界面处温度很高，固体不可能伸入液相中生长，固体表面在宏观上看起来是平滑的，固、液界面呈平面状向液相推进。晶体在负温度梯度条件下长大时，固体表面上的偶然凸起便可深入液体中过冷度较大的地方生长。这时，固、液界面的平面状就变得不稳定，固体表面也不再是光滑的，而会形成一些伸长的晶柱，在界面上产生凸起。若这时界面附近过冷度很大，固相表面上的凸出部分就会很快地伸向过冷液体中长大并在侧面产生分枝，形成树枝状晶体。这种生长情况即为树枝状生长；若过冷度较小，固相表面某些偶然凸出的部分可能会伸入过冷区长大，但不能向纵深发展，使生长的界面介于平面和树枝状之间，形成一种凹凸不平的类似胞状的结构，称为胞状组织，这种生长情况为胞状生长。当其他条件相同时，选分结晶合金更利于树枝晶的生长。对实际钢液来说，由于杂质元素及合金元素的作用，结晶时多为树枝状的生长方式。

2) Dendrite growth. Figure 2-4 shows the dendrite growth. Generally, crystallization always grows preferentially in the place where solute segregation is the smallest and heat dissipation is the fastest. Because the thermal conductivity of edges and corners is better than other directions, the growth speed of edges and corners is faster than other directions. Iron is a cubic lattice with hexahedral structure. It grows from eight angles to the tip of rhombic cone, forming the main axis of dendrite (primary axis). Then it grows a bifurcate called secondary axis on the side of the main axis, regenerates the tertiary axis, and develops successively until the dendrites meet each other and form a dendrite grain. The dendrites with uniform development in all directions are called equiaxed grain; only the dendrites with prominent development in one direction are called columnar grain.

2）树枝晶长大。图 2-4 为树枝晶生长示意图。一般结晶总是在溶质偏析最小和散热最快的地方优先生长。由于棱角比其他方向导热性好，且距离未被溶质富集的液体最近，因此棱角方向的长大速度比其他方向要快。铁为立方晶格，呈正六面体结构，从八个角长成为菱锥体的尖端，就构成了树枝晶主轴（一次轴），然后在主轴侧面长出分叉称为二次轴，再生出三次轴，依次发展下去，直到晶枝彼此相遇，形成一个树枝状晶粒。各方向的主轴都得到较均匀发展的树枝状称为等轴晶；只有某一方向的主轴得到突出发展的树枝状称为柱状晶。

In actual ingots or billets, there are two kinds of crystal growth:

① Directional growth. When the molten steel is injected into the ingot mold, a large number

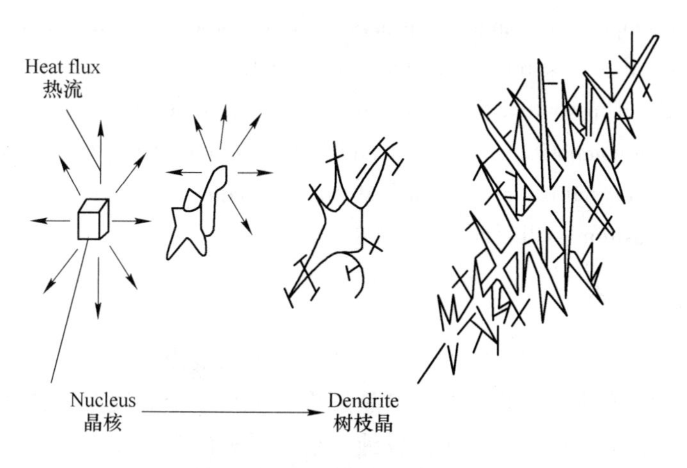

Figure 2-4 Dendrite growth
图 2-4 树枝晶生长示意图

of crystal cores are produced in the supercooled liquid contacting with the cold mold wall. At first, they can grow freely, but the crystals growing perpendicular to the mold wall quickly inhibit the development of crystals in other directions. The reason is that when the heat dissipation is fast perpendicular to the mold wall and the crystallization conditions are the best, so that the crystals growing perpendicular to the mold wall preferentially grow to the ingot center. Columnar grains with single growth direction perpendicular to the mold wall are formed.

② Equiaxial growth. After the columnar crystal grows to a certain length, the directional heat dissipation along the mold wall slows down, the temperature gradient gradually decreases, the columnar grain stops developing, and the liquid temperature in the ingot core drops without obvious temperature gradient. In this way, the crystal nucleus at the ingot center has similar growth conditions in all directions, so it grows into dendrites with basically the same crystal axis, i. e. equiaxed grain.

在实际钢锭或铸坯中，晶体有两种长大情况：

① 定向生长。钢液注入锭模或结晶器时，与冷模壁接触的过冷液体中产生大量结晶核心，开始它们可以自由生长，但垂直于模壁方向生长的晶体很快抑制了其他方向晶体的发展，其原因是垂直模壁方向散热快时，选分结晶条件最好，使垂直于模壁方向生长的晶体优先向锭心长大，从而形成了垂直于模壁的单方向生长的柱状晶。

② 等轴生长。在柱状晶长到一定长度后，沿模壁定向散热减慢，温度梯度逐渐减小，柱状晶停止发展，处于锭心的液体温度下降且无明显的温度梯度。这样锭心处晶核在各方向上具有相似的生长条件，因而长成各晶轴基本相同的树枝晶，即等轴晶。

2.1.2 Characteristics of Molten Steel Crystallization

Steel is an alloy containing carbon, silicon, manganese, phosphorus, sulfur and other elements. In fact, the solidification of steel is non-equilibrium crystallization, so the crystallization of liquid steel is different from that of pure metal. The crystallization process must be carried out and completed within a temperature range. The crystallization process is fractional crystallization. The crystal initially crystallized is relatively pure, the content of solute elements is low, and the melting point is high. The content of solute elements in the final crystal is high, and the melting point is also low. The composition of crystal and liquid is changed with the decrease of temperature. Only when the crystallization is finished and the equilibrium is reached, the crystal can reach the same composition as the original alloy.

2.1.2 钢液结晶的特点

钢是含有碳、硅、锰、磷、硫等元素的合金，钢的凝固在实际上属非平衡结晶，因此钢液的结晶具有不同于纯金属的特点。结晶过程必须在一个温度范围内进行并完成，结晶过程为选分结晶，最初结晶出的晶体比较纯，溶质元素的含量较低，熔点较高，最后生成的晶体溶质元素的含量较多，熔点也较低，无论是晶体或液体的成分，都随着温度的下降而不断地变化着，只有当结晶完毕后，并且达到平衡时，晶体才有可能达到和原始合金一样的成分。

The molten steel contains various alloy elements, so its crystallization temperature is not a point, but a temperature range, as shown in Figure 2-5. The liquid steel begins to crystallize at the liquid temperature and reaches the end of crystallization at the solid temperature. The difference between the liquid temperature and solid temperature is the crystallization temperature range. It can be expressed as:

$$\Delta T_C = T_L - T_S \tag{2-2}$$

The significance of determining the crystallization temperature range is as follows:

(1) It is the basis for determining the casting temperature and tapping temperature;

(2) The crystallization temperature range has an important influence on the crystal structure.

The effect on the solidified structure can be seen from the relationship between the crystallization temperature range and the width of the two-phase region. The crystallization of molten steel is completed in a temperature range, so the solid phase and liquid phase coexist in this temperature range.

There are three crystallization regions in the solidification process of molten steel: solid region, two-phase region and liquid region. In the two-phase region, the nucleation and the growth of the crystal nucleus are going on. The solidification of the steel billets is the process of the two-phase region moving from the surface of the slab to the center of the billet.

钢液含有各种合金元素，其结晶温度不是一个点，而是在一个温度区间，如图 2-5 所示。钢液在 $T_{液}$ 开始结晶，到达 $T_{固}$ 结晶完毕，结晶温度范围用 $\Delta T_{结晶}$ 表示：

$$\Delta T_{结晶} = T_{液} - T_{固} \tag{2-2}$$

确定结晶温度范围的意义在于：

(1) $T_{液}$ 是确定浇铸温度及出钢温度的基础；

(2) 结晶温度范围的大小对结晶组织有重要影响。

从结晶温度范围和两相区宽度的关系中可以看出对凝固组织的影响。由于钢液结晶是

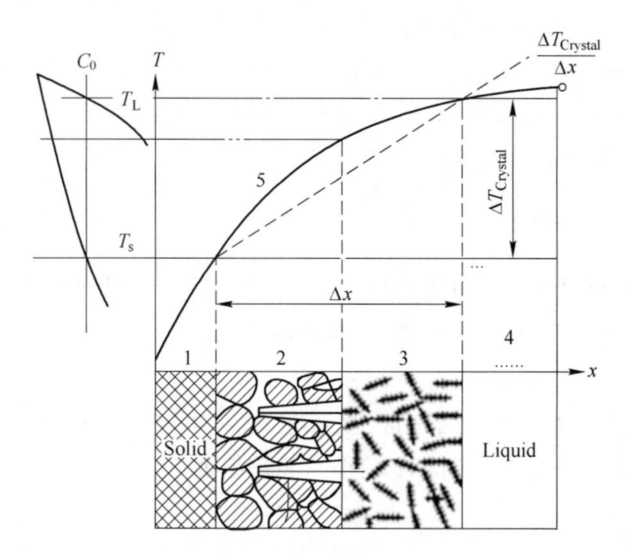

Figure 2-5　Crystallization temperature range and two-phase zone of molten steel

1—Solid area; 2, 3—Two-phase area; 4—Liquid area; 5—Temperature distribution

图 2-5　钢液的结晶温度范围和两相区的构成图

1—固相区; 2, 3—两相区; 4—液相区; 5—温度分布

在一个温度区间内完成的，因此在这个温度区间里固相与液相并存。

钢液凝固过程中一般出现 3 个结晶区域，即固相区、两相区和液相区。在两相区内进行着形核和晶核的长大过程，铸坯的凝固就是两相区由铸坯表面逐渐向铸坯中心推移的过程。

The width of the two-phase region is mainly determined by the crystallization temperature range of the molten steel and the temperature gradient in the melt at the solidification front. The wider the crystallization temperature range is, the lower the superheat of liquid steel and the smaller the cooling strength of billet surface are, the wider the two-phase zone is. A wide two-phase zone is not good for billet quality, so the width of the two-phase zone should be reduced properly. The specific process measures start from strengthening the cooling strength.

两相区的宽度主要取决于钢液的结晶温度范围和凝固前沿熔体中的温度梯度。结晶温度范围越宽，钢液过热度越低和铸坯表面冷却强度越小，则两相区越宽。较宽的两相区对铸坯质量不利，因此应适当减少两相区宽度。具体的工艺措施从加强冷却强度入手。

Task Implementation 任务实施

2.1.3　Grain Size Control after Crystallization

Grain size is one of the important indexes that affect the properties of steel. Generally speaking, the steel with fine grain structure has higher comprehensive mechanical properties, and its strength, hardness, toughness, plasticity, etc. are relatively good, so it is very important to control the grain size of steel in production.

After solidification, the grain size is determined by the nucleation rate N (nucleation number/$cm^3 \cdot s$) and growth rate v (cm/s). The larger the nucleation rate N is, the smaller the growth rate v is, and the finer the grain is.

2.1.3 结晶后晶粒大小控制

晶粒大小是影响钢材性能的重要指标之一，一般来说，细晶粒组织的钢材具有较高的综合力学性能，即其强度、硬度和韧性、塑性等都比较好，所以生产上相当重视控制钢材的晶粒尺寸。

金属凝固后，晶粒大小取决于晶核生成速率 N（成核数目/$cm^3 \cdot s$）和长大速率 v（cm/s），成核率 N 越大，长大速率 v 越小，晶粒越细。

2.1.4 Segregation

Segregation is the most common and basic phenomenon during solidification of any alloy. Solute elements are distributed between solid and liquid simply because the solubility of solute elements is different between the phases. In general, the phenomenon that nonuniform chemical composition of billet (or ingot) is called segregation. All kinds of elements, gases and non-metallic inclusions in steel have segregation phenomenon, but the degree of segregation is not the same.

When segregation occurs at the distance between dendrites (usually 50~400μm), the segregation of alloy elements in castings and ingots is called microsegregation. The formation of microsegregation is due to the different composition of liquid phase during solidification, additional solute precipitates from liquid phase into the narrow space between dendrites, mainly including composition segregation (dendrite segregation), cellular segregation and grain boundary segregation.

2.1.4 偏析控制

偏析是合金凝固过程中最常见和最基本的现象。由于溶质元素在各相中的溶解度不同导致元素在固相和液相的分布不同。通常把铸坯（或钢锭）中化学成分不均匀的现象称为偏析。钢中所含各种元素、气体和非金属夹杂物等均有偏析现象，但偏析程度并不一样。

当偏析发生在枝晶间距（通常为50~400μm）的距离时，把合金元素在铸件和铸锭中的偏析称为微观偏析。微观偏析的产生是由于凝固时液相成分的不同，额外的溶质从液相析出进入树枝晶之间的狭窄空间当中，主要包括晶内偏析（枝晶偏析）、胞状偏析和晶界偏析。

Besides the microscopic segregation, there exist macroscopic types of segregation often encountered in the actual casting. Figure 2-6 illustrates the typical types of segregation appearing in a continuously cast slab. There are three types of positive segregation and one type of negative segregation.

除微观偏析外，实际铸造会经常遇到宏观偏析。图 2-6 说明了连铸板坯中出现的典型偏析类型，包括 3 种正偏析和 1 种负偏析。

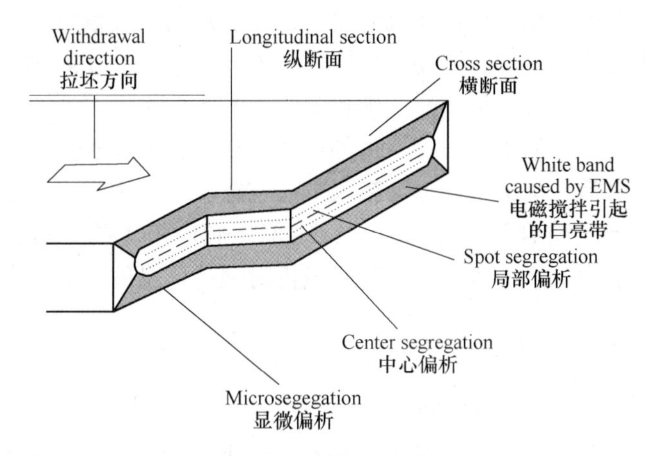

Figure 2-6　Typical examples of various segregation in slab

图 2-6　板坯中出现的典型偏析

Segregation has a great influence on the quality of billet, which may result in different properties of various parts of steel, and even reduce the yield of steel or even scrap the steel, so measures should be taken to reduce segregation. The following measures can be taken in the production process to control the segregation:

(1) Increase the condensation rate of liquid steel. The segregation can be reduced by inhibiting the diffusion of solute into the mother liquor.

(2) Suitable billet section. Small section can shorten solidification time and reduce segregation.

(3) Various methods are used to control the flow of molten steel. For example, using the suitable immersion nozzle, adding titanium, boron and other denaturants.

(4) Electromagnetic stirring. Stirring can break dendrites, refine grains and reduce segregation.

(5) Process factors. There are many measures in this respect, such as: properly reducing the casting temperature and the casting speed is conducive to reducing the segregation. Preventing the bulging deformation of the continuous casting slab can help eliminate the concentration of impurities mother liquor flowing into the center gap, so as to reduce the center segregation.

(6) Reduce the content of sulfur and phosphorus in molten steel. Sulfur and phosphorus are the elements with the greatest segregation tendency in steel, and they are also the most harmful to steel. Therefore, the influence of segregation on steel quality can be reduced by reducing the content of sulfur and phosphorus in molten steel.

偏析对铸坯质量有较大的影响，轻者造成钢材各部分性能不一，重者可能降低钢的成材率甚至使钢报废，因此应采取措施来减少偏析。生产工艺中可采取以下措施来控制偏析：

（1）增加钢液的冷凝速度。通过抑制选分结晶中溶质向母液深处的扩散来减小偏析。

（2）合适的铸坯断面。小断面可使凝固时间缩短，从而减轻偏析。

（3）采用各种方法控制钢液的流动。如适宜的浸入式水口、加入 Ti、B 等变性剂等。

（4）电磁搅拌。搅拌可打碎树枝晶、细化晶粒、减小偏析。

（5）工艺因素。这方面的措施很多，例如：适当降低铸温和拉速有利于减轻偏析；防止连铸坯鼓肚变形，可消除富集杂质母液流入中心空隙，以减小中心偏析等。

（6）降低钢液中硫、磷含量。硫、磷是钢中偏析倾向最大的元素，对钢的危害也最大，因此通过减少钢液中硫、磷含量也可减轻偏析对钢材质量的影响。

2.1.5 Formation and Discharge of Gas

The gas produced in the solidification process is mainly CO, H_2, etc. CO will be produced by the poor deoxidation of molten steel. More moisture in the material will dissolve into the molten steel and increase the hydrogen and oxygen content in the steel. The gas in the steel fails to float up, so it remains in the steel and form to solidified bubbles. If the solidification bubble is close to the surface of the billet, it will form the claw crack in the rolling process. After the liquid steel is completely solidified, hydrogen will still form very small bubbles, which is called flakes. The pressure inside the flakes defects is very high, which is enough to cause small cracks near the flakes and become a hidden danger of steel. Flakes can be eliminated by high temperature diffusion annealing or slow cooling.

2.1.5 气体的形成和排出

凝固过程中产生的气体主要是 CO、H_2 等，钢液脱氧不良会产生 CO，物料潮湿含较多的水分会溶入钢液，增加钢中氢和氧含量，存在于钢中的气体未能上浮，于是残存于钢中形成凝固气泡，若凝固气泡距铸坯表面很近即所谓皮下气泡，皮下气泡在铸坯轧制时会形成爪裂。在钢液完全凝固以后，氢依然会形成很细小的气泡析出，即所谓的白点。白点内压力非常高，足以使白点附近产生细小裂纹而成为钢材的隐患。白点可通过高温扩散退火或缓冷消除。

2.1.6 Formation and Discharge of Inclusions

Inclusions can destroy the continuity of steel and have some influence on the properties of steel. Inclusions can be controlled from two aspects: one is to float as much as possible; the other is to control the shape of inclusions. If the inclusion particles are small, spherical and uniform distributed, the harm is small. Therefore, the influence of inclusion on steel quality can be changed by controlling the composition, shape, quantity, size and distribution of inclusion.

2.1.6 夹杂物的形成和排出

夹杂物破坏钢基体的连续性，对钢的性能有一定的影响。可以从两方面来控制夹杂物：一是夹杂物尽量上浮；二是控制夹杂物形态。如果夹杂物颗粒很小，并呈球状且分布均匀，其危害较小，因此在夹杂物问题不能减少的情况下，可以通过控制夹杂物的成分、形态、数量、尺寸及分布来改变夹杂物对钢质量的影响。

2.1.7 Solidification Shrinkage

Shrinkage is the phenomenon that the volume and size of steel decrease during solidification and cooling. It has a great influence on the formation of defects such as shrinkage, porosity and internal cracks. The shrinkage of liquid steel can be divided into three stages with temperature drop and transformation, namely, liquid shrinkage, solidification shrinkage and solid shrinkage.

Shrinkage consists of two parts: one is volume shrinkage, which is mainly caused by liquid shrinkage and solidification shrinkage. It does not affect the shell size of the billet (ingot), but it can cause shrinkage, porosity or internal crack. The other part is solid shrinkage. The solid shrinkage has the largest shrinkage, which produces thermal stress in the process of temperature drop and tissue stress in the process of phase transformation. The formation of stress is the root of billet crack. Therefore, solid shrinkage has a great influence on the quality of billet.

2.1.7 凝固收缩

钢在凝固和冷却过程中所发生的体积和尺寸减小的现象称为收缩。它对铸坯（钢锭）的热裂、缩孔和疏松等缺陷的形成都有很大的影响。钢液的收缩随温降和相变可分为 3 个阶段，分别是液态收缩、凝固收缩和固态收缩。

收缩体现在铸坯（钢锭）上由两部分组成：一部分是体积收缩，主要由液态收缩和凝固收缩造成的，它不影响铸坯（钢锭）的外壳尺寸，但可以使内部出现缩孔、疏松或内裂。另一部分是固态收缩，固态收缩量最大，在温降过程中产生热应力，在相变过程中产生组织应力，应力的产生是铸坯裂纹的根源。因此，固态收缩对铸坯质量影响较大。

2.1.8 Stress Control during Solidification and Cooling

During the solidification and cooling process, the billet is mainly affected by three kinds of stresses: shrinkage stress, structural stress and mechanical stress, which will lead to billet cracks. Therefore, it is important to understand the stress in the cooling process of billet, to conduct reasonable cooling and to reduce the cooling stress for reducing the cracks in steel and improving the quality of billet.

2.1.8 凝固、冷却过程中应力控制

铸坯在凝固及冷却过程中主要受三种应力的作用，即收缩应力、组织应力和机械应力，这些应力会引发铸坯裂纹。因此，了解铸坯冷却过程的应力，进行合理的冷却，减小冷却应力，对于减少钢中裂纹、提高铸坯质量有着重要意义。

Cracks will be formed when the tensile stress of the billet exceeds the strength limit and the plastic allowable range of the steel in this position. Among the three kinds of stresses, thermal stress and structural stress undoubtedly play a key role, while mechanical stress increases the possibility of cracks.

The surface crack of billet increases the workload of finishing, affects the hot delivery and direct rolling of billet, and even causes the billet to be scrapped. The central crack will reduce the performance of steel and leave hidden dangers to steel, so it is necessary to try to reduce the cracks caused by stress.

(1) Reasonable water distribution and proper cooling system shall be adopted to keep the surface temperature of the billet away from the brittleness range under high temperature. The cooling shall be uniform to prevent the surface of the billet from overheating.

(2) For some alloy steel and steel with strong crack sensitivity, dry cooling can be used to make the surface and center temperature of the billet tend to be the same, and greatly reduce the generation of thermal stress.

(3) The hot cracking tendency of billet can be reduced by adjusting and controlling the composition of molten steel reasonably and reducing the content of harmful elements in steel.

(4) Keep strict roll shape and roll gap accuracy, and reduce roll bulging and centering error between rolls.

当铸坯所承受的拉应力超过该部位钢的强度极限和塑性允许的范围时，就会产生裂纹。在 3 种应力中，热应力、组织应力无疑起了关键作用，机械应力则加大了裂纹产生的可能性。

铸坯的表面裂纹增加了钢坯精整的工作量，影响铸坯的热送和直接轧制，严重时会使铸坯报废；而中心裂纹会降低钢材的性能，给钢材留下隐患，因此必须设法减少由于应力造成的裂纹，措施有：

(1) 采用合理的配水和合适的冷却制度，以使铸坯的表面温度避开高温下的脆性区间，冷却要均匀，防止铸坯表面回热过快。

(2) 对于某些合金钢、裂纹敏感性强的钢种可采用干式冷却，使铸坯表面、心部温度趋于一致，大大减少热应力的产生。

(3) 合理调节和控制钢液成分，降低钢中有害元素的含量，可确保减少铸坯的热裂倾向。

(4) 保持严格的辊子形状和辊缝精度，减少辊间铸坯鼓肚和对中误差。

Exercises:

(1) What are the thermodynamic and dynamic conditions of liquid steel crystallization?

(2) What changes will happen to the molten steel during solidification?

(3) Use the simulation software to check the solidification process of billet and compare the solidification of billet after the parameters change.

思考与习题：

(1) 钢液结晶的热力学条件和动力学条件是什么？

(2) 钢液在凝固过程中会发生哪些变化？

(3) 利用仿真软件完成查看钢坯的凝固过程，对比参数变化后钢坯的凝固情况。

Task 2.2 Solidification Process Control of Billet
任务 2.2 连铸坯凝固过程控制

Mission objectives:

任务目标:

(1) Master the solidification process of slab during continuous casting production.

(1) 掌握连铸生产过程中铸坯的凝固过程。

(2) Understand the influencing factors of solidification of molten steel in the mold and secondary cooling zone.

(2) 了解钢液在结晶器、二冷区凝固的影响因素。

(3) Understand the solidification structure of continuous casting slab, and be able to control the solidification structure of continuous casting slab through operation and parameter adjustment.

(3) 理解连铸坯的凝固结构,并能够通过操作和参数调整控制铸坯凝固结构。

Task Preparation 任务准备

2.2.1 Cognition of Continuous Casting Production Process

Some of the important phenomena which govern the continuous casting process and determine the quality of the product are illustrated in Figure 2-7. Steel flows into the mold through ports in the submerged entry nozzle, which is usually bifurcated. Argon gas is injected into the nozzle to prevent clogging. The resulting bubbles provide buoyancy that greatly affects the flow pattern, both in the nozzle and in the mold. They also collect inclusions and may become entrapped in the solidifying shell, leading to serious surface defects in the final product.

2.2.1 连铸生产过程认知

连铸控制和决定产品质量的过程如图 2-7 所示。钢水通过浸入式水口流入结晶器,吹氩以防止水口堵塞,产生的上浮气泡对喷嘴和结晶器中的钢水流动有很大影响。上浮过程能够收集夹杂物,夹杂物也可能包裹于坯壳中而导致铸坯出现严重的表面缺陷。

The jet leaving the nozzle flows across the mold and impinges against the shell solidifying at the narrow face. The jet carries superheat, which can erode the shell where it impinges on locally thin regions. In the extreme, this may cause a costly breakout, where molten steel bursts through the shell.

钢流离开水口穿过结晶器冲击坯壳表面。钢流具有较高过热度,冲击较薄的坯壳进而

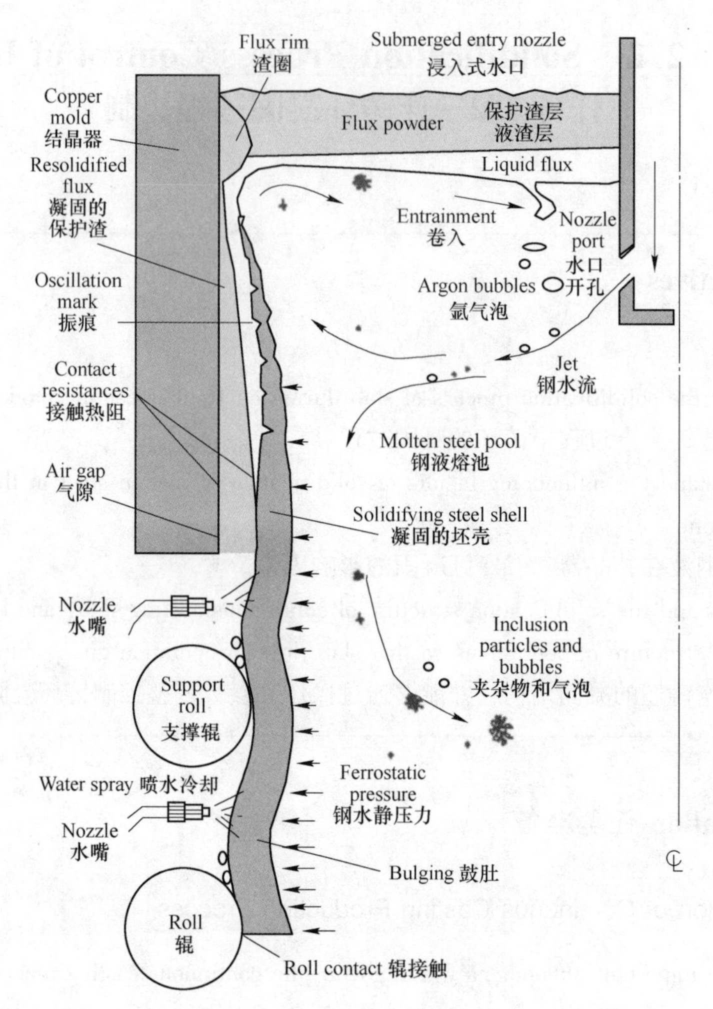

Figure 2-7 Schematic diagram of solidification in the mold

图 2-7 结晶器内凝固情况示意图

腐蚀坯壳。严重时会导致钢水从坯壳流出从而造成漏钢事故。

Typically, the jet impinging on the narrow face splits to flow upwards toward the top free surface and downwards toward the interior of the strand. Flow recirculation zones are created above and below each jet. This flow pattern changes radically with increasing argon injection rate or with the application of electromagnetic forces, which can either brake or stir the liquid. The flow pattern can fluctuate with time, leading to defects, so transient behavior is important.

通常，冲击窄面的钢流流向液面上部和内部，进而形成循环区域，当氩气流量增加或施加电磁感应时情况会发生变化，起到阻碍和搅拌钢水的效果。随着时间的变化，钢水发生波动导致缺陷的产生，因此钢水的瞬时状态是非常重要的。

Liquid flow along the top free surface of the mold is very important to steel quality. The horizontal velocity along the interface induces flow and controls heat transfer in the liquid and solid flux layers, which float on the top free surface. Inadequate liquid flux coverage leads to nonuniform initial solidification and a variety of surface defects.

结晶器上部钢水的流动对钢坯质量有着很大的影响，表面流动的钢水引发液面波动，控制钢水和浮在液面的保护渣之间的传热过程，不足的液态保护渣导致初始坯壳凝固不均和各种表面缺陷。

If the horizontal surface velocity is too large, the shear flow and possible accompanying vortices may entrain liquid flux into the steel. This phenomenon depends greatly on the composition dependent surface tension of the interface and possible presence of gas bubbles, which collect at the interface and may even create a foam. The flux globules then circulate with the steel flow and may later be entrapped into the solidifying shell lower in the caster to form internal solid inclusions.

如果水平方向流速过大，产生的涡流会将覆盖剂带入钢中。这种现象在很大程度上取决于液面的表面张力和其中可能存在的气泡，气泡会在界面处聚集，甚至可能产生泡沫。覆盖剂随着钢液流动被卷入连铸机下方的坯壳中，在钢坯内部形成夹杂物。

The vertical momentum of the steel jet lifts up the interface where it impinges the top free surface. This typically raises the narrow face meniscus, and creates a variation in interface level, or ‘standing wave’, across the mold width. The liquid flux layer tends to become thinner at the high points, with detrimental consequences.

垂直方向运动的钢水冲击表面引起液面上升，通常会使窄面的弯月面高度上升，使结晶器液面振动或穿过结晶器的波纹，液渣层逐渐变薄并产生有害影响。

The molten steel contains solid inclusions, such as alumina. These particles have various shapes and sizes and move through the flow field while colliding to form larger clusters and may attach to bubbles. They either circulate up into the mold flux at the top surface, or are entrapped in the solidifying shell to form embrittling internal defects in the final product.

钢水中含有夹杂物，这些夹杂物具有不同的形状和大小，通过流动碰撞聚集，也可能被气体吸附。一部分随着钢水流动到上部保护渣中，另一部分被坯壳包裹进入铸坯中形成内部脆性缺陷。

Mold powder is added to the top surface to provide thermal and chemical insulation for the molten steel. This oxide-based powder sinters and melts into the top liquid layer that floats on the top free interface of the steel. The melting rate of the powder and the ability of the molten flux to flow and to absorb detrimental alumina inclusions from the steel depends on its composition, governed by time-dependent thermodynamics. Some liquid flux resolidifies against the cold mold wall, creating a solid flux rim which inhibits heat transfer at the meniscus. Other flux is consumed into the gap between the shall and mold by the downward motion of the steel shell, where it encourages uniform heat transfer and helps to prevent sticking.

在结晶器液面加入保护渣可以为钢水提供隔热和成分稳定的作用。以氧化物为主体的粉末经过烧结并熔化，漂浮在钢水上部。保护渣的熔化速度、流动能力和吸收氧化铝有害夹杂的能力取决于其化学成分，并受相关热力学条件的控制。部分覆盖剂在结晶器壁上再次凝固，形成的固态区域抑制了弯月面处的传热。部分覆盖剂进入坯壳和结晶器之间的缝隙而被消耗，其有助于保证传热均匀并减少黏结漏钢。

Periodic oscillation of the mold is needed to prevent sticking of the solidifying shell to the mold walls, and to encourage uniform infiltration of the mold flux into the gap. This oscillation affects the level fluctuations and associated defects. It also creates periodic depressions in the shell surface, called 'oscillation marks', which affect heat transfer and act as initiation sites for cracks.

结晶器通过周期的振动防止坯壳粘在结晶器壁上，帮助保护渣均匀进入气隙中，但振动会影响缺陷的产生，在坯壳表面产生“振痕”，它是坯壳表面产生的周期性凹陷，振痕影响传热过程，也是裂纹产生的初始位置。

Initial solidification occurs at the meniscus and is responsible for the surface quality of the final product. It depends on the time-dependent shape of the meniscus, liquid flux infiltration into the gap, local superheat contained in the flowing steel, conduction of heat through the mold, liquid mold flux and resolidified flux rim, and latent heat evolution. Heat flow is complicated by thermal stresses which bend the shell to create contact resistance, and nucleation undercooling, which accompanies the rapid solidification and controls the initial microstructure.

钢水在弯月面处开始凝固，是影响钢坯表面质量的主要原因。凝固过程受到弯月面的形状、渗入间隙的保护渣、钢水的局部过热、结晶器的热传导、液态和重新凝固的保护渣、潜热释放等因素的影响。随着钢坯的快速凝固和控制初始组织结构，热应力使坯壳弯曲产生接触的阻力和形核过冷使热传递变得复杂。

Further solidification is governed mainly by conduction and radiation across the interfacial gap between the solidifying steel shell and the mold. This gap consists mainly of mold flux layers, which move down the mold at different speeds. It is greatly affected by contact resistances, which depend on the flux properties and shrinkage and bending of the steel shell, which may create an air gap. The gap size is controlled by the amount of taper of the mold walls, which is altered by thermal distortion. In addition to controlling shell growth, these phenomena are important to crack formation in the mold due to thermal stress and mold friction.

钢水之后的凝固过程主要由坯壳和结晶器间隙的热传导和热辐射控制，保护渣在间隙中以不同的速度向下移动。气隙的形成主要取决于保护渣的性能、坯壳收缩和弯曲的过程。结晶器锥度影响气隙的大小，除了控制坯壳生长之外，受到热阻和结晶器摩擦的影响，这些现象还对结晶器内裂纹的形成起着重要的作用。

As solidification progresses, microsegregation of alloying elements occurs between the dendrites as they grow outward to form columnar grains. The rejected solute lowers the local solidification temperature, leaving a thin layer of liquid steel along the grain boundaries, which may later form embrittling precipitates. When liquid feeding cannot compensate for the shrinkage due to solidification, thermal contraction, phase transformations and mechanical forces, then tensile stresses are generated. When the tensile stresses concentrated on the liquid films are high enough to nucleate an interface from the dissolved gases, then a crack will form.

随着凝固过程的进行，向外生长树枝晶之间发生了合金元素的微观偏析。未溶解的成分降低了局部的凝固温度，在晶界边上留下一层很薄的钢液，随后可能形成脆性沉淀物。当液态钢水不能及时补充由于凝固、热收缩、相变和机械力引起的收缩时，应力就会产生。当集中在液膜上的应力足够大时溶解的气体形成界面时，就会产生裂纹。

After the shell exits the mold and moves between successive rolls in the spray zones, it is subject to large surface temperature fluctuations, which cause phase transformations and other microstructural changes that affect its strength and ductility. It also experiences thermal strain and mechanical forces due to ferrostatic pressure, withdrawal, friction against rolls, bending and unbending. These lead to complex internal stress profiles which cause creep and deformation of the shell. This may lead to further depressions on the strand surface, crack formation and propagation.

在钢坯离开结晶器进入连续喷水的辊道移动后，由于表面温度波动会引起相变和影响强度和塑性的微观结构变化。由于钢水静压力、传热、辊道摩擦、弯曲和矫直产生的热应力和机械作用，钢坯内部应力分布不均，坯壳产生蠕变和变形，这些都会导致钢坯表面进一步凹陷造成裂纹的形成和扩展。

Lower in the caster, fluid flow is driven by thermal and solutal buoyancy effects, caused by density differences between the different compositions created by the microsegregation. This flow leads to macrosegregation and associated defects, such as centerline porosity, cracks and undesired property variations.

钢坯高度降低后，显微偏析造成的不同成分的密度不同，导致钢水的流动受到热应力和溶质浮力的影响。宏观偏析及中心疏松、裂纹和其他导致性能变差的缺陷都是由于这种流动而产生的。

2.2.2 Low Power Structure Characteristics of Billet

The solidification of continuous casting billet is equivalent to the solidification of ingot with high aspect ratio, and the billet solidifies while running in the continuous casting machine, forming a long liquid cavity. There are three stages of billet solidification:

2.2.2 铸坯凝固的过程认知

连铸坯凝固相当于高宽比特别大的钢锭凝固，且铸坯在连铸机内边运行边凝固，内部具有很长的液相区域，铸坯凝固分以下 3 个阶段：

(1) The primary shell is formed in the mold. When the molten steel is poured into the water-cooled mold, the cooling speed is very fast (100℃/s) on the meniscus, and a certain thickness of primary shell is formed near the mold wall. With the decrease of temperature, high temperature shell transforms from δ to γ. The shell is in the state of dynamic balance of contraction and bulging, and a stable air gap is formed at the lower part of the mold, the heat transfer is slowed down, and the growth speed of shell is slowed down. At the outlet of the mold, the shell should have a certain thickness (billet) to resist the static pressure of the molten steel and strength to resist the friction.

(1) 结晶器内形成初生坯壳。钢水浇入水冷结晶器，在弯月面冷却速度很快（100℃/s），在器壁附近形成一定厚度的初生坯壳。随着温度下降，高温坯壳发生 $\delta\rightarrow\gamma$ 相变，坯壳处于收缩和鼓肚的动平衡状态，到结晶器下部形成稳定气隙，传热减慢，坯壳生长速度减慢。在结晶器出口，坯壳应具有一定抵抗钢液静压力的厚度（小方坯）和抗摩擦力的强度。

(2) The shell in the secondary cooling zone grows steadily. The casting slab with liquid core is pulled out from the crystallizer and enters the secondary cooling zone to receive water spray cool-

ing. The spray water drops away the heat from the surface of the billet, resulting in a large temperature gradient on the surface and center of the slab. The most rapid heat dissipation is perpendicular to the surface of the slab, which makes the dendrites grow parallel to the surface and form columnar crystals.

（2）二冷区坯壳稳定生长。带液芯的铸坯从结晶器拉出来进入二冷区接受喷水冷却，喷雾水滴从铸坯表面带走热量，使铸坯表面和中心形成了大的温度梯度。垂直于铸坯表面方向散热最快，使树枝晶平行于生长面而形成了柱状晶。

（3）Accelerated solidification at the end of the liquid phase cavity. The dendrites at the solid-liquid interface of the liquid phase cavity are broken by the forced convection of the liquid. The dendrites participate in the circulation of the liquid steel to the middle of the slab. Some of the dendrites melt again, accelerating the disappearance of superheat, and the rest are the core of the equiaxed crystals. In this way, the columnar crystal is connected with the equiaxed crystal deposited in the paste area at the bottom of the liquid cavity, and the liquid core is completely solidified to form the low power structure of the continuous casting billet.

（3）液相穴末端的加速凝固。液相穴的固液交界面的树枝晶被液体的强制对流折断，枝晶片参与钢液循环至铸坯中部，一部分枝晶重新熔化，加速过热度的消失，剩下的作为等轴晶的核心。这样，生长的柱状晶与沉积在液相穴底部糊状区的等轴晶相连接，液芯完全凝固构成了连铸坯低倍结构。

2.2.3 Cognition of Billet Solidification Structure

Generally speaking, the continuous casting billet is composed of chill crystal zone, columnar crystal grain zone and central equiaxed crystal grain zone from the edge to the center (shown in Figure 2-8), which has no essential difference with ingot. Because of the high cooling strength, the chill crystal zone of continuous casting billet is usually thicker than ingot, the columnar crystal is more developed but not as thick as ingot, and the central equiaxed crystal zone is also composed of coarse equiaxed crystal. The whole structure of continuous casting billet is denser than that of ingot, and the grains are finer. The solidification of billet with long liquid phase cavity makes it difficult for feeding molten steel and easy to form center porosity.

2.2.3 铸坯凝固结构认知

一般情况下，连铸坯从边缘到中心是由激冷层、柱状晶带和中心等轴晶带组成，如图2-8所示，与钢锭无本质区别。只是由于冷却强度大，连铸坯的激冷层往往要厚些，柱状晶较发达但不如钢锭那么粗大，中心等轴晶带也是由粗大等轴晶组成。连铸坯整个结构比钢锭致密，晶粒也要细一些，且液相穴很长的铸坯凝固，钢水补缩困难易产生中心疏松。

（1）Chill crystal zone. The surface of billet is composed of fine and free equiaxed grains, also known as fine equiaxed zones. The width of fine equiaxed crystal band is generally 2~5mm, which is formed under the condition of the highest cooling rate at the meniscus of the mold. Its thickness mainly depends on the superheat of molten steel. The higher the casting temperature is, the thinner the chill crystal zone is; the lower the casting temperature is, the thicker the chill crys-

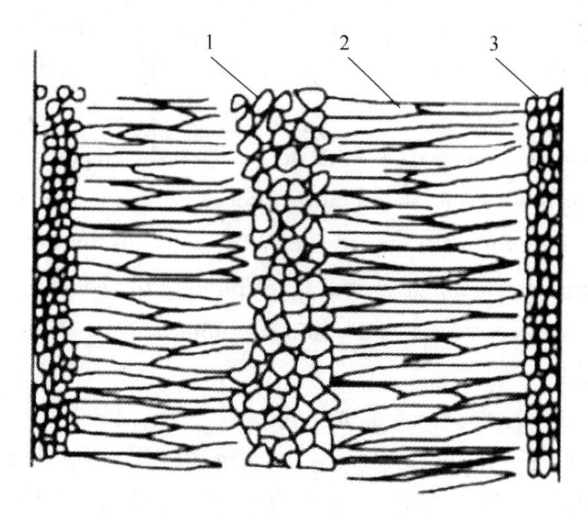

Figure 2-8　Solidification structure of continuous casting billet

1—Central equiaxed crystal grain zone; 2—Columnar crystal grain zone; 3—Chill crystal zone

图 2-8　连铸坯凝固结构示意图

1—中心等轴晶带；2—柱状晶区；3—激冷层

tal zone is.

The molten steel injected into the mold contacts with the wall of the water-cooled copper mold at the meniscus, the cooling rate can reach 100℃/s, the surface of molten steel is strongly cooled, the temperature drops rapidly below the liquidus, and a large degree of supercooling is obtained, so that the nucleation rate of the molten steel greatly exceeds the growth rate of the crystal core. At the same time, the impurities in the mold wall and the undercooled melt provide good conditions for nucleation. A large number of nuclei are formed in the undercooled melt almost at the same time. In the continuous downward movement of the billet, they hinder each other's growth. Therefore, fine equiaxed crystals with different orientations are obtained on the surface of the billet.

（1）激冷层。铸坯表皮由细小等轴晶组成，也称细小等轴晶带。细小等轴晶带宽度一般为2~5mm，它是在结晶器弯月面处冷却速度最高的条件下形成的。其厚度主要取决于钢水过热度，浇铸温度越高，激冷层就越薄；浇铸温度越低，激冷层就厚一些。

注入结晶器内的钢液，在弯月面处与水冷铜结晶器壁相接触，冷却速度可达100℃/s，表层钢液被强烈冷却，温度迅速降到液相线以下，获得了较大的过冷度，使钢液的形核速率大大超过了晶核的长大速率。同时，结晶器壁和过冷熔体中的杂质为形核提供了良好的条件，过冷熔体内几乎同时形成了大量的晶核，在铸坯连续向下的运动中，彼此间妨碍各自的长大，因而，铸坯表层得到不同取向的细小等轴晶。

（2）Columnar crystalgrain zone. The shrinkage in the process of the formation of the chill crystal zone makes the air gap appear in the mold wall of 100~150mm below the meniscus, which reduces the heat transfer speed. At the same time, the internal and external heat dissipation of the molten steel makes the temperature of the chill zone rise and no new crystal nucleus is formed. With the development of directional heat transfer of molten steel, the columnar crystal zone begins to form. The columnar crystal near the chill zone is very fine, and basically does not bifurcate. From the vertical section, the columnar crystal is not completely perpendicular to the surface, but inclines upward at a certain angle (about 10°), from the outer edge to the center, and the number of columnar crystal changes from variable to less. The development of columnar crystal is irregular, and it may form transgranular structure through the center of billet in some parts. For arc caster, the low power structure of billet is asymmetric. Because of the gravity, the crystal sinks and the growth of the outer arc columnar crystal is restrained, the columnar crystal on the inner arc side is longer than that on the outer arc side, and the inner cracks of the billet are often concentrated on the inner arc side.

(2) 柱状晶区。铸坯激冷层形成过程中的收缩，使结晶器弯月面以下约100~150mm的器壁处产生了气隙，降低了传热速度。同时，钢液内部向外散热使激冷层温度升高，不再产生新的晶核。在钢液定向传热得到发展的条件下，柱状晶带开始形成。靠近激冷层的柱状晶很细，基本上不分叉。从纵断面看，柱状晶并不完全垂直于表面而是向上倾斜一定角度（约10°），从外缘向中心，柱状晶个数由多变少呈竹林状。柱状晶的发展是不规则的，在某些部位可能会贯穿铸坯中心形成穿晶结构。对于弧形连铸机，铸坯低倍结构具有不对称性。由于重力作用，晶体下沉，抑制了外弧柱状晶生长，故内弧侧柱状晶比外弧侧要长些，且铸坯内裂纹也常常集中在内弧侧。

Casting temperature and cooling condition have influence on the growth of columnar crystal. If the casting temperature is high, the columnar crystal zone will be wide; if the cooling strength of the secondary cooling zone is increased, the temperature gradient will be increased, and the columnar crystal will be promoted; if the casting section is increased, the temperature gradient will be reduced, and the width of columnar crystal will be reduced.

浇铸温度、冷却条件等对柱状晶生长均有影响。浇铸温度高，柱状晶带就宽；二冷区冷却强度加大，将增加温度梯度，也促进柱状晶发展；铸坯断面加大，则减小温度梯度，从而减小柱状晶的宽度。

(3) Central equiaxed crystal grain zone. With the development of solidification front, the temperature gradient of solidification layer and solidification front decreases gradually, the width of two-phase area expands continuously, and the temperature of molten steel in the center of billet drops to the liquidus temperature, which provides supercooling condition for the crystallization of molten steel in the center. A large number of equiaxed crystals are formed and grow rapidly, forming irregular equiaxed crystal bands, and the grains are larger than those in the chill crystal zone.

(3) 中心等轴晶区。随着凝固前沿的推移，凝固层和凝固前沿的温度梯度逐渐减小，两相区宽度不断扩大，铸坯芯部钢水温度降至液相线温度后，为芯部钢液的结晶提供了过冷度条件，大量等轴晶产生并迅速长大，形成无规则排列的等轴晶带，且晶粒比激冷层的粗大。

Task Implementation 任务实施

2.2.4 Control of Secondary Cooling on Slab Quality

The quality of the secondary cooling zone has a great influence on the output of the caster and the quality of the billet. When the other process conditions are fixed, it can improve the casting speed if the secondary cooling strength increases. The secondary cooling strength is closely related to the defects of the billet (such as internal crack, surface crack, bulging of the slab and rhomboidity deformation).

(1) The uneven cooling results in the rise of shell temperature, and the billet is easy to form intermediate cracks or subcutaneous cracks.

(2) When the surface temperature is too low (for example, less than 900℃), it is easy to form surface transverse cracks.

(3) When the cooling strength is not enough, the straightening crack will be easily formed when the billet is straightened with liquid core.

(4) When the surface temperature of the slab is too high in the secondary cooling zone, the bulging deformation of the slab will easily occur and the center segregation will be aggravated.

(5) The secondary cooling strength is too high, which makes the columnar crystal develop easily to form transgranular structure, and makes the center porosity and segregation aggravate.

Therefore, the cooling requirements for the secondary cooling zone are: high cooling efficiency to accelerate the heat transfer; appropriate water spray to make the surface temperature distribution of the billet uniform; the billet should be completely solidified before straightening; the surface temperature of the billet should be greater than 900℃ during straightening.

2.2.4 二次冷却对铸坯质量的控制

二冷区冷却的好坏对铸机产量和铸坯质量都有很大的影响。当其他工艺条件一定时，二冷强度增加，可提高拉速；而二冷强度又与铸坯缺陷（如内裂纹、表面裂纹、铸坯鼓肚和菱变等）密切相关。

（1）冷却不均匀，导致坯壳温度回升，铸坯易产生中间裂纹或皮下裂纹。

（2）铸坯矫直时表面温度过低（如小于900℃），易产生表面横裂纹。

（3）冷却强度不够，铸坯带液芯矫直易产生矫直裂纹。

（4）二冷区铸坯表面温度过高，板坯易产生鼓肚变形而使中心偏析加重。

（5）二冷强度过大，促使柱状晶发达易形成穿晶结构，使中心疏松和偏析加剧。

因此，对二冷区的冷却要求是：冷却效率要高，以加速热量的传递；喷水量合适，使铸坯表面温度分布均匀；铸坯在矫直前尽可能完全凝固；矫直时铸坯表面温度应大于900℃。

2.2.5 'Small ingot' Structure Control

After the billet enters the secondary cooling zone, the columnar crystal growth caused by the non-uniformity of cooling in the secondary cooling zone often results in a 'solidification bridge' with an interval of 5~10cm in the center of the longitudinal section of the slab, accompanied by porosity and shrinkage (especially small square billet). Because of its similar solidification structure with small ingots, it is called 'small ingot' structure, as shown in Figure 2-9.

2.2.5 “小钢锭”结构控制

铸坯进入二冷区后，二冷区冷却的不均匀性所导致的柱状晶不稳定生长，使铸坯纵断面中心常常出现间隔5~10cm的“凝固桥”，并伴随有疏松和缩孔（尤其是小方坯）。因其与小钢锭的凝固结构相似，故称为“小钢锭”结构，如图2-9所示。

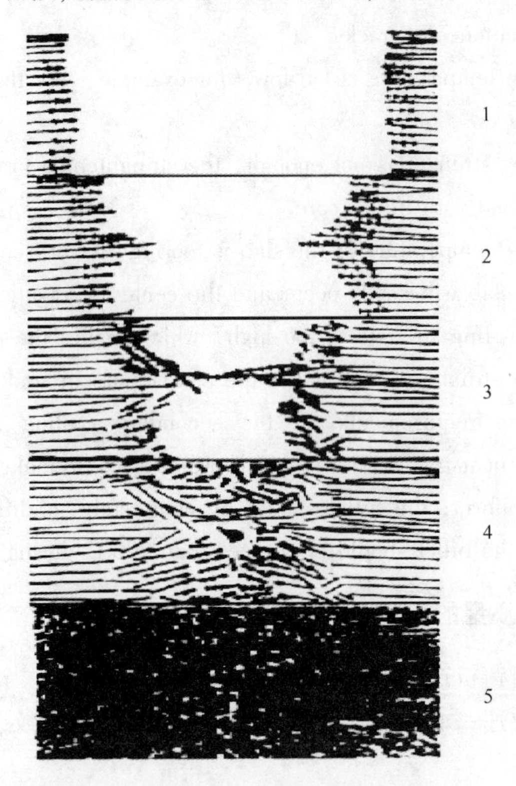

Figure 2-9 Structure formation of ‘small ingot’

1—Uniform growth of columnar crystals; 2—Preferential growth of some columnar crystals; 3—Overlapping bridge of columnar dendrites; 4— ‘Small ingot’ solidifies and forms shrinkage cavity; 5—Macrostructure of actual billet

图2-9 “小钢锭”结构形成示意图

1—柱状晶均匀生长；2—某些柱状晶优先生长；3—柱状树枝晶搭接桥；4—“小钢锭”凝固并产生缩孔；5—实际铸坯的宏观结构

In actual production, measures such as uniform cooling, low superheat casting and electromagnetic stirring can be adopted to reduce or avoid the ‘small ingot’ structure of continuous casting slab.

实际生产中，可采用二冷区铸坯均匀冷却、低过热度浇铸、电磁搅拌等措施来减轻或避免连铸坯的“小钢锭”结构。

2.2.6 Control of Billet Structure

The low times structure of continuous casting billet, especially the excessive development of columnar crystal, not only affects the processing performance of steel, but also affects the mechanical performance and service performance of steel (such as corrosion, welding, etc.). About 75% of the industrial steel is used after hot processing. After hot rolling (forging) of the billet, the shape of the steel has changed during hot processing. There is a large extension along the billet axis. Gasholes and porosity are reduced and dendrites are crushed. At the same time, the distribution of regional segregation in the billet has changed. The plastic inclusions are extended to different degrees, and the brittle inclusions are formed point chain distribution, these changes have an impact on the performance of steel. The influence of equiaxed crystal and columnar crystal on the properties of steel is different. The equiaxed crystal structure is dense, each equiaxed crystal is embedded with each other, the combination is relatively strong, the processing performance is good, and the mechanical properties of steel are isotropic.

2.2.6 铸坯结构的控制

连铸坯的低倍结构，尤其是柱状晶过分发展，既影响钢的加工性能，也影响钢的力学性能和使用性能（如腐蚀、焊接等）。工业用钢大约有75%是经过热加工后使用的，铸坯经过热轧（锻）成材，在热加工时钢的外形发生变化，沿钢坯轴向有较大的延伸，焊合铸坯中的气孔和疏松，压碎树枝晶；与此同时，铸坯中区域偏析的分布有所改变，塑性夹杂物得到不同程度的延伸，脆性夹杂物成点链状分布，这些变化都对钢的性能带来影响。而等轴晶和柱状晶这两种结构对钢性能的影响不同，等轴晶结构致密，各个等轴晶体彼此相互嵌入，结合比较牢，加工性能好，钢材的力学性能呈各向同性。

The relative size between columnar and equiaxed crystal area of the continuous casting billet is mainly determined by casting temperature. When the casting temperature is high, the columnar crystal area will be wide, as shown in Figure 2-10. When the superheat of low carbon steel is greater than 20℃, the width of columnar crystal will increase sharply. Therefore, the liquidus temperature casting close to the steel grade is an effective way to expand the equiaxed crystal region. However, the superheat of molten steel is controlled very low (such as less than 10℃), which is difficult to operate, and it is easy to freeze the nozzle and increase the inclusions in the billet. Therefore, in order to keep the molten steel casting under certain superheat (such as 30℃), the following measures should be taken to expand the equiaxed crystal zone:

连铸坯中柱状晶区和等轴晶区的相对大小主要取决于浇铸温度。浇铸温度高，柱状晶区就宽，如图2-10所示。低碳钢过热度大于20℃，柱状晶宽度就急剧增加，因此，接近钢种的液相线温度浇铸是扩大等轴晶区的有效手段。但是，钢水过热度控制得很低（如小于10℃），在操作上有一定难度，易使水口冻结，且使铸坯夹杂物增加。因此，保持钢水在一定过热度下浇铸（如30℃），为扩大等轴晶区就需采取以下措施：

(1) Micro coolant is added to the mold. Steel strip or micro steel block is added into the mold to eliminate the superheat of molten steel, so that it can solidify rapidly at liquidus temperature and accelerate solidification. Its defect is uneven melting of coolant and easy to contaminate molten steel. For this reason, the Italian Metallurgical Research Center put forward a new technology of accelerating solidification. This method is to feed the core wire into the mold through the core hole of the tundish stopper. The core wire is composed of aluminum, titanium, iron and other

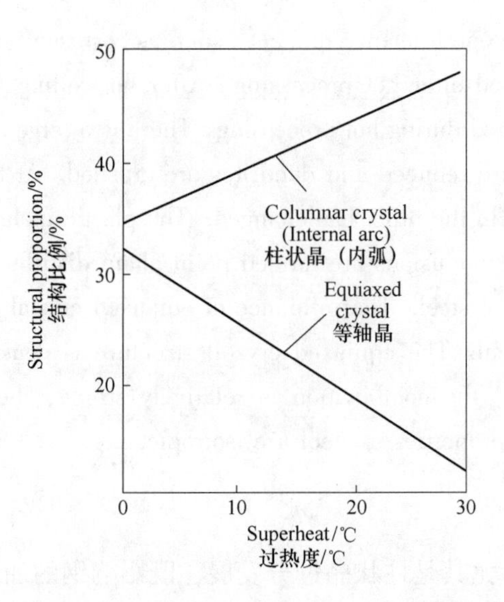

Figure 2-10 Relationship between superheat and solidification structure

图 2-10 过热度与凝固结构的关系

powder, which is used to reduce the superheat of the liquid steel, thus accelerating the solidification and improving the equiaxed crystal rate.

（1）结晶器加入微型冷却剂。结晶器内加入钢带或微型钢块，消除钢水过热度，使其迅速地在液相线温度凝固。其缺陷是冷却剂熔化不均，易污染钢液。为此，意大利冶金研究中心提出了加速凝固的新工艺，此法是通过中间包塞棒芯孔向结晶器喂入包芯线，包芯线成分为铝、钛、铁等粉剂，用以降低钢液的过热度，从而加速凝固，提高等轴晶率。

（2）Spray metal powder. Metal powder of different sizes is injected into the mold to absorb overheating and provide crystal core, enlarge equiaxed crystal area and improve product performance.

（2）喷吹金属粉末。在结晶器内喷入不同尺寸的金属粉，以吸收过热和提供结晶核心，扩大等轴晶区，改善产品性能。

（3）Control the cooling water quantity in the secondary cooling area. Large amount of secondary cooling water, low surface temperature and large temperature gradient inside and outside the billet are conducive to the growth of columnar crystal, and the columnar crystal area is wide. When the amount of secondary cooling water is reduced, the width of columnar crystal area is reduced, and the equiaxed crystal area is increased. Therefore, reducing the amount of secondary cooling water is a factor to inhibit the growth of columnar crystals, but its effect is limited.

（3）控制二冷区冷却水量。二冷水量大，铸坯表面温度低，铸坯内外温度梯度大，有利于柱状晶生长，柱状晶区就宽。而降低二冷水量，可使柱状晶区宽度减少，等轴晶区有所增加。因此，减小二冷水量是抑制柱状晶生长的一个因素，但其效果是有限的。

（4）Add nucleating agent. The nucleating agent was added into the mold to increase the number of crystallized cores and enlarge the equiaxed region. Requirements for nucleating agent:

it is solid at liquid steel temperature; it does not decompose into elements and enter steel at liquid steel temperature; it does not float up and exists at the solidification front; the nucleating agent is wetted with liquid steel as much as possible, and the lattice is close to each other, so that there is adhesion between nucleating agent and liquid steel. Commonly used nucleating agents are Al_2O_3, ZrO_2, TiO_2, V_2O_5, AlN, VN, ZrN, etc.

（4）加入形核剂。在结晶器内加入形核剂，以增加结晶核心数量，扩大等轴晶区。对形核剂的要求有：在钢液温度下为固体；在钢液温度下不分解为元素而进入钢中；不上浮而存在于凝固前沿；形核剂尽可能与钢水润湿，晶格彼此接近，使形核剂与钢液间有黏附作用。常用的形核剂物质有 Al_2O_3、ZrO_2、TiO_2、V_2O_5、AlN、VN、ZrN 等。

（5）Electromagnetic stirring is adopted. The function of electromagnetic stirring is to make the molten steel produce forced convection circulation flow under the action of electromagnetic force, so that the dendrite at the solidification front is fused or broken, and the dendrite fragments, as the core of equiaxed crystal, grow up and expand the equiaxed crystal area, at the same time, the dendrite bridging is eliminated, the porosity and shrinkage in the center of billet are improved, and the center segregation is reduced. Electromagnetic stirring can also eliminate the surface blowholes and inclusions, improve the surface quality and the purity of the steel, reduce the temperature gradient of the liquid steel at the solidification front, that is, allow the liquid steel to have higher superheat, and promote the uniformity of the shell thickness, which will be conducive to prevent the billet from cracks and steel leakage accidents, and improve the casting speed.

（5）采用电磁搅拌。电磁搅拌的作用是在电磁力作用下使钢水产生强制对流循环流动，使凝固前沿的树枝晶熔断或折断，枝晶碎片作为等轴晶核心长大进而扩大等轴晶区，同时消除了树枝晶搭桥，改善中心疏松和缩孔，减轻中心偏析。电磁搅拌还可消除皮下气孔和皮下夹杂，改善表面质量，提高纯净度，减小凝固前沿钢液温度梯度，即允许钢液有更高的过热度，促进坯壳厚度的均匀性，这将有利于防止铸坯产生裂纹和发生漏钢事故，可提高拉速。

Electromagnetic stirrer can be installed in mold（M-EMS），secondary cooling zone（S-EMS）and final stage of solidification（F-EMS）. In addition to improving the center porosity and center segregation, M-EMS almost includes all the functions of electromagnetic stirring, which has the most obvious effect and the best effect. The role of S-EMS is to restir the two-phase region composed of equiaxed crystal and liquid steel near the end of solidification, which can further reduce the central segregation of billet including high carbon steel. According to different quality requirements, it can be used to mix separately or in combination.

电磁搅拌器可以安装在结晶器（M-EMS）、二冷区（S-EMS）和凝固末端（F-EMS）。M-EMS 除了改善铸坯中心疏松、中心偏析外几乎包括了电磁搅拌的所有功能，作用最明显，效果最好。S-EMS 是在接近凝固终点再次搅拌由等轴晶和钢液组成的两相区，可以进一步减轻包括高碳钢在内的中心偏析。根据质量要求的不同，可以采用单独搅拌，也可以是联合搅拌。

Exercises:

(1) Describe the solidification process of billet and analyze how to control the billet structure.

(2) How is the shell formed in the mold?

(3) Use the simulation software to check the solidification process of billet and observe the production process of billet.

思考与习题:

(1) 简述钢坯的凝固过程，分析钢坯结构的控制方法。

(2) 结晶器内坯壳是如何形成的?

(3) 利用仿真软件完成查看钢坯的凝固过程，观察钢坯的生产过程。

Project 3　Operation of Continuous Casting Equipment
项目3　连铸设备操作

Task 3.1　Operation of Ladle and Ladle Turret
任务3.1　钢包及钢包回转台操作

Mission objectives:

任务目标:

(1) Master the structure and main parameters of ladle and ladle turret.

(1) 掌握钢包及钢包回转台的结构和主要参数。

(2) Be able to operate and maintain relevant equipment.

(2) 能够对相关设备进行操作和维护。

(3) Be able to complete the installation of ladle nozzle, ladle baking and other operations.

(3) 能够完成安装长水口、钢包烘烤等操作。

Task Preparation 任务准备

3.1.1　Cognition of Ladle Equipment

Ladle is a device for holding and carrying molten steel and for casting. It is also a container for steel refining. In the process of casting, the steel flow can be controlled by opening the size of the ladle nozzle, and it can also be used for the refining outside the furnace to improve the quality of the molten steel. With the development of metallurgy, the function of ladle can be simply summarized as: holding, carrying, refining, casting molten steel, tipping, slag pouring and landing.

3.1.1　钢包设备认知

钢包又称盛钢桶、钢水包、大包等；它是用于盛装、运载钢液并进行浇铸的设备，也是钢液炉外精炼的容器。在浇铸过程中可以通过开启水口的大小来控制钢流量，还可以用于炉外精炼改善钢水质量。随着冶金的进步，钢包的作用可以简洁地总结为盛放、运载、精炼、浇铸钢水、倾翻、倒渣和落地放置等功能。

3.1.1.1 Capacity of Ladle

The ladle capacity is determined by the tapping of the steelmaking furnace or the lifting capacity of the tapping crane. The determination of the capacity of the ladle shall meet the following requirements:

(1) The ladle can hold rated molten steel volume of the steelmaking furnace;

(2) The ladle can hold 10% of the excess capacity (including a certain amount of slag liquid);

(3) When the ladle is filled with steel, the slag liquid surface to the ladle edge shall keep 100~200mm safe space;

(4) The ladle used for refining should be designed according to the needs of the secondary refining.

3.1.1.1 钢包的容量

钢包容量是根据炼钢炉的出钢量或者出钢跨起重机的起重量来确定。一般钢包容量的确定应满足以下要求：

(1) 钢包能容纳炼钢炉的额定钢水量；

(2) 钢包还能容纳10%的超装量（其中包括一定量的渣液）；

(3) 当钢包盛装钢水后，其渣液面到钢包上沿应留有100~200mm的安全空间。

(4) 精炼用的钢包应根据两次精炼的需要来设计。

3.1.1.2 Shape of Ladle

Ladle is a barrel shaped container with a circular cross section, its shape and size are shown in Table 3-1, and the determination shall meet the following requirements:

(1) The ratio of the diameter to the height of the ladle. When the ladle capacity is fixed, in order to reduce the heat loss of the ladle, the area of the inner surface of the ladle should be reduced. Therefore, the ratio of the average inner diameter to the height of the ladle is generally 0.9~1.1.

(2) Taper. In order to facilitate the pouring of residual steel and residue from the ladle and take out the bottom agglomerate, the ladle is generally made with a taper of 10%~15%.

(3) Ladle shape. In order to facilitate gas and non metallic inclusions floating and exclusion, and reduce impact force of steel flow when casting, which requires that the shape of ladle cannot be made into thin and high shape.

3.1.1.2 钢包的形状

钢包是一个具有圆形截面的桶状容器，其形状与尺寸见表3-1，应满足以下要求：

(1) 钢包的直径与高度之比。钢包容量一定时，为了减少钢包的散热损失，应使钢包的内表面面积缩小，因此钢包的平均内径与高度之比，一般选择0.9~1.1。

（2）锥度。为了便于钢水浇铸后能从钢包内倒出残钢、残渣以及取出包底凝块，一般钢包内部制成上大下小，并具有10%~15%的锥度。

（3）钢包外形。为了便于钢水中气体和非金属夹杂物的上浮和排除，并降低开浇时钢流的冲击力，要求钢包的外形不能做成细高形状。

Table 3-1　Main parameter of ladle

表 3-1　钢包主要数据

Capacity 容量 /t	Volume 容积 $/m^3 \cdot t^{-1}$	Metal part weight 金属部分质量/t	Lining quality 包衬质量 /t	Total weight 总重/t	Diameter to height ratio 直径与高度之比	Taper 锥度 /%	Thickness of steel plate of wall 包壁钢板厚度/mm	Thickness of steel plate of bottom 包底钢板厚度/mm	Ladle height 钢包高度 /mm
25	4. 65	6. 64	4. 89	11. 53	0. 98	10. 0	18	22	2316
50	9. 16	15. 47	6. 75	22. 22	1. 01	10. 0	22	26	2652
90	15. 32	18. 69	16. 40	35. 09	0. 96	10. 0	24	32	3228
130	20. 50	29. 00	16. 50	45. 50	0. 956	7. 5	26	34	3860
200	30. 80	40. 60	29. 00	69. 60	0. 845	8. 2	28	38	4659
260	40. 20	47. 50	32. 00	79. 50	0. 935	6. 5	28	38	4750

3. 1. 1. 3　Structure of Ladle

The ladle is mainly composed of ladle body and refractory lining.

(1) Ladle body. The ladle body is composed of shell, reinforced hoop, trunnion, slag overflow hole, steel injection hole, ventilation hole and tipping device, as shown in Figure 3-1.

(2) Ladle refractory. The ladle lining is in contact with high temperature molten steel and slag for a long time, which is eroded by the casting flow and slag, especially for the ladle used for refining outside the furnace. The erosion of the lining not only reduces the life of the ladle, but also increases the number of inclusions in the molten steel. Therefore, it is of great significance to select proper refractory for ladle lining to improve steel quality, stable operation and production efficiency.

3. 1. 1. 3　钢包的结构

钢包主要由钢包本体和耐火材料组成。

（1）钢包本体。钢包本体由外壳、加强箍、耳轴（作用是吊运钢包）、溢渣口、注钢口、透气口、倾翻装置部件组成，如图 3-1 所示。

（2）钢包耐火材料。钢包内衬与高温钢水、炉渣长时间接触，受到铸流冲刷和炉渣侵蚀，尤其是用于炉外精炼的钢包，侵蚀更为严重；内衬的侵蚀不仅会降低钢包的寿命，还会增加钢液中的夹杂物数量。因此，钢包的内衬选用合适的耐火材料对改善钢的质量、稳定操作、提高生产效率有着重要的意义。

Ladle refractories are working layer, permanent layer and thermal insulation layer from inside

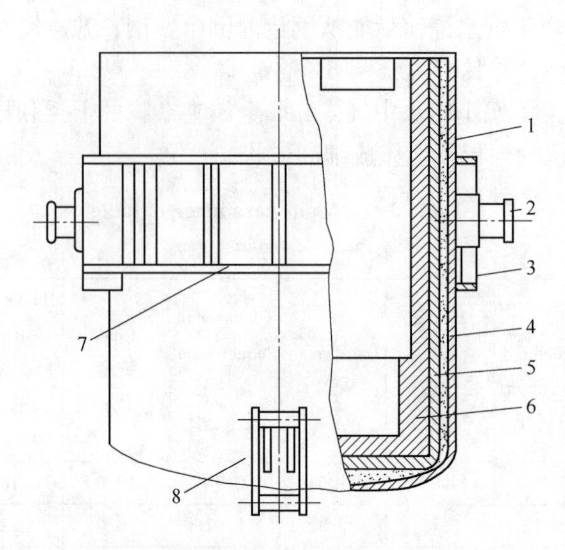

Figure 3-1　Ladle structure

1—Ladle shell; 2—Trunnion; 3—Supporting pedestal; 4—Insulation layer; 5—Permanent layer; 6—Working layer; 7—Reinforced hoop; 8—Tipping rings

图 3-1　钢包结构

1—包壳；2—耳轴；3—支撑座；4—保温层；5—永久层；6—工作层；7—腰箍；8—倾翻吊环

to outside. Because the working conditions of each layer of ladle lining are different, the selection and masonry of refractory materials are also different.

钢包耐火材料从内到外依次为工作层、永久层、保温层。由于钢包内衬各层的工作条件不同，因此耐火材料的选择及砌筑也不同。

1) The function of thermal insulation layer is to keep warm, so as to reduce heat loss. It is close to the steel plate of shell, with a thickness of about 10～15mm. It is usually built with asbestos board or polycrystalline fire-resistant fiber board.

2) The permanent layer is also called non-working layer, with a thickness of 30～60mm. In order to prevent the ladle burn through accident, it is generally built by clay brick with certain thermal insulation performance or high alumina brick.

3) The working layer is in direct contact with molten steel and slag, which is subject to chemical erosion, mechanical erosion, quench and heat, and spalling caused by it. Use magnesia carbon brick and high aluminum brick to build comprehensively.

It should be noted that the ladle must be fully baked before used.

1）保温层的作用是保温，以减少热损失。它紧贴外壳钢板，厚约 10～15mm，常用石棉板或多晶耐火纤维板砌筑。

2）永久层也称非工作层，永久层厚为 30～60mm。为了防止钢包烧穿事故，一般由有一定保温性能的黏土砖或高铝砖砌筑。

3）工作层直接与钢液、炉渣接触，受到化学侵蚀、机械冲刷和急冷急热作用及由其引起的剥落。使用镁碳砖、高铝砖等综合砌包。

需要注意的是，钢包必须经过充分烘烤才可使用。

3.1.1.4 Ladle nozzle

(1) Ladle slide gate. Set a ladle nozzle in the ladle bottom, which known as steel injection hole can pour out the molten steel. It is mainly composed of nozzle bricks installed around. As shown in Figure 3-2. It is through the relative movement between two upper and lower slide plates with nozzle holes, so as to achieve the purpose of opening and closing and adjusting the flow of molten steel. Because of the high temperature steel, high pressure and erosion, the quality of the slide plate brick is very strict.

(2) Ladle long nozzle. Long nozzle, also known as protective tube, is used between ladle and tundish to protect steel flow from secondary oxidation, prevent steel flow splashing and open pouring slag rolling, which has obvious effect on improving steel quality. The use of long nozzle can also reduce the temperature drop of tundish steel, which is beneficial to control the superheat of molten steel, improve the macrostructure and operation conditions of billet.

Reoxidation: When the air contact with the liquid steel in the ladle and tundish, injection flow and mold, the molten steel will be reoxidated and form oxide film or oxide slag scale on the surface of the molten steel during continuous casting.

3.1.1.4 钢包水口

(1) 钢包滑动水口。在钢包底部一侧设置一个钢包水口，它可使钢水流出，又称注钢口。主要由周围安装的水口砖组成，如图 3-2 所示。它是通过两块带水口孔的上、下滑板砖之间相对移动，从而达到开闭、调节钢水流量大小的目的。由于滑板砖承受高温钢水、高压力及冲蚀作用，故对质量要求十分严格。

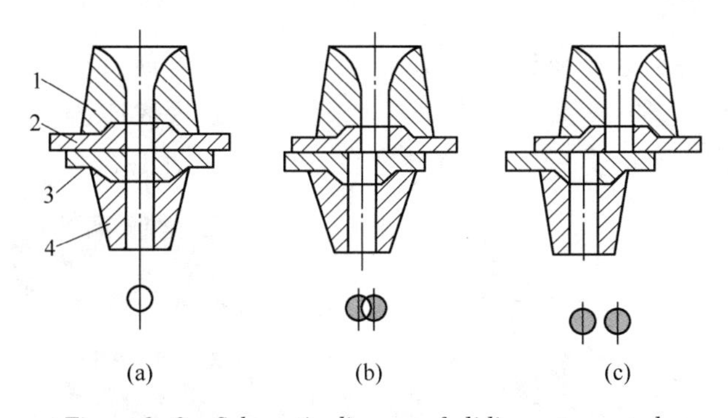

Figure 3-2 Schematic diagram of sliding gate control

(a) Fully open; (b) Half open; (c) Fully closed

1—Upper nozzle; 2—Upper slide plate; 3—Lower slide plate; 4—Lower nozzle

图 3-2 滑动水口控制原理示意图

(a) 全开; (b) 半开; (c) 全闭

1—上水口; 2—上滑板; 3—下滑板; 4—下水口

(2) 钢包长水口。长水口又称保护套管，长水口用于钢包与中间包之间，保护钢流不受二次氧化，防止钢流飞溅以及敞开浇铸的卷渣问题对提高钢质量有明显效果。使用长水口还可以减少中间包钢水温降，对合理控制钢水过热度、改善铸坯低倍组织和操作条件都有利。

二次氧化是指连铸过程中，钢包和中间包的钢液和铸流以及结晶器内的钢液与空气接触发生氧化，在钢液表面产生氧化膜或氧化渣皮的过程。

At present, there are two kinds of long nozzle: fused quartz and aluminum carbon.

1) Fused quartz long nozzle. The main component is SiO_2. It can be used for pouring general steel grades without baking. Steel grades with high manganese content should not be used.

2) Aluminum carbon long nozzle. This kind of nozzle is mainly made of corundum and graphite, and its main component is Al_2O_3. It is suitable for casting special steel with good corrosion resistance, little pollution to molten steel and needs baking before use.

目前常用的长水口有熔融石英质和铝碳质两种。

1）熔融石英质长水口。主要成分是 SiO_2。使用前可以不烘烤，用于浇铸一般钢种，含锰高的钢种不宜使用。

2）铝碳质长水口。这类水口主要是由刚玉和石墨为主要原料制成的产品，主要成分为 Al_2O_3。适合浇铸特殊钢种，耐侵蚀性能好，对钢液污染小，使用前烘烤。

3.1.2 Cognition of Ladle Turret

The function of ladle turret is to store and support ladle. In the casting process, the ladle can be changed and transferred to the top of the tundish through rotation, which creates conditions for continuous casting of multiple furnaces sequence casting. It is common equipment for carrying and casting in modern continuous casting, which is usually set between the receiving span and the casting span column line, as shown in Figure 3-3.

3.1.2 钢包回转台认知

钢包回转台的作用是存放、支撑钢包。它在浇铸过程可通过转动，实现钢包之间的更换、并转送至中间包的上方，为多炉连浇创造条件。它是现代连铸中应用较普遍的运载和浇铸的设备，通常设置于接受跨与浇铸跨柱列线之间，如图 3-3 所示。

Figure 3-3 Ladle turret

图 3-3 钢包回转台

3.1.2.1 Structure of Ladle Turret

The ladle turret is mainly composed of steel structure, rotary drive device, rotary clamping device, lifting device, weighing device, lubricating device and accident driving device.

3.1.2.1 钢包回转台的结构

钢包回转台主要由钢结构部分、回转驱动装置、回转夹紧装置、升降装置、称量装置、润滑装置及事故驱动装置等部件组成。

3.1.2.2 Main Parameters of Ladle Turret

The main parameters of ladle turret are: bearing capacity, rotation speed (generally, the rotation speed of ladle turret is 1r/min), rotation radius, ladle lifting stroke (generally 600~800mm) and ladle lifting speed (generally 1.2~1.8m/min).

3.1.2.2 钢包回转台的主要参数

钢包回转台的主要参数有：承载能力、回转速度（一般钢包回转台的回转转速为1r/min）、回转半径、钢包升降行程（通常为600~800mm）和钢包升降速度（一般为1.2~1.8m/min）。

Task Implementation 任务实施

3.1.3 Installation of Ladle Long Nozzle

At present, the installation of long nozzle mainly adopts lever type fixed device. Before baking the tundish, put the long nozzle into the support ring of the lever mechanism, and then bake with the tundish. As shown in Figure 3-4, after the ladle drainage and starting casting are normal, the long nozzle is rotated to be connected with the lower nozzle of the ladle slide gate, and make the long nozzle and the lower ladle nozzle in close contact. Connect the argon blowing sealing sleeve and supply argon to seal the connection of the long nozzle.

3.1.3 钢包长水口的安装

目前长水口的安装主要采用杠杆式固定装置。在中间包烘烤前，先将长水口放入杠杆机构的托圈内，然后与中间包一起烘烤。如图3-4所示，在钢包引流开浇正常后，旋转长水口与钢包滑动水口的下水口对接，使长水口与钢包下水口紧密接触。连接吹氩密封套，并供氩气对长水口连接处进行氩封。

3.1.4 Judge Whether the Ladle is Out of Service

(1) After finishing last furnace casting, pour out the remaining steel residue.

(2) Observe whether there is red part around and at the bottom of the ladle shell.

(3) Observe whether there are local cavities and serious corrosion in ladle lining.

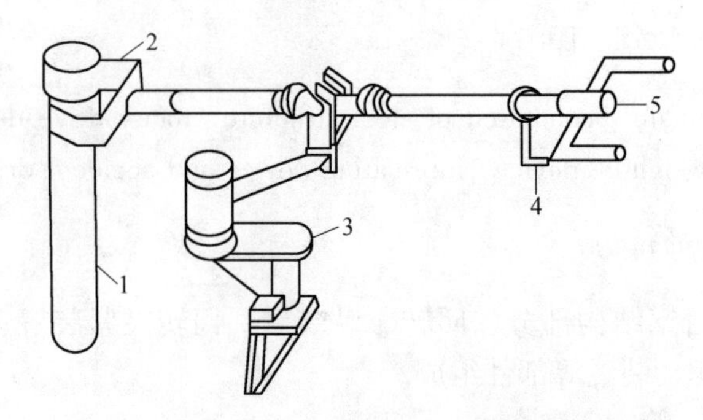

Figure 3-4 Installation diagram of ladle long nozzle

1—Long nozzle; 2—Supporting ring; 3—Support; 4—Counterweight; 5—Operating rod

图 3-4 长水口的安装示意图

1—长水口；2—托圈；3—支座；4—配重；5—操作杆

(4) Use visual inspection or simple measuring tools to detect the depth and diameter of the cavity and erosion.

(5) According to the standard of ladle out of service, make a comprehensive judgment to decide whether to make hot repair, cold repair or out of service.

3.1.4 钢包是否停用的判断

(1) 上炉浇钢完毕，倒尽余钢残渣。

(2) 观察钢包外壳四周及底部有无发红部分。

(3) 观察钢包内衬有否局部空洞及严重蚀损处。

(4) 采用目测或简易测量工具探测空洞及蚀损处的深度和直径。

(5) 与钢包停用标准相对照，综合进行判断，决定热补、冷补或是停用。

Criteria for ladle disuse: it is necessary to make a comprehensive judgment on whether the ladle is disused at a later stage. The thickness of the lining after erosion is usually estimated by tapping and observing the redness of the shell. The corresponding safety residual lining thickness should be determined for ladles of different volumes. At the same time, the cracks of the slag line, lower working layer, bottom bricks and base bricks should be checked. In addition, if some parts are seriously eroded, they should also be stopped according to the specific situation.

钢包判断停用的标准：钢包用到后期是否停用，必须经过综合判断。通常采用敲击法和观察包壳发红程度来估计侵蚀后的包衬厚度，对不同容量的钢包应确定相应的安全残衬厚度，同时应检查渣线、下部工作层、包底及座砖的砖缝，然后根据其中一个或几个部位的侵蚀情况来决定。如果某些部位侵蚀严重，也应视具体情况而停用。

3.1.5 Ladle Cleaning and Maintain

The residual steel residue inside the used ladle must be completely removed, otherwise it will seriously affect the quality requirements of the next furnace steel, so the ladle shall be cleaned.

After the residual steel is removed, the severely eroded parts shall be repaired. In order to improve the service life of ladle and prevent steel leakage, hot repair should be carried out in time. The residual steel and residue shall be cleaned before repair.

(1) After finishing casting in the last furnace, pour out the residual steel residue in the ladle as soon as possible.

(2) Clean up cold steel residue in the hole and nozzle in time.

(3) If there is cold steel in the bottom of the ladle, the ladle must be recumbent and the cold steel must be melted and removed with oxygen.

(4) Check the damage of ladle slag line, bottom, wall and base, repair and maintain in time.

(5) Lift the ladle to the baking position to bake or set aside.

3.1.5 钢包清理与维护

使用过的钢包，其内部的残钢残渣必须要彻底清除，否则将严重影响下一炉钢的质量要求，因此要对钢包进行清理。

清除残钢残渣后，应修补侵蚀严重的部位。为了提高钢包的使用寿命及防止漏钢，应该及时进行热修。修补前应将该处的残钢、残渣清理干净。

(1) 上一炉浇铸完毕，尽快将钢包内余钢残渣倒清。

(2) 及时清理包口冷钢残渣。

(3) 若包底有冷钢则必须将钢包横卧，用氧气将冷钢进行熔化清除。

(4) 检查钢包渣线、包底、包壁、座砖损坏情况，及时进行修补及维护。

(5) 将钢包吊至烘烤位置进行烘烤或待用。

3.1.6 Use and Maintenance of Ladle Sliding Nozzle

By changing the relative position of the upper and lower slide plate holes, we can adjust the flow of molten steel. The steps of replacing the ladle sliding nozzle as follows:

(1) Clean up the residual steel and residue in the upper and lower sliding plate grooves and the upper and lower nozzle.

(2) Install the upper slide plate.

(3) Install the lower nozzle and lower sliding plate.

(4) Adjust the clearance between upper and lower sliding plates.

(5) Repair the ladle slide plate.

3.1.6 钢包滑动水口的使用与维护

钢包滑动水口通过改变上、下滑板孔的相对位置，调节浇铸钢水流量。更换钢包滑动水口如下：

(1) 清理上下滑板槽内及上下水口内残钢、残渣。

(2) 安装上滑板。

(3) 安装下水口、下滑板。

(4) 调节上、下滑板间隙。

(5) 修复钢包滑板。

3.1.7 Operation of Ladle Baking

Before receiving steel, the ladle must be baked in the position of ladle baking device according to the specified ladle baking system. Through baking treatment, it can not only remove the moisture in the refractory lining of the ladle, improve the scour and erosion resistance of the refractory lining, but also improve the service temperature of the refractory lining of the ladle, reduce the damage of the refractory lining, and reduce the temperature drop of the molten steel of the ladle.

3.1.7 钢包烘烤操作

钢包在受钢前必须在钢包烘烤装置的工位上，按规定的钢包烘烤制度进行烘烤作业。通过烘烤处理，既能够去除钢包耐火衬中的水分，提高耐火衬的耐冲刷、耐侵蚀性能；又能够提高钢包耐火衬的使用温度，减少耐火衬的破损，减少钢包的钢水温降。

3.1.7.1 Baking Medium

The fuel medium mainly includes oil fuel medium and gas fuel medium.

(1) Oil fuel medium mainly includes heavy oil and light diesel oil.

(2) Gas fuel medium mainly includes coal gas and natural gas.

Combustion supporting medium mainly includes compressed air and oxygen.

3.1.7.1 烘烤介质

燃料介质主要有燃油类燃料介质和燃气类燃料介质。

(1) 燃油类燃料介质主要有重油和轻柴油。

(2) 燃气类燃料介质主要有煤气和天然气。

助燃介质主要有压缩空气和氧气。

3.1.7.2 Ladle Baking Operation Steps

(1) Lift the ladle to the baking position.

(2) Ignite the open fire and place it at the nozzle of the toaster.

(3) Slowly open the gas valve to ignite the gas, then adjust the gas valve to make the flame reach the required level, and then open the air for combustion.

(4) After the new ladle is baked as required, close the baking equipment. Close the air valve first, then close the gas valve and stop baking.

3.1.7.2 钢包烘烤操作步骤

(1) 把钢包吊至烘包位置。

(2) 点燃明火，并将明火放置于烘烤器的喷口处。

(3) 缓慢开启煤气阀门使煤气点燃，随后调节煤气阀使火焰达到所需程度，再开空气进行助燃。

(4) 待新钢包按要求进行烘烤完毕后，关闭烘烤器。关闭烘烤器时先关闭空气，随后关闭煤气阀，停止烘烤。

3.1.7.3 Precautions

(1) Strictly control the opening degree of gas and air valve to make the temperature rise slowly, and the flame temperature is appropriate to prevent the lining from cracking and peeling off.

(2) Baking requirements are shown in Table 3-2.

(3) Gas should be turned on first when turning on and turned off later when turning off. When using heavy oil as fuel, pay attention to whether the heavy oil is fully atomized, and whether the mixture of fuel and air is reasonable, so that the heavy oil can be fully burned.

3.1.7.3 注意事项

(1) 严格控制煤气和空气阀开启度，使温度缓慢上升，火焰温度适宜，防止包衬开裂剥落。

(2) 烘烤要求见表3-2。

(3) 煤气使用开启时要先开，关闭时要后关。用重油做燃料时，应注意重油是否充分雾化，燃料与空气的混合要合理，使重油充分燃烧为宜。

Table 3-2 Ladle baking requirements

表3-2 钢包烘烤要求

Fire intensity 火势	Distance between flame and bottom 火焰与包底距离	Baking time/h 烘烤时间/h
Small fire 小火	600mm	8
Medium fire 中火	200~300mm	8
Large fire 大火	The flame bounces back to the ladle wall after it hits the ladle bottom directly 火焰直射包底后反弹到钢包壁	20

3.1.8 Operation of Ladle Turret

(1) The rotary table can rotate 360° forward and reverse at will, but it can only start to rotate when the ladle reaches a certain height.

(2) When the rotation of the turntable in one direction is not completely stopped, the operation in the opposite direction is not allowed.

(3) When sitting in the ladle, the operator should be careful to avoid excessive impact on the rotary table.

(4) Regularly check whether the lubrication of each lubricating point is normal.

(5) Check all steel structures irregularly, such as fork arm rotating disk, tower base and rotating ring, etc., and handle in time in case of cracks or deformation. The main welds shall be subject to ultrasonic or radiographic inspection every year, and the defective welds shall be subject to follow-up inspection to pay close attention to the expansion trend.

(6) Regularly check whether the bolts of each fastener are loose, especially the prestressed

anchor bolts shall be inspected at random every year, and any problem found shall be handled in time.

(7) Regularly check whether the lifting hydraulic cylinder and hydraulic joint leak oil, whether the action is normal, and whether the spherical thrust bearing is seriously worn and damaged.

(8) Regularly check whether the operation of each transmission part and each moving part is flexible and normal.

(9) Regularly test run the accident drive device, and check the operation and air pressure of the pneumatic motor.

(10) Check whether the pneumatic clamping device is worn or damaged and whether the action is flexible and normal regularly.

3.1.8 钢包回转台操作

(1) 回转台可以正反360°任意旋转，但必须在钢包升到一定高度时，才能开始旋转。

(2) 当回转台朝一个方向旋转未完全停止时，不允许反方向操作。

(3) 在坐包时，应该小心操作避免对回转台产生过大的冲击。

(4) 定期检查各润滑点的润滑是否正常。

(5) 不定期检查各钢结构，如叉臂旋转盘、塔座和回转环等，发现有开裂或变形等缺陷时，要及时处理。对主要焊缝应每年进行超声波或射线探伤，对有缺陷的焊缝应进行跟踪检查，密切注意其是否有扩展趋势。

(6) 定期检查各紧固件的螺栓有无松动现象，特别是预应力地脚螺栓要每年进行抽检，发现问题及时处理。

(7) 定期检查升降液压缸及液压接头是否漏油，动作是否正常，其球面推力轴承是否严重磨损和损坏。

(8) 定期检查各传动部位以及各活动部位运作是否灵活正常。

(9) 定期试运转事故驱动装置，检查气动马达的运转及气压等情况。

(10) 要定期检查气动夹紧装置有无磨损、损坏现象，动作是否灵活、正常。

Exercises:

(1) What is the structure and function of the ladle?

(2) How to check and maintain the ladle after use?

(3) Use simulation software to complete the preparation of ladle, rotate ladle turret and install long nozzle for casting.

思考与习题:

(1) 钢包的结构和作用是什么?

(2) 钢包使用后应该如何检查和维护?

(3) 利用仿真软件完成钢包的准备工作，旋转钢包回转台并安装长水口准备浇铸。

Task 3.2 Operation of Tundish and Tundish Car
任务3.2 中间包及中间包车操作

Mission objectives:

任务目标:

(1) Master the structure and main parameters of tundish and tundish car.

(1) 掌握中间包及中间包车的结构和主要参数。

(2) Be able to operate and maintain relevant equipment.

(2) 能够对相关设备进行操作和维护。

(3) Be able to complete tundish baking, replacement of submerged entry nozzle, assembly of stopper rod, etc.

(3) 能够完成中间包烘烤、更换浸入式水口、装配塞棒等操作。

Task Preparation 任务准备

3.2.1 Cognition of Tundish

Tundish is a transition device for molten steel casting between the ladle and the mold, which is used to receive molten steel in the ladle and inject molten steel into the mold. The main functions of tundish are reducing pressure, stabilizing flow, inclusions removal, separating flow, storing steel and tundish metallurgy.

3.2.1 中间包认知

中间包也称为中包、中间罐。中间包是位于钢包与结晶器之间用于钢液浇铸的过渡装置,用于接受钢包钢水及向结晶器内注入钢水。中间包具有减压、稳流、去夹杂、分流、贮钢和中间包冶金等重要作用。

3.2.1.1 Main Parameters of Tundish

The main technical parameters of tundish are tundish capacity, tundish height, tundish length and width, inner wall slope of tundish, tundish nozzle diameter and nozzle spacing, etc.

3.2.1.1 中间包的主要参数

中间包的主要技术参数有中间包容量、中间包高度、中间包长度和宽度、中间包内壁斜度、中间包水口直径及水口间距等。

(1) Tundish capacity. The capacity of tundish is mainly determined according to the capacity of ladle, section size of billet, strand number of tundish, casting speed, time of replacing ladle during sequence casting, residence time of molten steel in tundish and other factors. Generally, the capacity of tundish is 20%~40% of the ladle capacity.

（1）中间包容量。中间包容量主要根据钢包容量、铸坯断面尺寸、中间包流数、浇铸速度、多炉连浇时更换钢包的时间、钢水在中间包内的停留时间等因素来确定。一般中间包的容量为钢包容量的 20%~40%。

(2) Tundish height. The height of tundish mainly depends on the depth requirement of molten steel in the tundish. Generally, the depth of molten steel in the tundish is 600~1200mm. During sequence continuous casting, the minimum liquid level of molten steel in the tundish shall not be less than 300mm, so as to avoid vortex of molten steel and its involvement in slag liquid. In addition, a distance of 100~200mm shall be left between the liquid level of molten steel and the upper entrance of the tundish.

（2）中间包高度。中间包高度主要取决于钢水在包内的深度要求，一般中间包内钢水的深度为 600~1200mm。在多炉连浇时中间包内的最低钢水液面深度不应小于 300mm，以免钢水产生旋涡，并卷入渣液；另外钢水液面至中间包上口之间应留 100~200mm 的距离。

(3) Tundish length. The length of tundish mainly depends on the distance between the tundish nozzles.

（3）中间包长度。中间包长度主要取决于中间包的水口位置距离。

(4) Tundish width. The tundish width can be determined according to the amount of molten steel that should be stored in the tundish.

（4）中间包宽度。中间包宽度可根据中间包应存放的钢水量来确定。

(5) Slope of inner wall of tundish. The inner wall of the tundish has a certain slope, which is beneficial to clean up the residual steel and residue in the tundish. Generally, the slope of inner wall of tundish is 10% ~ 20%.

（5）中间包内壁斜度。中间包内壁有一定的斜度，其作用是有利于清理中间包内的残钢、残渣。一般中间包内壁斜度为 10%~20%。

(6) Diameter and spacing of tundish nozzle. The diameter of tundish nozzle shall be determined according to the liquid steel flow required by the maximum casting speed of continuous casting machine.

The distance between nozzles is a special technical parameter of the tundish of the multi strands casting machine. The distance of spacing nozzles is the center distance between adjacent molds in the multi strands casting machine.

（6）中间包水口直径及水口间距。中间包水口直径应根据连铸机的最大拉速所需要的钢液流量来确定。

水口间距是多流连铸机中间包特有的技术参数，水口间距是指多流连铸机中相邻的各个晶器之间的中心距离。

3.2.1.2 Tundish Structure

The structure and shape of the tundish shall have the minimum heat dissipation area and good heat preservation performance. The commonly used section shapes of tundish are circle, ellipse, triangle, rectangle and T shape.

The shape of tundish shall be simple to facilitate hoisting, storage, masonry, cleaning and other operations. According to the number of the nozzle flow, it can be divided into single flow, multi flow, etc., the number of the nozzle flow of the tundish is generally 1~5 flow.

3.2.1.2 中间包结构

中间包的结构、形状应具有最小的散热面积，良好的保温性能。一般常用的中间包断面形状为圆形、椭圆形、三角形、矩形和T形等。

中间包的形状力求简单，以便于吊装、存放、砌筑、清理等操作。按其水口流数可分单流、多流等，中间包的水口流数一般为1~5流。

The three-dimensional structure of the tundish is shown in Figure 3-5. The shell is made of welded steel plate, with refractory lining inside, hooks andtrunnions on both sides of the tundish for lifting, and cushion under the trunnion for stable sitting on the tundish car.

中间包的立体结构如图3-5所示。外壳是由钢板焊接而成，其内衬砌有耐火材料，中间包的两侧有吊钩和耳轴，便于吊运；耳轴下面还有坐垫，以稳定地坐在中间包车上。

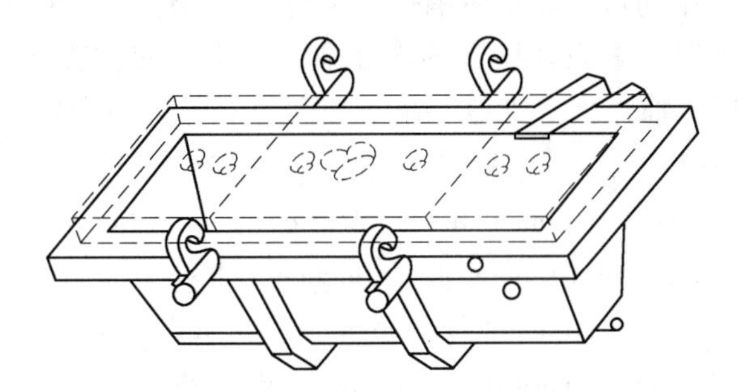

Figure 3-5 Three dimensional structure of tundish

图3-5 中间包立体结构示意图

The tundish structure is mainly composed of shell, cover, lining, nozzle and nozzle control mechanism (sliding nozzle mechanism, stopper mechanism), as shown in Figure 3-6.

中间包结构主要由包壳、包盖、内衬、水口及水口控制机构（滑动水口机构、塞棒机构）等装置组成，如图3-6所示。

(1) Tundish shell. The tundish shell is a box structure made of welded steel plate. In order to make the tundish shell has enough rigidity and can work in high temperature and heavy load environment without deformation for many times, reinforcing hoops and ribs should be welded outside the shell. Weld trunnions on both sides or around to support and lift tundish. In addition, there are overflow holes and tapping holes for molten steel, and many vent holes are drilled on the shell steel plate.

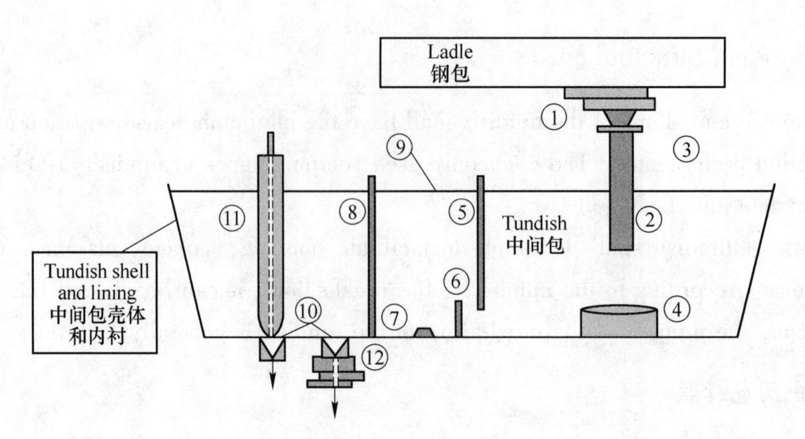

Figure 3-6　Tundish structure

1—Two-plate ladle slidegate system; 2—Ladle shroud (submerged); 3—Extraneous gas shrouding ring; 4—Impact pad (turbulance reducing); 5—Weir; 6—Dam; 7—Porous plug ('bubble curtain'); 8—Baffle; 9—Tundish flux; 10—Inner tundish well nozzle; 11—Stopper rod (Ar purged); 12—Three-plate slidegate system

图 3-6　中间包结构示意图

1—双滑板水口；2—钢包保护水口；3—气体保护环；4—湍流抑制器；5—挡堰；6—挡坝；7—多孔塞（气幕）；8—隔板；9—中间包覆盖剂；10—中间包内水口；11—塞棒（氩气吹扫）；12—三滑板滑动水口

（1）中间包壳体。中间包壳体是由钢板焊接而成的箱形结构件，为了使中间包壳体具有足够的刚度，能在高温、重载环境下多次作业不变形，应在壳体外部焊加固箍和加强筋。在两侧或四周焊接耳轴，用来支撑和吊运中间包。另外还设置钢水溢流孔、出钢孔，在壳体钢板上钻削许多排气孔。

(2) Tundish cover. The function of the tundish cover is to keep warm, prevent the molten steel splashing in the tundish, and reduce the influence of high temperature radiation and baking of molten steel in the tundish on adjacent equipment (such as the bottom of the ladle, the rotating arm of the ladle turret, the manipulator device of the long nozzle, etc.). The tundish cover is welded with steel plate and is lined with masonry refractory. The tundish cover is provided with a molten steel injection hole, a stopper hole, a tundish baking hole, a temperature measuring hole and a lifting ring for hoisting.

（2）中间包盖。中间包盖的作用是保温、防止中间包内的钢水飞溅，减少邻近设备（如钢包底部、钢包回转台转臂、长水口机械手装置等）受到包内钢水高温辐射、烘烤的影响。中间包盖用钢板焊接而成，其内衬砌筑耐火材料。在中间包盖上设置钢水注入孔、塞棒孔、中间包烘烤孔、测温孔及吊装用吊环。

(3) Refractory lining of tundish. The refractory lining of tundish consists of working layer, permanent layer and thermal insulation layer. The thermal insulation layer is made of asbestos board and thermal insulation brick, and the permanent layer is made of clay brick, or cast a whole; the working layer is in direct contact with the liquid steel, mainly made of high aluminum brick, magnesia brick or thermal insulation plate, and the surface of the working layer is sprayed

with 10~30mm thick fire-resistant mud to improve the protection effect.

The refractory lining of the large capacity tundish is also provided with dam or weir to isolate the disturbance of the ladle casting flow to the molten steel in the tundish, so that the flow of the molten steel in the tundish is more reasonable, which is more conducive to the floating of non-metallic inclusions in the molten steel, so as to improve the purity.

（3）中间包耐火衬。中间包耐火衬由工作层、永久层和绝热层等组成。其中绝热层用石棉板、保温砖砌筑而成，永久层用黏土砖砌筑，或用浇铸料整体浇铸成形；工作层与钢液直接接触，主要使用高铝砖、镁质砖或绝热板进行砌筑，还可以在工作层砌砖表面喷涂10~30mm 厚的耐火泥浆以提高保护效果。

大容量中间包的耐火衬中还设置挡墙，隔离钢包的铸流对中间包内钢水的扰动，使中间包内钢水的流动更趋合理，更有利于钢水中非金属夹杂物的上浮，从而提高纯净度。

3.2.2 Tundish Nozzle and Nozzle Control Mechanism

The tundish nozzle can control the flow of steel from tundish to mold, which is usually controlled by stopper rod, sliding nozzle, sizing nozzle, submerged entry nozzle, etc. It is very important to select the proper control mode of steel flow according to the steel grades.

3.2.2 中间包水口及水口控制机构

中间包水口可以控制从中间包流入结晶器的钢流大小，通常采用塞棒、滑动水口、定径水口、浸入式水口等来进行控制。根据所浇钢种，选择合适的钢流控制方式，对于高效生产连铸坯是非常重要的。

3.2.2.1 Tundish Stopper Rod

Stopper rod is a refractory bar installed in the tundish to control the opening and closing of the nozzle and the flow of molten steel by the lifting displacement. It is composed of rod core, sleeve brick and plug brick. The rod core is usually made of plain carbon steel round steel with a diameter of 30~60mm. The upper end is connected with the cross arm of the lifting mechanism by bolt, the lower end is connected with the plug brick by thread or pin, and the middle sleeve brick. Figure 3-7 is the schematic diagram of stopper structure and lifting device. The stopper rod shall be laid carefully and used after baking and drying for more than 48h to avoid steel leakage caused by rupture of refractory material.

3.2.2.1 中间包塞棒

塞棒是装在中间包内靠升降位移控制水口开闭及钢水流量的耐火材料棒。它由棒芯、袖砖和塞头砖组成。棒芯通常由直径为 30~60mm 的普碳钢圆钢加工而成，上端靠螺栓与升降机构的横臂连接，下端靠螺纹或销钉与塞头砖连接，中间套袖砖，如图 3-7 所示。塞棒需仔细砌筑，并经 48h 以上的烘烤干燥后使用，以避免耐火材料炸裂造成漏钢事故。

The stopper control mechanism can be either manual or automatic. In the manual mode, the liquid steel flow is controlled by raising or lowering the stopper rod position by the handle. The au-

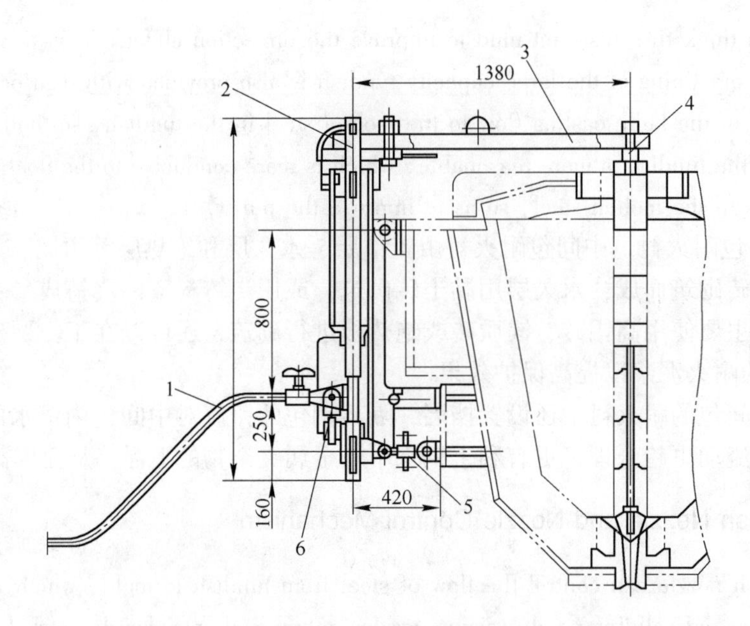

Figure 3-7 Diagram of tundish stopper mechanism

1—Control handle; 2—Lifting slide rod; 3—Cross beam; 4—Stopper core rod; 5—Bracket adjusting device; 6—Sector gear

图 3-7 中间包塞棒机构简图

1—操纵手柄；2—升降滑杆；3—横梁；4—塞棒芯杆；5—支架调整装置；6—扇形齿轮

tomatic control adopts hydraulic cylinder or motor, which can be linked with the mold liquid level control to achieve the purpose of controlling the liquid steel flow.

The stoppers in tundish are easy to soften, deform and even break due to immersion in molten steel for a long time. For this reason, zirconium carbon or aluminum carbon refractories are generally used, which are integrally formed under high pressure.

In order to improve the service life of the stopper rod, the steel pipe can be inserted into the stopper rod center to introduce compressed air or argon for cooling, as shown in Figure 3-8, which can not only control the injection flow, but also prevent the nozzle from blocking and pure steel.

塞棒控制机构可以为手动，也可以自动控制。手动模式通过手柄提高或降低塞棒位置来控制钢水流量大小。自动控制采用液压缸或电机，可以与结晶器液面控制联动，以达到控制钢水流量的目的。

中间包塞棒，由于长时间在钢水中浸泡，容易软化变形，甚至断裂造成事故。为此，一般用锆-碳或铝碳质耐火材料，经过高压整体成形。

为提高塞棒使用寿命，可在塞棒中心插入钢管引入压缩空气或氩气进行冷却，如图3-8，这样不仅可以控制注流，还可以在一定程度上起到防止水口堵塞、纯净钢液的作用。

3.2.2.2 Tundish Sliding Gate

The sliding gate controls the steel flow through the relative position between the sliding plates. Compared with the stopper rod, it is safe and reliable, can accurately control the steel flow, and is

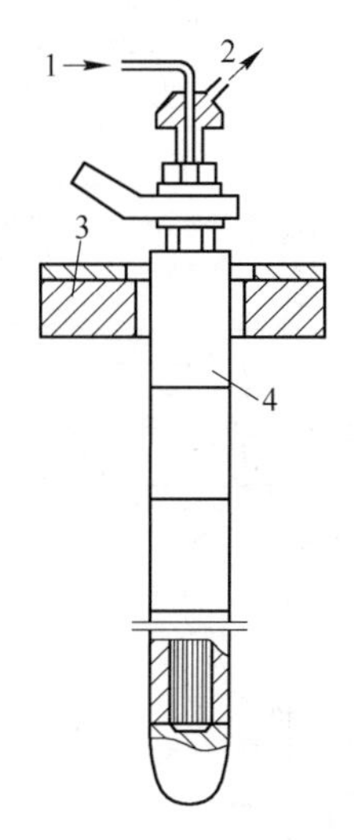

Figure 3-8　Air cooling diagram of stopper

1—Air inlet; 2—Air outlet; 3—Tundish cover; 4—Stopper

图 3-8　塞棒空气冷却图

1—空气入口；2—空气出口；3—中间包盖；4—塞棒

conducive to automation. The tundish mainly adopts three-layer sliding plate to control the steel flow, as shown in Figure 3-9. The sliding gate is mainly composed of upper nozzle and sliding plate, lower nozzle and sliding plate and intermediate sliding plate. When working, the upper and lower slide plates are fixed, and the middle slide plate is used to control the injection flow.

3. 2. 2. 2　中间包滑动水口

滑动水口是通过滑板之间相对位置来控制钢流的，相比塞棒而言安全可靠，能精确控制钢流，有利于实现自动化。中间包主要采用三层式滑板控制钢流，如图 3-9 所示，其结构主要由上水口和上滑板、下水口和下滑板及中间滑板等组成。工作时上下滑板固定不动，利用中间的滑板活动来控制注流。

3. 2. 2. 3　Submerged Entry Nozzle

The submerged entry nozzle is located between the tundish and the mold, the upper end of the nozzle is connected with the tundish, and the lower end is inserted into the mold steel water. It insulates the contact between the casting flow and the air, prevents the splashing caused by the impact of the casting flow to the steel surface, and prevents the reoxidation. Through the selection of nozzle shape, the flow state of molten steel in the mold can be adjusted to promote the separation of inclusions and improve the quality of steel.

3. 2. 2. 3　浸入式水口

浸入式水口位于中间包和结晶器之间，水口上端与中间包相连，下端插入结晶器钢水中。它隔绝了铸流与空气的接触，防止铸流冲击到钢液面引起飞溅，杜绝二次氧化。通过水

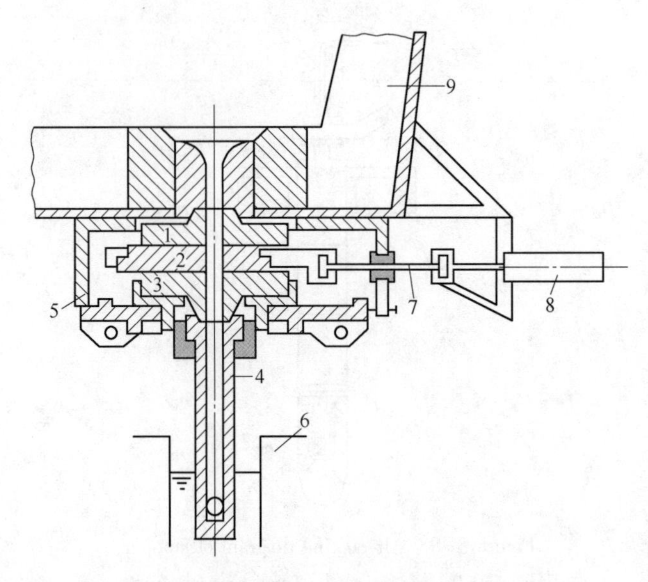

Figure 3-9 Reciprocating sliding gate

1—Upper fixed sliding plate; 2—Movable sliding plate; 3—Lower fixed sliding plate; 4—Submerged nozzle; 5—Sliding nozzle box; 6—Mold; 7—Connecting rod; 8—Oil cylinder; 9—Tundish

图 3-9 往复式滑动水口

1—上固定滑板；2—活动滑板；3—下固定滑扳；4—浸入式水口；5—滑动水口箱体；6—结晶器；7—连杆；8—油缸；9—中间包

口形状的选择，可以调整钢液在结晶器内的流动状态，以促进夹杂物的分离，提高钢的质量。

(1) Parameters of submerged entry nozzle. The type and parameters of submerged nozzle mainly depend on the size and shape of casting section, casting speed and steel grade. For all kinds of casters, the submerged nozzle is in the center of the mold section. The change of the center position and immersion depth of the submerged nozzle will cause the change of the molten steel flow pattern in the mold. Generally, immersion depth of submerged nozzle is (125±25) mm.

(1) 浸入式水口的参数。浸入式水口类型及参数的确定主要取决于浇铸断面的大小、形状、铸坯拉速以及钢种等。对于各种连铸机，浸入式水口都处于结晶器断面的中心位置。浸入式水口的对中位置和浸入深度的变化将引起结晶器内钢液流态的变化。一般浸入式水口浸入深度为 (125±25) mm。

If the position of the submerged nozzle is not right, the flow state of the molten steel in the mold will be asymmetric, and the billet will be prone to longitudinal cracks. Generally, the centering deviation is not more than ±1mm.

浸入式水口位置不对中，会使结晶器内钢液的流动状态不对称，热中心偏离，易使铸坯产生纵裂纹。通常要求对中偏差不大于±1mm。

If the immersion depth of the nozzle is too shallow, the heat center will move up and the steel level will be active, which will easily cause slag entrapment; if the immersion depth of the nozzle is too deep, the heat center will move down, which will cause bad slag formation and even damage the solidified shell of the casting billet, which will cause steel leakage.

水口浸入深度过浅，使热中心上移，钢液面活跃，容易引起卷渣；水口浸入深度过深，使热中心下移，会引起化渣不良，甚至会破坏铸坯凝固壳引起漏钢。

(2) Classification of submerged entry nozzles. According to the material, the submerged entry nozzle can be divided into two types:

(2) 浸入式水口的分类。按照材质可以将浸入式水口分为两类：

1) Fused silica submerged entry nozzle. The features of the long nozzle are: good thermal shock resistance, high mechanical strength and chemical stability, and acid resistance slag erosion resistance. Do not need baking before use. This kind of long nozzle is suitable for casting general steel grades, not suitable for casting steel grades with high manganese content.

1) 熔融石英质浸入式水口。这种长水口的特点是：抗热冲击性好，有较高的机械强度和化学稳定性好，耐酸性渣侵蚀性。在使用前不必烘烤。这种长水口适用于浇铸一般钢种，不适宜浇铸锰含量较高的钢种。

2) Aluminum carbon submerged entry nozzle. This kind of nozzle is mainly made of corundum and graphite. It is characterized by strong adaptability to steel grades, especially suitable for casting special steel, and little pollution to molten steel. The material of the nozzle can also be adjusted according to the casting time and steel grade. Aluminum carbon submerged entry nozzle can only be used after baking, otherwise there is a risk of cracking. In addition, there is an aluminum carbon submerged nozzle that can be used without baking.

2) 铝碳质浸入式水口。这类水口主要是由刚玉和石墨为主要原料制成的产品。它的特点是对钢种的适应性强，特别适合于浇铸特殊钢，对钢水污染小。该水口的材质，还可以根据浇铸时间的长短和钢种进行调整。铝碳质浸入式水口，一般在使用前需烘烤后才能使用，否则有开裂的危险。另外，还有一种不经烘烤即能使用的铝碳质浸入式水口。

At present, the most commonly used submerged nozzles are single hole straight barrel type and double side hole type. As shown in Figure 3-10, the straight hole type nozzle is generally used for square billets or small slabs; the side hole type nozzle is mainly used for square billets or slabs, and its side hole inclination angle has three kinds of upward, downward or horizontal.

目前，使用最多的浸入式水口有单孔直筒型和双侧孔式两种，如图 3-10 所示，直孔式水口一般用于方坯或小板坯；侧孔式水口主要用于大方坯或板坯，其侧孔倾角有向上倾斜、向下倾斜或呈水平状三种。

3.2.3 Cognition of Tundish Car

The tundish car is used to support, transport and replace the tundish equipment, as shown in Figure 3-11. The car structure shall be conducive to casting, slag bailing, oxygen burning and other operations; at the same time, it shall also have the functions of traverse, lifting adjustment and weighing.

3.2.3 中间包车认知

中间包车是用来支撑、运输、更换中间包的设备，如图 3-11 所示，车的结构要有利

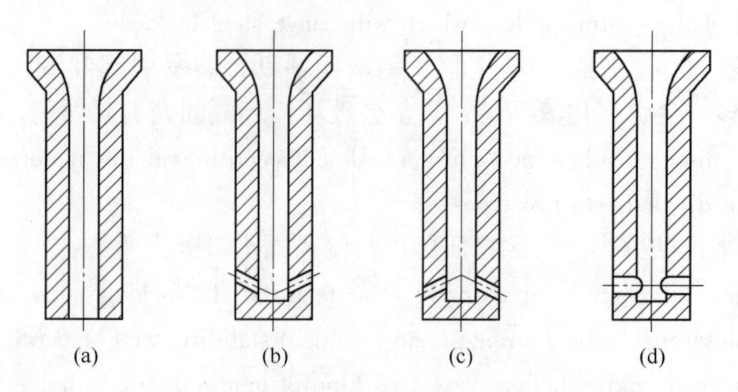

Figure 3-10 Basic types of submerged entry nozzle

(a) Single hole straight cylindrical nozzle; (b) Side hole upward inclined nozzle; (c) Side hole downward inclined nozzle; (d) The side hole horizontal nozzle

图 3-10 浸入式水口基本类型

(a) 单孔直筒形水口;(b) 侧孔向上倾斜状水口;(c) 侧孔向下倾斜呈倒 Y 形水口;(d) 侧孔呈水平状水口

于浇铸、捞渣和烧氧等操作;同时还应具有横移、升降调节和称量功能。

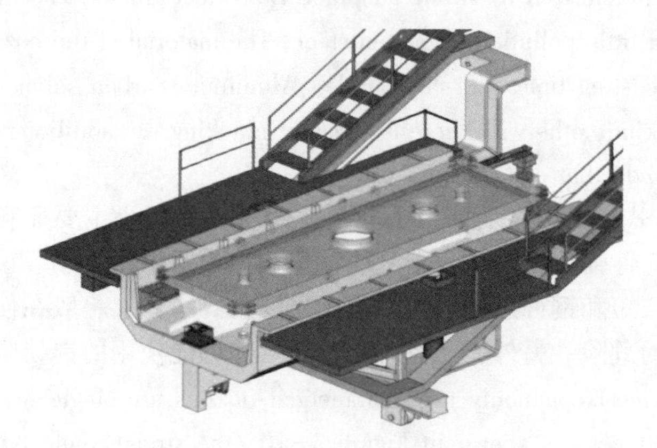

Figure 3-11 Tundish car structure

图 3-11 中间包车结构图

According to the location of tundish nozzle, main beam and track, tundish car can be divided into two types: suspended type and gate type.

中间包车按中间包水口、中间包车的主梁和轨道的位置,分为悬吊式和门式两种类型。

The frame is a saddle frame welded by steel plate, which makes enough space around the submerged nozzle of tundish, so that the operator can observe, sample, add protective slag and remove the residue of steel level in the mold near the mold.

The lifting device can make the tundish rise and fall. The lifting frame for placing the tundish is supported by four screw jacks and driven by two motors through two universal joint shafts. The two groups of motor drive system are connected by bevel gear box and connecting shaft, which has good synchronization and self-locking.

车架是钢板焊接的鞍形框架,这种结构使得中间包浸入式水口周围具有足够的空间,便于操作人员靠近结晶器进行观察、取样、加保护渣及去除结晶器内钢液面残渣。

升降装置能使中间包上升、下降。放置中间包的升降框架由四台丝杆千斤顶支撑，由两台电机通过两根万向接轴驱动。两组电动机驱动系统用锥齿轮箱和连接轴连接起来，具有良好的同步性和自锁性。

Task Implementation 任务实施

3.2.4 Installation of Tundish Stopper Rod

3.2.4 中间包塞棒的安装

Operation steps:

(1) Clean up. Remove the steel and slag on the collet (cross beam), check whether there are any sundries in the hole and edge of the nozzle, and use compressed air to clean the area around the nozzle brick.

操作步骤：

(1) 清理。清除夹头（横梁）上的粘钢粘渣，检查水口砖孔内及其边缘有无杂物，用压缩空气将水口砖周围吹扫干净。

(2) Check and adjust the lifting mechanism.

(2) 检查并调节升降机构。

(3) Assembling:

1) In the process of handling and assembly, the stopper shall be handled with care and shall not be collided. Take out the stopper from the drying room, and fix the stopper on the lifting mechanism chuck with nut washer.

2) During assembly, the graphite sleeve with thread in the threaded hole of the stopper head shall be installed flat, and it shall be closely contacted after being inserted into the bolt without loosening. After the bolt is installed, a layer of cement shall be evenly applied on the stopper rod head, the metal round plate shall be covered, the nut shall be tightened, and it shall be placed naturally for 4~8h. The assembly shall be carried out in a special place, vertical assembly shall be adopted, and the stopper rod shall not be placed horizontally.

3) Special tools shall be used for stopper crane. When it is hoisted to the installation position, it shall be aligned with the installation hole. Only the bolt at the upper part of stopper rod shall be close to the support arm, then the fixing nut shall be tightened and the argon gas pipe shall be connected.

4) Check. Check whether the stopper rod is installed accurately and whether the stopper head brick on the stopper rod is tightly connected with the nozzle.

5) Knock down the bottom pin of pillow. When the stopper rod head is closely matched with the nozzle, the pin at the bottom of the pillow can be knocked tightly.

(3) 装配：

1) 塞棒在搬运组装过程中，应轻拿轻放，不得碰撞。从烘房内取出塞棒，并用螺帽垫圈将塞棒固定在升降机构夹头上。

2）组装时塞棒头部螺纹孔中的带丝口石墨套要装平整，套入螺栓后应紧密接触，不得松动，上完螺栓在塞棒头部均匀涂抹一层胶泥，盖上金属圆板，拧紧螺帽，自然放置4~8h。组装在专用地点进行，采用立式组装，塞棒不允许平放。

3）塞棒吊运用专用工具，吊至安装位置要对准安装孔，将塞棒上部螺栓与支撑臂靠紧，然后拧紧固定螺帽，接上氩气管。

4）检查。检查塞棒位置安装是否准确，塞棒上的塞头砖与水口是否密合。

5）敲紧枕底销。当塞头与水口配合严密后，方可将枕底销敲紧。

Matters needing attention:

(1) The stopper rod shall not be installed vertically in the center of the nozzle brick.

(2) Thestopper head brick shall be closely matched with the nozzle.

(3) Do not use too much force when installing the stopper to avoid breaking.

(4) Before use, the stopper and the submerged nozzle should be baked together, and the rapid temperature rise is required to reach about 1000℃. Before baking, put fire-resistant fiber cylinder on the stopper rod to make the temperature uniform.

注意事项：

(1) 塞杆安装不要垂直对准水口砖中心。

(2) 塞头砖与水口要配合严密。

(3) 塞棒安装时不要用力过猛，以免断裂。

(4) 塞棒在使用前与浸入式水口要一起烘烤，要求快速升温达到1000℃左右。烘烤前在塞棒外套上耐火纤维筒以使温度均匀。

3.2.5 Installation of Tundish Submerged Entry Nozzle

3.2.5 中间包浸入式水口的安装

3.2.5.1 Installation of Integral Submerged Nozzle

The installation of the integral submerged nozzle is carried out in the preparation process before the tundish baking:

(1) Select the appropriate material according to the steel type, and check the size of submerged nozzle according to the drawing.

(2) Apply mud evenly on the conical surface combined with the nozzle sleeve brick.

(3) Install the nozzle vertically (without skewing) into the nozzle sleeve brick.

(4) Use mallet to knock the nozzle brick into the sleeve brick, so that the upper plane of the nozzle brick and the upper plane of the nozzle sleeve brick meet the process requirements.

(5) Remove the excess mud on the nozzle brick and sleeve brick.

(6) For the multi strand caster, the distance between the lower end of the integral nozzle and the mold should be adjusted when installing the integral nozzle.

3.2.5.1 整体式浸入式水口的安装

整体式浸入式水口的安装在中间包烘烤前的准备过程中进行：

（1）按钢种选择合适材质，按图纸检查浸入式水口尺寸。

（2）在与水口套筒砖结合的锥面上均匀涂抹泥料。

（3）将水口垂直（不能歪斜）装入水口套筒砖内。

（4）用木槌将水口砖敲入套筒砖内，使水口砖上平面与水口套筒砖上平面位置符合工艺要求。

（5）清除水口砖、套筒砖上多余泥料。

（6）多流连铸机，安装整体水口需校正整体水口下端间距，要与结晶器间距相一致。

3.2.5.2 Installation of Split Type Submerged Nozzle for Billet Continuous Casting

The split submerged nozzle can be installed in production or installed before casting. Its operation procedure is the same as that of the whole submerged nozzle except that there is one more set of nozzle.

3.2.5.2 小方坯连铸分体式浸入式水口安装

分体式浸入式水口可以在生产中进行安装，也可先安装好再开浇，其操作程序比整体浸入式水口多一个套水口的工序外，其余相同。

3.2.6 Tundish Immersion Nozzle Position Adjustment

Operation steps:

(1) After the tundish stops baking, check the tundish steel flow control device and quickly drive the tundish truck to the casting position.

(2) Adjust the left and right positions of the tundish nozzle through the tundish traveling mechanism; adjust the internal and external arc positions of the nozzle through the traverse device on the tundish car.

(3) Adjust the immersion depth of long nozzle through tundish lifting device or other lifting device.

(4) During the start-up casting, the submerged nozzle is at the required start-up position, and the submerged nozzle will be lowered to the casting position after the start-up casting is normal.

(5) In the casting process, the immersion depth can be adjusted by adjusting the height of the submerged nozzle within the allowable immersion depth range, so as to improve the service life of the submerged nozzle.

3.2.6 中间包浸入式水口位置调整

操作步骤：

（1）中间包停烘后，检查中间包钢流控制装置，迅速将中间包车开至浇铸位置。

（2）通过中间包车行走机构调整中间包水口的左右位置；通过中间包车上的横移装置

调整水口的内、外弧位置。

(3) 通过中间包升降装置或其他起重装置调整长水口浸入深度。

(4) 开浇时，浸入式水口处于要求的开浇位置，待开浇正常后将浸入式水口降至浇铸位。

(5) 浇铸过程中，在允许的浸入深度范围内，可通过调整浸入式水口高低来调整浸入深度，以提高浸入式水口寿命。

3.2.7 Tundish Baking

Baking the tundish can improve the temperature of the refractory lining in the tundish, remove the moisture and reduce the temperature drop and heat loss of the molten steel in the tundish. Generally installed on the continuous casting platform, the refractory lining of the tundish is preheated to 1200℃ before receiving the steel.

3.2.7 中间包烘烤

烘烤中间包，可以提高中间包内耐火衬的温度；去除其中水分，减少中间包内钢水的温降和热损耗。一般安装在连铸平台上，在受钢前，将中间包的耐火衬预热到1200℃。

The baking device of tundish and nozzle relies on flame jet to bake the refractory lining of tundish to reach the specified temperature.

The function of the baking burner is to mix the fuel medium and combustion supporting medium, ignite and spray out the baking flame.

The baking device of tundish and nozzle generally uses gas as the fuel medium, and the combustion supporting medium mainly includes compressed air, blowing air and oxygen, etc.

中间包、水口烘烤装置依靠火焰喷射，烘烤中间包的耐火衬，达到规定的温度。

烘烤烧嘴的作用是混合燃料介质和助燃介质，点火喷射出烘烤火焰。

中间包、水口的烘烤装置一般以煤气作为燃料介质，助燃介质主要有压缩空气、鼓风气和氧气等。

3.2.7.1 Inspection before Tundish Baking

Operation steps:

(1) Check whether the gas pipeline leaks and whether the combustion supporting fan is normal.

(2) Check whether the gas pressure is normal.

(3) Check the nozzle for blockage.

(4) Check whether the opening and closing of gas switch and air switch are flexible and whether there is air leakage.

(5) Check whether the measuring instrument is under control.

(6) Check whether the baking equipment (such as nozzle lifting system) can act as required.

3.2.7.1 中间包烘烤前的检查

操作步骤：

（1）检查煤气管道是否漏气，检查助燃风机是否正常。
（2）检查煤气压力是否正常。
（3）检查喷嘴有没有堵塞。
（4）检查煤气开关及空气开关开启关闭是否灵活、有否漏气。
（5）检查计量仪器是否处于受控状态。
（6）检查烘烤设备（如喷嘴升降系统等）能否按要求动作。

3.2.7.2 Baking Operation of Tundish and Nozzle

（1）The tundish shall be accurately positioned to ensure that the baking hole, nozzle and baking nozzle of the tundish are aligned.

（2）If the ladle long nozzle is used for casting, the ladle long nozzle shall be placed in the baking position in the tundish.

（3）After the inspection of tundish steel flow control mechanism, open the stopper rod or sliding plate for baking.

（4）Align the nozzle baking oven with the nozzle at the bottom of the tundish or cover the long nozzle of the tundish.

（5）Open the fuel valve, ignite and bake, and adjust the gas or fuel flow according to the baking requirements.

（6）The heating curve of tundish baking can be divided into two stages, the first is medium fire, then is high fire, and the time distribution is about half to half.

（7）Generally, the total baking time of tundish is controlled at 60～90min, and the final baking temperature is not less than 1100℃.

（8）When the baking time and temperature are reached, the baking can be stopped according to the production schedule, and immediately enter the state to be casted.

（9）The baking condition of tundish can be judged by baking time: it must be longer than the shortest time; it can be judged by temperature. A few casters are judged by continuous temperature measurement of tundish, and most casters are judged by observing the lining color of tundish with naked eyes, and it must be bright red.

（10）Some casters use cold tundish made of heat insulation plate, but the nozzle must be baked.

（11）The split type of tundish long nozzle needs to be baked in a long nozzle baking furnace, and it is taken out and installed on the tundish nozzle before tundish casting.

3.2.7.2 中间包及水口的烘烤操作

（1）中间包准确定位，保证中间包烘烤孔、水口与烘烤嘴对中。
（2）若采用钢包长水口浇铸，将钢包长水口放置于中间包内的烘烤位置。
（3）检查中间包钢流控制机构后，打开塞杆或滑板烘烤。
（4）将水口烘烤炉对准中间包底部的水口或套住中间包长水口。
（5）打开燃料阀，点火烘烤，并根据烘烤要求调整燃气或燃油流量。
（6）中间包烘烤升温曲线分两个阶段，先是中火，后大火升温，时间分配大致各一半。

（7）一般中间包总烘烤时间控制在60～90min，最终烘烤温度不低于1100℃。

（8）达到烘烤时间和烘烤温度后可根据生产安排停止烘烤，并立即进入待浇铸状态。

（9）中间包烘烤情况一是可从烘烤时间来判断：必须大于最短时间；二是从温度来判断：少数铸机由中间包连续测温来判断，大多数铸机用肉眼观察中间包内衬颜色来判断，必须达到亮红色。

（10）有些铸机采用绝热板砌的冷中间包，但水口必须烘烤。

（11）分离式中间包长水口需长水口烘烤炉烘烤，在中间包开浇前装在中间包水口上。

Precautions:

(1) The baking device must be used correctly. For gas baking, the gas safety operation regulations shall be strictly implemented. When the gas is baked, the gas must be turned on first, and then the air. When the baking is finished, the air must be turned off first and then the gas.

(2) During baking, observe the baking situation at any time to prevent flameout or abnormal flame.

(3) The baking time must meet the process requirements, but it should not be too long.

注意事项：

（1）必须正确使用烘烤装置，对于煤气烘烤，应严格执行煤气安全操作规程。煤气烘烤时，必须先开煤气，后开空气，烘烤结束时，必须先关空气后关煤气。

（2）烘烤过程中，应随时对烘烤情况进行观察，防止熄火或火焰异常。

（3）烘烤时间必须符合工艺要求，但也不宜过长。

3.2.8 Inspection of Tundish Car

(1) Before taking the tundish, the tundish car should be in the down position. When placing the tundish on the tundish car with a crane, do not put it directly in the position. When it is placed at a certain height from the tundish car, operate the tundish car to lift and catch the tundish.

(2) When the tundish is running in one direction and not fully stopped, the reverse operation is not allowed.

(3) For thin oil lubricated parts, regularly check the oil level and cleanliness.

(4) The rolling bearing of each operation part shall be inspected regularly, and it shall be replaced or repaired in time in case of any abnormality.

(5) Regularly check whether the screws at the connection of each transmission part are loose. If any abnormality is found, tighten or replace them in time

(6) Regularly check whether the limit travel switch of the lifting device acts accurately, and adjust or replace it in time in case of any abnormality.

(7) Check whether the pneumatic and hydraulic components and pipelines of the long nozzle manipulator are leaking, whether the movable parts are blocked, and whether the connecting partsare loose. If any abnormality is found, handle it in time and clean the pneumatic filter in time.

3.2.8 中间包车检查

（1）坐中间包之前，中间包车应处于下降位。当用吊车往中间包车上放中间包时，不要直接放到位，应在放到离中间包车一定高度时，操作中间包车上升接住中间包。

（2）当中间包车朝一个方向运行未完全停车时，不允许反方向操作。

（3）对稀油润滑的部位，要定期检查油位高度及检验油质清洁度。

（4）各运行部位的滚动轴承要定期检查，发现异常应及时更换或修理。

（5）定期检查各传动部位连接处螺丝是否松动，如发现异常情况应及时拧紧或更换

（6）定期检查升降装置的限位行程开关动作是否准确，如发现异常应及时调整或更换。

（7）检查长水口机械手各气动和液压元件及管线是否有泄漏、各活动部位是否有卡阻现象，连接部位有无螺栓松动现象，如发现异常应及时处理，并及时清洗气动过滤器。

Exercises:

(1) What is the structure and function of tundish?

(2) How to check and maintain the tundish after use?

(3) Use the simulation software to operate the tundish and related equipment, drive the baked tundish to the casting position, install the submerged nozzle and align it, and prepare for casting.

思考与习题：

（1）中间包的结构和作用是什么？

（2）中间包使用后应该如何检查和维护？

（3）利用仿真软件操作中间包及相关设备，将烘烤后中间包开到浇铸位，安装浸入式水口并对中，做好浇铸准备。

Task 3.3 Operation of Mold and Mold Oscillation Device
任务 3.3 结晶器及结晶器振动装置操作

Mission objectives:

任务目标：

(1) Master the structure and main parameters of mold and mold oscillation device.

（1）掌握结晶器及结晶器振动装置的结构和主要参数。

(2) Be able to operate and maintain relevant equipment.

（2）能够对相关设备进行操作和维护。

(3) Be able to complete mold inspection and maintenance, oscillation device detection, liquid level control and other operations.

（3）能够完成结晶器检查和维护、振动装置检测、液面控制等操作。

Task Preparation 任务准备

3.3.1 Cognition of Mold

The mold is a water-cooled bottomless ingot mold, which is the core component of the continuous casting machine, called the heart of the continuous casting equipment. The function of the mold is to cool the molten steel in the mold, preliminarily solidify and form the shell with a certain thickness. Through the oscillation of the mold, the shell is separated from the mold wall without breaking or steel leakage. The mold is cooled by cooling water, usually called primary cooling.

3.3.1 结晶器认知

结晶器是一个水冷的无底钢锭模，是连铸机的核心部件，称为连铸设备的心脏。结晶器的作用是钢液在结晶器内冷却、初步凝固成型，且形成具有一定厚度的坯壳，通过结晶器的振动，使坯壳脱离结晶器壁而不被拉断和漏钢。结晶器采用冷却水冷却，通常称为一次冷却。

3.3.1.1 Requirements of Mold

The solidification process is carried out under the continuous and relative motion between the shell and the mold wall. Therefore, the mold should have good thermal conductivity and rigidity, not easy to deform; the weight should be light to reduce the inertia force during oscillation; the wear resistance of the inner surface should be good to improve the service life; the mold structure should be simple, easy to manufacture and maintain.

3.3.1.1 结晶器的要求

凝固过程是在坯壳与结晶器壁连续、相对运动下进行的。为此，结晶器应具有良好的导热性和刚性，不易变形；重量要轻，以减少振动时的惯性力；内表面耐磨性要好，以提高寿命；结晶器结构要简单，便于制造和维护。

3.3.1.2 Classification and Structure of Mold

According to the shape of mold, it can be divided into vertical mold and arc mold. The vertical mold is used for vertical, vertical bending and straight arc casters, while the arc mold is used for full arc and elliptical casters. In terms of its structure, there are tube mold and combined mold; tube mold is mostly used for billet, round billet and rectangular billet, while combined mold is mostly used for bloom, large rectangular billet and slab. According to the specification and shape of the billet, it can be divided into round billet, rectangular billet, square billet, slab and shaped billet mold.

3.3.1.2 结晶器的分类和结构

按结晶器的外形可分为直结晶器和弧形结晶器。直结晶器用于立式、立弯式及直弧形连铸机；而弧形结晶器用在全弧形和椭圆形连铸机上。从其结构来看，有管式结晶器和组合式结晶器；小方坯、圆坯及矩形坯多采用管式结晶器，而大方坯、大矩形坯和板坯多采用组合式结晶器。按照铸坯规格和形状可分为圆坯、矩形坯、方坯、板坯及异型坯结晶器等。

（1）Tube mold. The structure of tube mold is shown in Figure 3-12. The inner tube is cold drawn special-shaped seamless copper tube, the outer sleeve is covered with steel shell, and there is a gap of about 7mm between the copper tube and the steel sleeve for cooling water, i. e. cooling water gap. Copper pipe and steel sleeve can be made into arc or straight shape. The upper outlet of the copper pipe is fixed on the steel shell with a screw through the flange, and the lower outlet of the copper pipe is generally the free end, allowing thermal expansion and cold contraction; however, the upper and lower outlet must be sealed without water leakage. The mold jacket is cylindrical. The bottom foot plate is arranged in the middle part of the outer sleeve to fix the mold on the oscillation frame.

（1）管式结晶器。管式结晶器的结构如图 3-12 所示。其内管为冷拔异型无缝铜管，外面套有钢质外壳，铜管与钢套之间留有约 7mm 的缝隙通冷却水，即冷却水缝。铜管和钢套可以制成弧形或直形。铜管的上口通过法兰用螺钉固定在钢质的外壳上，铜管的下口一般为自由端，允许热胀冷缩；但上下口都必须密封，不能漏水。结晶器外套是圆筒形的。外套中部有底脚板，将结晶器固定在振动框架上。

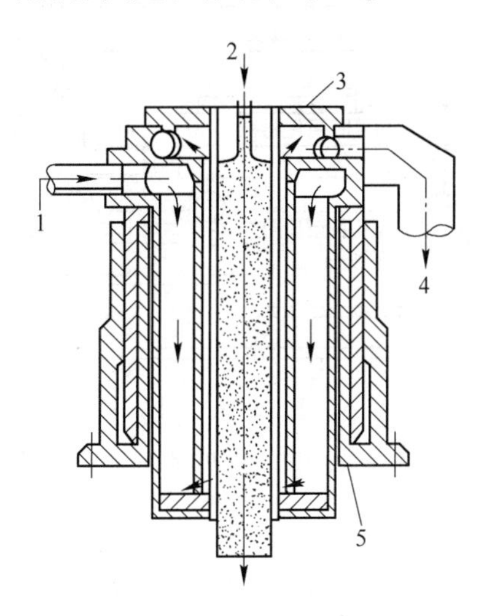

Figure 3-12 Tube mold

1—Cooling water inlet; 2—Liquid steel; 3—Collet; 4—Cooling water outlet; 5—Oil pressure cylinder

图 3-12 管式结晶器

1—冷却水入口；2—钢液；3—夹头；4—冷却水出口；5—油压缸

The tubular mold is simple in structure, easy to manufacture and maintain. It is widely used in the casting of small and medium section billets. The maximum casting section is 180mm×180mm. In addition, some tube molds cancel the cooling water gap and directly spray cooling with cooling water, that is, spray tube mold.

管式结晶器结构简单，易于制造、维修，广泛应用于中小断面铸坯的浇铸，最大浇铸断面为 180mm×180mm。另外，有的管式结晶器取消冷却水缝，直接用冷却水喷淋冷却，即为喷淋式管式结晶器。

（2）Combined mold. The combined mold is composed of 4 composite wall plates. Each composite panel is composed of copper inner wall and steel shell. A lot of grooves are milled on the copper plate surface contacting with the steel shell to form a middle water seam. The composite wall panel is connected and fixed with double bolts, as shown in Figure 3-13. The cooling water enters from the lower part and discharges from the upper part after passing through the water joint. The four panels have their own water cooling system. At the corner of the inner wall of 4 composite panels, 3~5mm thick copper strip with 45° chamfer shall be padded to prevent corner crack of slab.

（2）组合式结晶器。组合式结晶器由 4 块复合壁板组合而成。每块复合壁板都由铜质内壁和钢质外壳组成。在与钢壳接触的铜板面上铣出许多沟槽形成中间水缝。复合壁板用双螺栓连接固定，如图 3-13 所示，冷却水从下部进入，流经水缝后从上部排出。4 块壁板有各自独立的水冷却系统。在 4 块复合壁板内壁相结合的角部，垫上厚 3~5mm 并带 45°倒角的铜片，以防止铸坯角裂。

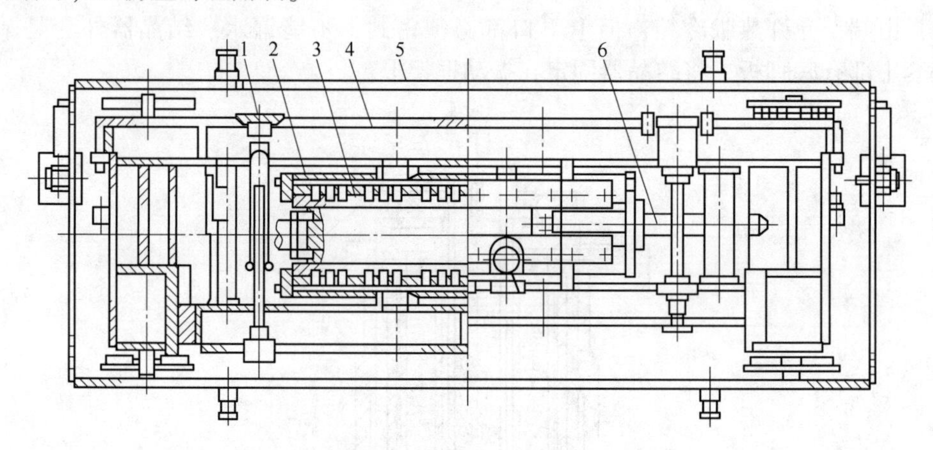

Figure 3-13 Structure of combined mold

1—Thickness adjustment and clamping mechanism; 2—Inner wall of narrow surface; 3—Inner wall of wide surface; 4—Mold outer frame; 5—Oscillation frame; 6—Width adjusting mechanism

图 3-13 组合式结晶器结构图

1—调厚与夹紧机构；2—窄面内壁；3—宽面内壁；4—结晶器外框架；5—振动框架；6—调宽机构

The combined mold can realize on-line automatic adjustment in the casting process, and the slab mold with adjustable width has been widely used. The mold width can be adjusted by manual, electric or hydraulic drive. The thickness of the inner wall plate is 20~50mm, which can be processed and repaired after abrasion, but the thinnest thickness shall not be less than 10mm.

For arc mold, the two side composite plates are flat, and the inner and outer arc composite plates are arc-shaped. The four sides of the mold are flat.

组合式结晶器可以实现浇铸过程中在线自动调整，现已广泛采用宽度可调的板坯结晶器。可用手动、电动或液压驱动调节结晶器的宽度。内壁板厚度在20~50mm，磨损后可加工修复，但最薄不能小于10mm。

对弧形结晶器来说，两块侧面复合板是平的，内外弧复合板做成弧形的。而直结晶器四面壁板都是平面状的。

(3) Multistage mold. With the increase of the casting speed, the thickness of the shell at the bottom of the mold becomes thinner and thinner. In order to prevent the deformation of the slab or steel leakage, the multi-stage mold technology is adopted. It can also reduce corner crack and rhombic deformation of billet. The multi-stage mold is to install foot roller, copper plate or cooling grid at the lower outlet of the mold.

(3) 多级结晶器。随着连铸机拉坯速度的提高，出结晶器下口的铸坯坯壳厚度越来越薄；为了防止铸坯变形或出现漏钢事故，采用多级结晶器技术。它还可以减少小方坯的角部裂纹和菱变。多级结晶器即在结晶器下口安装足辊、铜板或冷却格栅。

1) Foot roller. Multiple pairs of closely arranged clamping rollers are installed on the four sides of the lower outlet of the mold, with small diameter and sufficient rigidity. Nozzles are installed between the rollers to spray water for cooling. These small rollers are called foot rollers, as shown in Figure 3-14 (a). In order to prevent the foot rollers from causing transverse stress to the slab, the installation position of the foot rollers shall be aligned with the mold.

1) 足辊。在结晶器的下口四面装有多对密排夹辊，其直径较小且具有足够的刚度，辊间安有喷嘴，这些小辊称为足辊，如图3-14 (a) 所示，为了防止足辊对铸坯造成横向应力，足辊的安装位置应与结晶器对中。

2) Cooling plate. A piece of copper plate is installed on each side of the lower outlet of the mold, and the corner of the billet is cooled by water spray. The copper plate is supported on the surface of the shell by spring, which ensures the uniform cooling of the billet. As shown in Figure 3-14 (b), this device has a slightly larger drawing resistance, but the cooling effect is very good, which is mainly used on the tubular mold of the billet caster.

2) 冷却板。在结晶器下口每面安装一块铜板，且在铸坯角部喷水冷却，铜板靠弹簧支撑紧贴在坯壳表面，保证了铸坯的均匀冷却，如图3-14 (b) 所示，这种装置拉坯阻力稍大，但冷却效果很好，主要用在小方坯连铸机管式结晶器上。

3) Cooling grid. Cooling grid is a kind of copper plate with many square holes, also called grid or grating. Cooling water is directly sprayed on the surface of the slab through square holes. A cooling grid is installed on each side of the lower outlet of the mold, and the back of the grid is provided with stiffener plate, as shown in Figure 3-14 (c). The cooling effect of this device is good, the resistance of blank drawing is slightly large, and it is difficult to clean after steel leakage.

3) 冷却格栅。冷却格栅是一种带有许多扁方孔的铜板，也称格栅或格板。冷却水通过方孔直接喷射在铸坯表面。结晶器下口每面安装一块冷却格板，格板背面有加强筋板，如图3-14 (c) 所示，这种装置冷却效果较好，拉坯阻力略大，发生漏钢后清理困难。

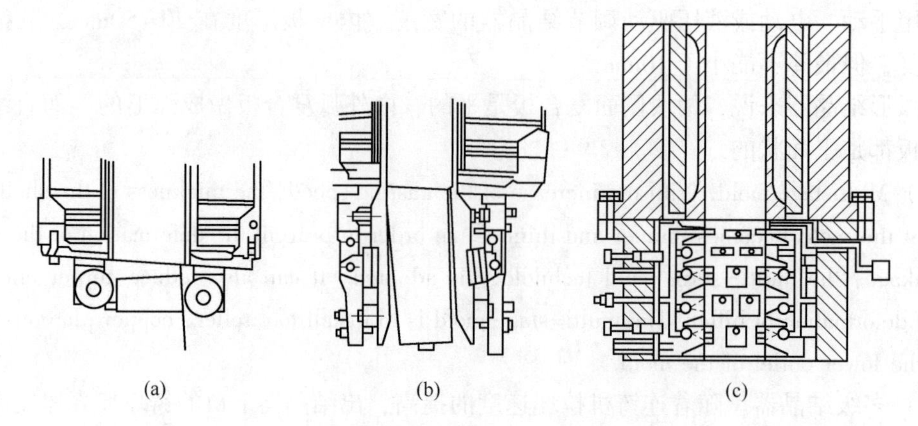

Figure 3-14 Structure diagram of multistage mold

(a) Foot roller; (b) Cooling plate; (c) Cooling grid

图 3-14 多级结晶器结构示意图

(a) 足辊；(b) 冷却板；(c) 冷却格栅

3.3.1.3 Main Parameters of Mold

(1) Section size of mold. The section size of mold shall be determined according to the nominal section size of continuous casting billet. However, the deformation of the semi-finished billet will be caused by the gradual shrinkage and straightening of the billet during the cooling and solidification process. For this reason, it is required that the section size of the mold should be larger than the nominal section size of the continuous casting billet, usually about 2%~3% larger. According to the section shape of billet, the mold of billet, slab, rectangle slab, round billet and beam blank can be used.

3.3.1.3 结晶器的主要参数

(1) 结晶器断面尺寸。结晶器的断面尺寸应根据连铸坯的公称断面尺寸来确定。但由于连铸坯在冷却凝固过程中逐渐收缩以及矫直时都将引起半成品铸坯的变形。为此，要求结晶器的断面尺寸应当比连铸坯断面公称尺寸大一些，通常大 2%~3%左右。根据铸坯的断面形状可采用方坯、板坯、矩形坯、圆坯及异型坯结晶器。

(2) Length of mold. The shell must have a certain thickness when casting billetl eaving the mold. If the shell thickness is small, the billet is prone to bulge and even steel leakage. The length of the mold is an important factor for the thickness of the shell. The length of the mold shall ensure that the thickness of the shell when the billet is cast out of the lower outlet of the mold is greater than or equal to 10~25mm. The lower limit can be taken in the production of small section billets, while the upper limit should be taken in the production of large section billets.

(2) 结晶器长度。铸坯出结晶器时坯壳要有一定的厚度。若坯壳厚度较小，铸坯就容易出现鼓肚，甚至漏钢。结晶器的长度是能否具有足够坯壳厚度的重要因素。结晶器的长度应保证铸坯出结晶器下口时的坯壳厚度大于或等于 10~25mm。生产小断面铸坯时可取下限，而生产大断面铸坯时则应取上限。

（3）Taper of mold. The molten steel is cooled in the mold to form a shell. As the cooling shell shrinks, air gap is formed between the shell and the copper wall. In order to reduce the air gap, improve the thermal conductivity of the mold and accelerate the growth of the shell, the mold should have an inverted taper. Generally, the lower outlet of the mold is slightly smaller than the upper inlet.

The selection of back taper is very important. If the selection is too small, the shell will break away from the inner wall of the mold too early, which will seriously affect the cooling effect and cause the shell to bulge under the static pressure of the molten steel, even steel leakage. If the selection is too large, it will increase the drawing resistance and accelerate the wear of the mold inner wall.

According to the practice, the back taper of the general billet tube mold is 0.4%/m~0.9%/m according to the different steel types. The recommended value of the back taper of the billet mold is shown in Table 3-3. The back taper of the wide surface of the slab mold is 0.9%/m~1.1%/m, and that of the narrow surface is 0~0.6%/m. The back taper is usually 1.2%/m for the round billet mold which is cast with mold flux.

（3）结晶器倒锥度。钢水在结晶器中冷却生成坯壳。由于冷却坯壳收缩，在坯壳与铜壁之间形成气隙。为了减少气隙，提高结晶器的导热性能，加速坯壳生长，结晶器要有倒锥度。一般结晶器下口比结晶器上略小。

倒锥度的选择十分重要，选择过小，坯壳会过早脱离结晶器内壁，严重影响冷却效果，使坯壳在钢水静压力作用下产生鼓肚变形，甚至发生漏钢。选择过大，会增加拉坯阻力，加速结晶器内壁的磨损。

根据实践，一般方坯管式结晶器的倒锥度，依据钢种不同，取0.4%/m~0.9%/m，方坯结晶器倒锥度的推荐数值见表3-3，板坯结晶器宽面倒锥度在0.9%/m~1.1%/m，窄面则为0~0.6%/m。采用保护渣浇铸的圆坯结晶器，倒锥度通常是1.2%/m。

Table 3-3 Back taper of billet mold

表3-3 方坯结晶器的倒锥度

Section length 断面边长 /mm	Back taper 倒锥度 /% · m^{-1}
80~100	0.4
110~140	0.6
140~200	0.9

（4）Water seam area of mold. In the process of forming billet shell in mold, 96% of heat released by molten steel is taken away by cooling water through heat conduction. In unit time, the heat taken away by the unit surface area is called cooling strength. The main factors that affect the cooling strength of the mold are the heat conductivity of the inner wall of the mold and the flow rate of the cooling water in the mold. It is necessary to reasonably determine the total area A of the water seam of the mold, which is as follows:

$$A = \frac{10000}{36} \times \frac{QL}{V} \qquad (3-1)$$

Where Q——Water consumption per meter around the mold, $m^3/(h \cdot m)$;

L——Length around the mold, m;

V——Cooling water flow rate, m/s.

Generally, the water consumption in the mold is determined by experience. If the cooling water volume in the mold is too large, the slab will crack, and if it is too small, it is easy to cause bulging deformation or steel leakage. The waterseam type of the mold is shown in Figure 3-15.

（4）结晶器的水缝面积。钢水在结晶器内形成坯壳的过程中，其放出的热量96%是通过热传导由冷却水带走的。在单位时间内，单位表面积铸坯被带走的热量称为冷却强度。影响结晶器冷却强度的因素，主要是结晶器内壁的导热性能和结晶器内冷却水的流速和流量。必须合理确定结晶器的水缝总面积A，其公式为：

$$A = \frac{10000}{36} \times \frac{QL}{V} \qquad (3-1)$$

式中 Q——结晶器每米周边长耗水量，$m^3/(h \cdot m)$；

L——结晶器周边长度，m；

V——冷却水流速，m/s。

通常结晶器内的耗水量根据经验确定。结晶器内冷却水量过大，铸坯会产生裂纹，过小又易造成鼓肚变形或漏钢，结晶器的冷却水缝形式如图3-15所示。

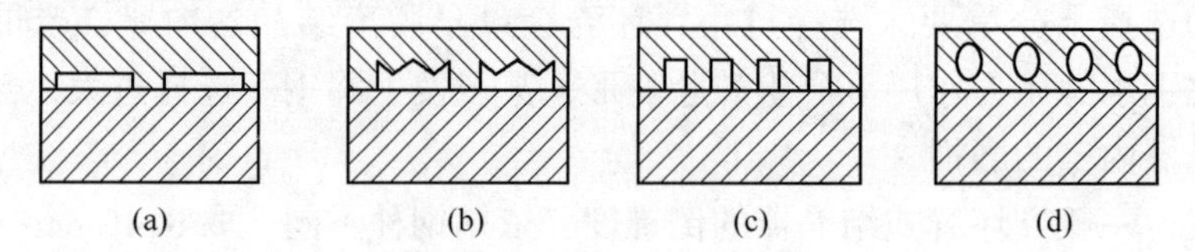

Figure 3-15 Cooling water tank type of mold

（a）Straight type;（b）Epsilon type;（c）Grooved（15mm×5mm）;（d）Drilled type

图3-15 结晶器的冷却水槽形式

（a）一字形；（b）山字形；（c）沟槽式（15mm×5mm）；（d）钻孔式

（5）Mold life, The service life of the mold actually refers to the time when the mold cavity keeps the original design size and shape. Only by keeping the original design size and shape can the quality of the billet be guaranteed. The life of mold is generally expressed by the length of casting billet. Under normal operating conditions, one mold can cast slab with length of 10000~150000m. The service life of the mold can also be expressed by the number of heats cast from the beginning to the repair of the mold, with the range of 100~150 ladles.

The measures to improve the service life of the mold include: improving the cooling water quality of the mold; ensuring the arc alignment accuracy of the mold foot roll and the secondary cooling area; regularly maintaining the mold; reasonably selecting the material and design parameters of the inner wall of the mold, etc.

（5）结晶器寿命。结晶器使用寿命实际上是指结晶器内腔保持原设计尺寸、形状的时间长短。只有保持原设计尺寸、形状、才能保证铸坯质量。结晶器寿命一般用结晶器浇铸铸坯的长度来表示。在一般操作条件下，一个结晶器可浇铸板坯 10000~15000m。结晶器寿命也可用结晶器从开始使用到修理前所浇铸的炉数来表示，其范围为 100~150 炉。

提高结晶器寿命的措施有：提高结晶器冷却水水质；保证结晶器足辊、二次冷却区的对弧精度；定期检修结晶器；合理选择结晶器内壁材质及设计参数等。

3. 3. 1. 4 Material of Mold

As the inner wall of the mold is directly in contact with high temperature steel, the inner wall material should have the following properties: good thermal conductivity, sufficient strength, wear resistance, plasticity andmachinability. The materials are mainly as follows:

(1) Copper. The inner wall material of the mold is generally made of red copper and brass, because it has the advantages of good heat conductivity, easy processing and low price, but it has poor wear resistance and short service life.

(2) Copper alloy. The inner wall of the mold is made of copper alloy, which can improve the strength, wear resistance and service life of the mold. If we use copper chromium zirconium arsenic alloy or copper zirconium magnesium alloy to make the inner wall of mold, the effect is good.

(3) Copper plate plating. Plating 0. 1~0. 15mm thick on the copper plate of the mold can improve the wear resistance. At present, the plating is mainly alloy of chromium, nickel, tungsten and other elements.

3. 3. 1. 4 结晶器的材质

由于结晶器内壁直接与高温钢水接触，所以内壁材料应具有以下性能：导热性好，足够的强度、耐磨性、塑性及可加工性。材质主要有以下几种：

(1) 铜。结晶器的内壁材料一般由紫铜、黄铜制作，因为它具有导热性好，易加工，价格便宜等优点，但耐磨性差，使用寿命较短。

(2) 铜合金。结晶器内壁采用铜合金材料，可以提高结晶器的强度、耐磨性，延长使用寿命。如使用铜-铬-锆-砷合金或铜-锆-镁合金制作结晶器内壁，效果都不错。

(3) 铜板镀层。在结晶器的铜板上镀 0. 1~0. 15mm 厚的镀层，能提高耐磨性，目前镀层主要为铬、镍、钨等元素的合金。

3. 3. 1. 5 Steel Leakage Detection Device

In order to predict the breakout accident of the mold, several sets of thermocouples are embedded through the evenly distributed bolts outside the copper walls of the mold; the temperature data measured by the thermocouples are input into the computer or the instrument, if the temperature at a certain point suddenly rises, it indicates that there is breakout near this point. The more sets of thermocouples, the more accurate the detection.

According to the friction between the inner wall of the mold and the shell, the steel leakage of the shell in the mold can also be determined.

3.3.1.5 漏钢检测装置

为了能够预报结晶器漏钢事故，在结晶器四面铜壁外通过均布的螺栓埋入多套热电偶；热电偶测到的温度数据输入计算机或仪表上显示，若某一点温度突然升高，说明这一点附近出现了漏钢。热电偶的套数越多，检测也越精确。

也有根据结晶器内壁与铸坯坯壳间摩擦力的大小来测定结晶器内坯壳是否漏钢的。

3.3.2 Cognition of Mold Lubrication Device

In order to prevent the shell from sticking to the inner wall of the mold, reduce the resistance of drawing and the wear of the inner wall of the mold, and improve the surface quality of the billet, the mold must be lubricated. At present, the main means of lubrication are lubricating oil and protective slag lubrication.

(1) Oil lubrication. A layer of 0.025~0.05mm thick uniform oil film and oil-gas film are formed between the shell and the inner wall of the mold by the lubricating oil to achieve the purpose of lubrication. This device is mainly used in open stream casting mold.

(2) Flux lubrication. At present, the mold powder is usually used to lubricate the continuous casting machine. The protective slag can be added manually or by vibrating feeder, and the device is shown in Figure 3-16. This improves the working conditions, and the control of the amount of additives is accurate.

3.3.2 结晶器润滑装置认知

为防止铸坯坯壳与结晶器内壁黏结，减少拉坯阻力和结晶器内壁的磨损，改善铸坯表面质量，结晶器必须进行润滑。目前的润滑手段主要有润滑油和保护渣润滑。

(1) 润滑油润滑。润滑油在坯壳与结晶器内壁之间形成一层厚 0.025~0.05mm 的均匀油膜和油气膜，达到润滑的目的。这种装置主要应用在敞开浇铸的结晶器上。

(2) 保护渣润滑。目前连铸机通常采用保护渣达到润滑的目的。保护渣可人工加入，也可用振动给料器加入，其装置如图 3-16 所示。这改善了劳动条件，且加入量控制准确。

3.3.3 Cognition of Mold Oscillation Device

3.3.3.1 Effect of Mold Oscillation

The functions of the mold oscillation device are as follows:

(1) The inner wall is well lubricated to prevent the bonding between the primary shell and the inner wall of the mold;

(2) The oscillation are helpful to improve the surface quality of the slab and form the slab with smooth surface;

(3) When the bond occurs, the oscillation can force the demolding and eliminate the bond;

(4) When the shell in the mold is broken, the pressing is obtained by the synchronous oscillation of the mold and the billet.

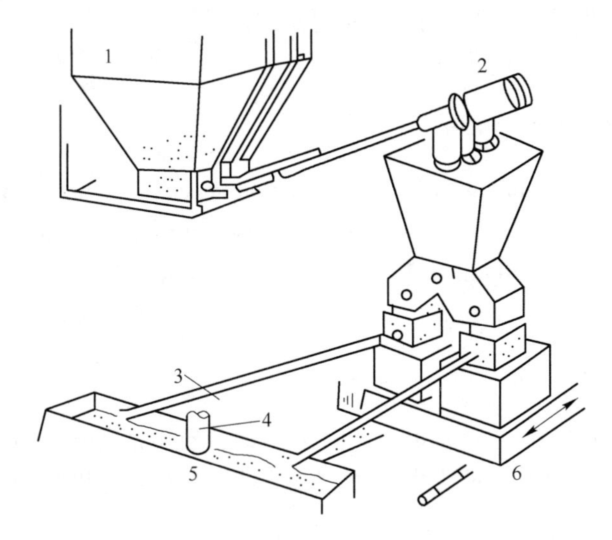

Figure 3-16　Automatic adding device of mold powder

1—Protective slag tank；2—Vibrating tube conveyor；3—Feed pipe；4—Submerged nozzle；5—Mold；6—Silo car

图 3-16　结晶器保护渣自动加入装置

1—保护渣罐；2—振动管式输送机；3—给料管；4—浸入式水口；5—结晶器；6—料仓车

3.3.3　结晶器振动装置认知

3.3.3.1　结晶器振动的作用

结晶器振动装置的作用如下：

（1）内壁获得良好的润滑、防止初生坯壳与结晶器内壁的黏结；

（2）振动有利于改善铸坯表面质量，形成表面光滑的铸坯；

（3）当发生黏结时，通过振动能强制脱模，消除黏结；

（4）当结晶器内的坯壳被拉断，通过结晶器和铸坯的同步振动得到压合。

3.3.3.2　Requirements for Mold Oscillation Device

（1）The oscillation device shall move in strict accordance with the required oscillation curve. The four corner positions of the whole oscillation frame shall rise to the top dead center or descend to the bottom dead center at the same time. During the oscillation, it is not allowed to shift and shake in front and back, left and right directions;

（2）The oscillation device shall be stable, soft and elastic during oscillation;

（3）The equipment is easy to manufacture, install and maintain.

3.3.3.2　结晶器振动装置的要求

（1）振动装置应当严格按照所需求的振动曲线运动，整个振动框架的 4 个角部位置均应同时上升到达上止点或同时下降到达下止点，在振动时不允许出现前后、左右方向的偏移与晃动现象；

（2）振动装置在振动时应保持平稳、柔和、有弹性；

（3）设备的制造、安装和维护方便。

3. 3. 3. 3 Mold Oscillation Mode

According to the characteristics of mold oscillation speed, it can be divided into four types: synchronous oscillation, negative strip oscillation, sinusoidal oscillation and non sinusoidal oscillation. The oscillation curve is shown in Figure 3-17, which is mainly reflected by the change between mold movement speed and time.

3. 3. 3. 3 结晶器振动方式

按结晶器振动速度特征可分为 4 种：同步振动、负滑脱振动、正弦振动和非正弦振动，其振动曲线如图 3-17 所示，主要通过结晶器运动速度和时间之间的变化来进行体现。

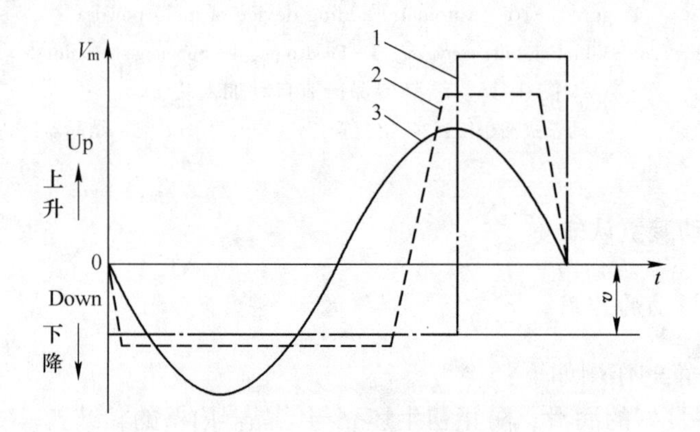

Figure 3-17 Oscillation characteristic curve

1—Synchronous oscillation; 2—Negative strip oscillation; 3—Sinusoidal oscillation

图 3-17 振动特性曲线

1—同步振动；2—负滑脱振动；3—正弦振动

(1) Synchronous oscillation. Synchronous oscillation is the earliest oscillation mode. When the oscillation device is working, the mold moves synchronously with the billet when it falls, and then the mold rises at three times of the casting speed. When the mold changes from falling to rising, the acceleration is very large, this will cause impact force and affect the stability of the mold.

（1）同步振动。同步振动是最早的一种振动方式，振动装置工作时，结晶器下降时与铸坯做同步运动，然后结晶器以 3 倍的拉速上升。由于结晶器在由下降转为上升时，加速度很大，会引起产生冲击力，影响结晶器的平稳性。

(2) Negative strip oscillation. Negative strip oscillation is an improved form of synchronous oscillation. When working, the mold first drops at a speed slightly higher than the casting speed, which is called negative strip, which is conducive to forced demolding and pressing of broken shell, and then rises at a higher speed. It is beneficial to the stability of oscillation and the thickening of billet shell, but a set of cam mechanism is needed, and the interlock between oscillation

mechanism and billet drawing mechanism must be ensured.

（2）负滑脱振动。负滑脱振动是同步振动的改进形式，工作时先是结晶器以稍大于拉速的速度下降，称为负滑动，这样有利于强制脱模及断裂坯壳的压合，然后再以较高的速度上升。它有利于振动的平稳和坯壳的增厚，但需用一套凸轮机构，且必须保证振动机构与拉坯机构联锁。

（3）Sinusoidal oscillation. The relationship between oscillation speed and time is a sine curve, as shown in Figure 3-18. The time of sinusoidal oscillation is equal, and the maximum oscillation speed of up and down is same. In the whole oscillation cycle, there is always relative movement between the slab and the mold, and in the process of mold falling, there is a small section of falling speed which is faster than the drawing speed, so it can prevent and eliminate the bond between the slab shell and the inner wall of the mold, and it can play a healing role for the slab shell which is pulled and cracked. In addition, since the velocity of the mold changes according to the sine principle, the acceleration must change according to the cosine principle, so the transition is relatively stable and the impact is small.

Sinusoidal oscillation can be realized by eccentric wheel linkage mechanism. It can improve the oscillation frequency, reduce the oscillation mark and improve the quality of slab, so sinusoidal oscillation is widely used.

（3）正弦振动。振动的速度与时间的关系为一条正弦曲线，如图 3-18 所示。正弦振动方式的上下振动时间相等，上下振动的最大速度也相同。在整个振动周期中，铸坯与结晶器之间始终存在相对运动，而且结晶器下降过程中，有一小段下降速度大于拉坯速度，因此可以防止和消除坯壳与结晶器内壁间的黏结，并能对被拉裂的坯壳起到愈合作用。另外，由于结晶器的运动速度是按正弦规律变化的，加速度则必然按余弦规律变化，所以过渡比较平稳，冲击较小。

正弦振动通过偏心轮连杆机构就能实现。利于提高振动频率、减小振痕，改善铸坯质量。因此，正弦振动方式应用广泛。

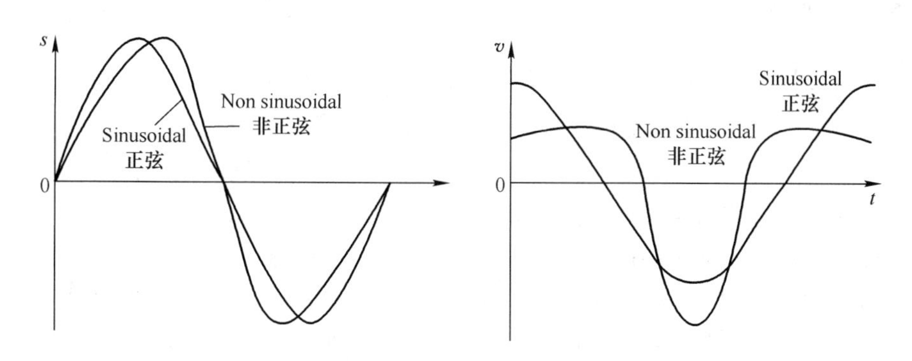

Figure 3-18　Comparison of sinusoidal and non sinusoidal oscillation curves

图 3-18　正弦振动与非正弦振动曲线比较图

（4）Nonsinusoidal vibration. In recent years, non sinusoidal oscillation technology has been applied in the field of continuous casting. The displacement waveform and velocity waveform of non sinusoidal oscillation and sinusoidal oscillation are shown in Figure 3-18. The non sinusoidal oscillation has the following characteristics: short negative strip time, long positive strip time, small difference between the mold upward velocity and the slab velocity. In use, it has the effect of little change in the surface quality of the billet, low leakage rate and stable operation of the equipment.

(4) 非正弦振动。近几年在连铸领域，非正弦振动技术得到应用。非正弦振动和正弦振动的位移波形和速度波形如图 3-18 所示。非正弦振动具有如下特点：负滑脱时间短、正滑脱时间长，结晶器向上运动速度与铸坯运动速度差较小。使用中具有铸坯表面质量变化不大、拉漏率降低、设备运行平稳故障率低的效果。

3.3.3.4 Mold Oscillation Parameters

The oscillation parameters of mold refer to amplitude, frequency, negative strip time, etc.

(1) Amplitude, the distance the mold moves from the horizontal position to the highest or lowest position, expressed as S, mm.

(2) Oscillation frequency, frequency of mold oscillation per minute, expressed as f, times/min.

(3) Negative strip time, the negative strip of the mold oscillation device means that when the oscillation speed of the mold drop is greater than the speed of drawing billet, the billet moves in opposite direction to that of drawing billet. The negative strip time is the time of negative strip for each vibration.

With the development of continuous casting technology, the oscillation frequency of mold is increasing constantly. At present, the mode of 'small amplitude and high frequency' is mainly used. At present, the frequency of 0~250 times/min is used. The oscillation amplitude is small, the liquid level fluctuation of mold steel is small, and the surface oscillation mark of billet is small. The longer the negative strip time is, the deeper the oscillation mark depth is, and the higher the probability of transverse crack is.

3.3.3.4 结晶器振动参数

结晶器振动参数主要是指振幅、频率、负滑动时间等。

(1) 振幅，结晶器从水平位置运动到最高或最低位置所移动的距离，用 S 表示，mm。

(2) 振动频率，结晶器每分钟振动的次数，用 f 表示，次/min。

(3) 负滑动时间，结晶器振动装置的负滑动是指当结晶器下降振动速度大于拉坯速度时，铸坯做与拉坯相反的运动。负滑动时间也就是每次振动进行负滑动的时间。

随着连铸技术的发展，结晶器的振动频率不断在增加，当前主要是用“小幅高频”的振动模式。结晶器振动频率高，对提高拉速和减轻振痕有利，目前采用 0~250 次/min 的频率。振幅小，结晶器钢水的液面波动小，铸坯表面振痕小。负滑动时间越长，铸坯振痕深度越深，相应产生铸坯的横裂纹几率也越高。

3.3.3.5 Mold Oscillation Mechanism

The oscillation mechanism of the mold is a device that makes the mold oscillate in a certain way, and its function is to make the mold oscillate regularly. Commonly used oscillation mechanisms are short arm four-bar oscillation mechanism, four eccentric wheel oscillation mechanism and hydraulic oscillation mechanism.

3.3.3.5 结晶器振动机构

结晶器的振动机构是使结晶器按照一定的振动方式进行振动的装置，其作用是使结晶器产生具有一定规律的振动。常用的振动机构有短臂四连杆式振动机构、四偏心轮式振动机构和液压振动机构等。

(1) Short arm four-bar oscillation mechanism. Short arm four-bar oscillation mechanism is widely used in small billet and large slab caster, but the oscillation mechanism of small billet caster is mostly installed on the inner arc side, while the oscillation mechanism of large slab caster is installed on the outer arc side, as shown in Figure 3-19 and Figure 3-20. The structure of the short arm four-bar oscillation mechanism is simple and easy to maintain. It can realize the arc movement of the mold accurately and improve the quality of the casting billet.

(1) 短臂四连杆式振动机构。短臂四连杆振动机构广泛应用于小方坯和大板坯连铸机上，只是小方坯连铸机振动机构多装在内弧侧，而大板坯连铸机振动机构安装在外弧侧，如图 3-19 和图 3-20 所示。短臂四连杆振动机构的结构简单，便于维修，能够较准确地实现结晶器的弧线运动，有利于铸坯质量的改善。

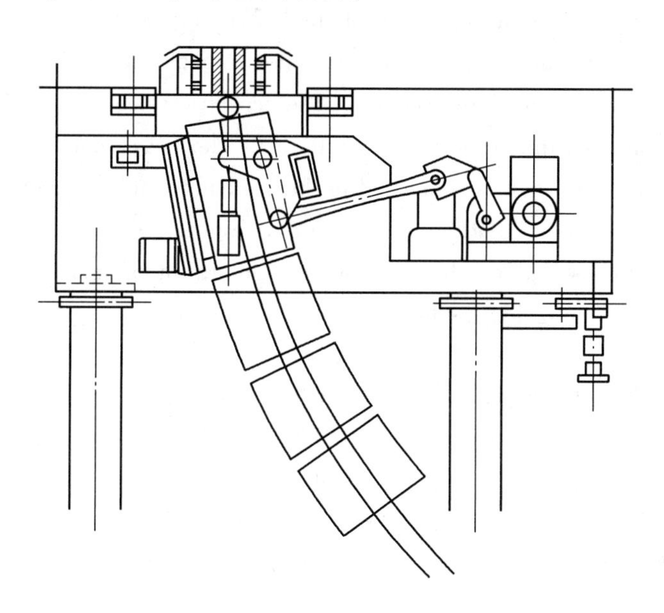

Figure 3-19 Short arm four-bar oscillation mechanism (inner arc side)
图 3-19 短臂四连杆振动机构（内弧侧）

The working principle of the four-bar mechanism can be seen in Figure 3-20. The motor is driven by the reducer through the eccentric wheel, the pull rod 3 makes reciprocating motion, and the pull rod 3 drives the swing of the connecting rod 4; the connecting rod 5 also swings with it, so that the oscillation frame 2 can oscillate according to the arc track.

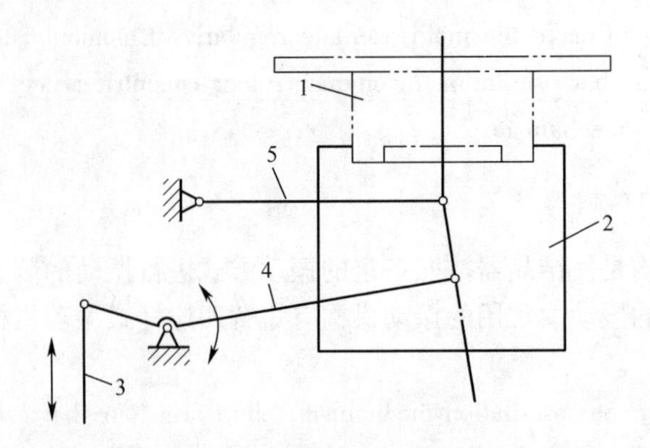

Figure 3-20 Short arm four-bar oscillation mechanism (outer arc side)

1—Mold; 2—Oscillation frame; 3—Pull rod; 4, 5—Connecting rod

图 3-20 短臂四连杆式振动机构（外弧侧）

1—结晶器；2—振动框架；3—拉杆；4，5—连杆

四连杆机构工作原理如图 3-20 所示，由电机通过减速机经偏心轮的传动，拉杆 3 做往复运动，拉杆 3 带动连杆 4 摆动；连杆 5 也随之摆动，使振动框架 2 能按弧线轨迹振动。

(2) Four - eccentric oscillation mechanism. Four - eccentric oscillation mechanism still belongs to sinusoidal oscillation, as shown in Figure 3-21. The motor 1 drives the central reducer 3, and drives the reducer 4 on the left and right sides through the cardan shaft. Each reducer drives the eccentric wheels 6 and 7 respectively. The eccentric wheels 6 and 7 have the same direction eccentric points, but the eccentricity is different. The arc operation of the mold uses two plate springs 8, one end of which is connected to the oscillation table frame 9, and the other end is connected to the proper position of the oscillation device to realize the arc oscillation. This kind of plate spring makes the oscillation table only swing in an arc without moving back and forth. Due to the small amplitude of the mold, the horizontal installation of two eccentric shafts will not cause obvious errors.

(2) 四偏心振动机构。四偏心振动机构仍属于正弦振动，如图 3-21 所示。电动机 1 带动中心减速机 3，通过万向轴带动左右两侧的减速机 4，每个减速机各自带动偏心轮 6 和 7；偏心轮 6 和 7 具有同向偏心点，但偏心距不同；结晶器弧线运行是利用两条板式弹簧 8，一端连接振动台框架 9，另一端连接在振动装置恰当的位置，实现弧形振动。这种板式弹簧使得振动台只能做弧线摆动，不会前后移动，由于结晶器振幅不大，两根偏心轴的水平安装，不会引起明显的误差。

The advantages of the four eccentric oscillation mechanism are: the mold oscillation is stable, no oscillation and jamming phenomenon, suitable for the application of high frequency and small amplitude technology, but the structure is complex.

四偏心振动机构的优点是：结晶器振动平稳，无摆动和卡阻现象，适合高频小振幅技术的应用，但结构较复杂。

(3) Mold hydraulic oscillation mechanism. Using hydraulic system as oscillation source, it

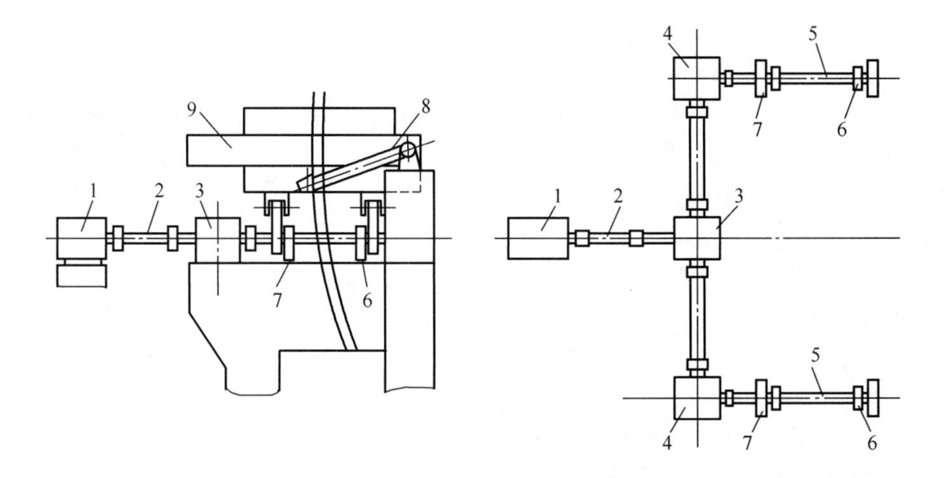

Figure 3-21　Four-eccentric oscillation mechanism

1—Motor; 2—Universal shaft; 3—Center reducer; 4—Angle reducer; 5—Eccentric shaft; 6, 7—Eccentric wheel; 8—Plate spring; 9—Oscillation table frame

图 3-21　四偏心振动机构

1—电动机；2—万向接轴；3—中心减速机；4—角部减速机；5—偏心轴；6，7—偏心轮；8—板式弹簧；9—振动台框架

has the characteristics of high control accuracy, flexible adjustment, small volume of online equipment, light weight and simple maintenance. It can not only meet the requirements of high frequency oscillation, but also eliminate the problems of motor burning and reducer damage caused by impact load in the transmission of motor and reducer. What's more, it can arbitrarily change the oscillation waveform according to the process requirements, control the negative strip speed and time, improve the lubrication and demolding between the mold and the continuous casting slab, and reduce the adhesive steel leakage accident; at the same time, it can reduce the oscillation frequency under the condition of high casting speed and the mechanism wear.

（3）结晶器液压振动机构。采用液压系统作为振动源，具有控制精度高、调整灵活、在线设备体积小、质量轻、维护简单等特点。它不仅能满足高频振动的要求，消除电机、减速器传动中由于冲击负荷造成电机烧损和减速器损坏等问题，更主要的是它可以根据工艺要求任意改变振动波形，控制负滑脱速度和负滑脱时间，改善结晶器与连铸坯之间的润滑与脱模，减少黏结性漏钢事故；同时降低了高拉速条件下的振动频率，减少机构磨损。

Task Implementation 任务实施

3.3.4　Inspection of Mold

3.3.4.1　Tool Preparation

（1）A steel ruler with millimeter scale long enough.

（2）A micrometer caliper matched with the mold size.

(3) One set of common internal and external card gauge.

(4) A set of taper meter.

(5) A pair of feeler gauge.

(6) There is one arc plate and one straight plate with enough length for mold alignment. Flatness and straight line must be verified.

(7) One set of low-voltage lighting.

(8) One set of water quality analyzer.

3.3.4 结晶器的检查

3.3.4.1 工具准备

(1) 足够长的带毫米刻度的钢尺一把。

(2) 与结晶器尺寸配套的千分卡尺一把。

(3) 普通内、外卡规一副。

(4) 锥度仪一套。

(5) 塞尺一副。

(6) 结晶器对中用的有足够长的弧度板、直板各一块。板度、直线必须经过校验。

(7) 低压照明灯一套。

(8) 水质分析仪一套。

3.3.4.2 Inspection Steps of Mold

(1) Inspect the inner wall of mold:

1) Visual inspection of the internal surface damage of the mold, focusing on the coating (or copper plate) wear, pits, cracks and other defects.

2) Use caliper, micrometer and ruler to check the dimension of upper and lower section of mold.

3) Check the mold side taper with a taper tester.

4) For the combined mold, the gap between the wide and narrow copper plates should be checked with a feeler gauge.

(2) Check the alignment between the mold and the secondary cooling section with arc plate and straight plate.

(3) After the cooling water of the mold is opened, check whether there is leakage or water leakage in the mold device.

(4) Check the temperature, pressure and flow of water in the mold, and observe the water temperature difference between the inlet and outlet of the mold during casting.

3.3.4.2 结晶器检查步骤

(1) 结晶器内壁检查:

1) 肉眼检查结晶器内表面损坏情况，重点在于镀层（或铜板）磨损、凹坑、裂纹等缺陷。

2）用卡规、千分卡、直尺检查结晶器上、下口断面尺寸。

3）用锥度仪检查结晶器侧面锥度。

4）对组合式结晶器，需用塞尺检查宽面和窄面铜板之间的缝隙。

（2）用弧度板、直板检查结晶器与二冷段的对中。

（3）结晶器冷却水开通后，检查结晶器装置是否有渗、漏水。

（4）检查结晶器进水温度、压力、流量，在浇铸过程中观察结晶器进出水温差。

3.3.5 Maintenance of Mold

(1) The damage to the inner wall of mold caused by improper operation should be avoided.

(2) The mold water tank shall be cleaned and decontaminated regularly, and the sealing parts shall be replaced regularly.

(3) Regularly and regularly analyze the cooling water quality of mold to ensure it meets the requirements.

(4) The inlet and outlet water pipes shall be flushed during the maintenance and replacement of the mold.

3.3.5 结晶器的维护

（1）使用中应避免各种不当操作对结晶器内壁的损坏。

（2）结晶器水槽应定期进行清理、除污，密封件应定期调换。

（3）定期、定时分析结晶器冷却水水质，保证符合要求。

（4）结晶器检修调换时应对进出水管路进行冲洗。

3.3.6 Adjustment of Mold Width and Taper

The adjustment and locking of mold width and taper of slab caster is a basic operation content that continuous casting operators often engage in. The online width adjustment and taper adjustment of the mold are all operated by electric coarse adjustment and manual fine adjustment. Generally, the following operation points shall be followed:

(1) The mold online shutdown adjustment and dimension adjustment device can only conduct width adjustment and taper adjustment in the preparation mode after shutdown, and is not allowed to operate in other modes.

(2) Before the width and taper adjustment of the mold, the wide panel clamping the narrow panel must be loosened first, and foreign matters such as slag and garbage in the gap shall be checked and removed to avoid scratches.

(3) According to the required width of the mold, start the drive motor to make the narrow panels on both sides of the mold move forward or backward as a whole.

(4) Based on the center line of the upper inlet of the mold, use a ruler to measure the width of the upper inlet of the narrow panel on the left and right sides of the mold respectively to check whether the width of the mold meets the requirements.

(5) After the width and size of the mold are roughly adjusted by electric motor, then the taper and width of the mold are adjusted manually and precisely.

(6) The taper state of the narrow mold panel is measured by using the mold taper meter. Then, according to the difference between the set taper value and the actual measured value, the manual adjustment of the hand wheel is used to fine tune and make it reach the set taper position state.

(7) After the taper state of the narrow panels on the left and right sides of the mold is manually adjusted in place, it is necessary to retest and adjust the width of the upper inlet of the mold.

(8) After the manual taper adjustment and width adjustment are completed, the manual adjusting hand wheel can be pulled out and recycled.

(9) Finally, the mold wide panel in the loose state is tightened again to clamp the narrow panel to lock it.

3.3.6 结晶器宽度及锥度的调整

对板坯连铸机进行结晶器宽度及锥度的调整、锁定是连铸操作人员经常从事的一项基本的操作内容。结晶器在线调宽、调锥度装置的调整方式都是采用电动粗调、手动精调等操作，通常应遵循以下操作要点：

(1) 结晶器在线停机调整、调维度装置只能在停机后的准备模式状态下进行调宽、调锥度操作，在其他模式状态下不允许操作。

(2) 在实施结晶器调宽、调锥度前，必须先将夹紧窄面板的宽面板松开，并检查和清除积缝隙内的黏渣、垃圾等异物，以避免划伤。

(3) 根据结晶器所需调整宽度的尺寸，启动驱动电动机，使结晶器两侧的窄面板分别整体向前或向后移动。

(4) 以结晶器上口中心线为基准，使用直尺分别测量结晶器左右两侧窄面板上口的宽度尺寸，以检查结晶器的宽度尺寸是否达到要求。

(5) 结晶器宽度尺寸进行电动粗调后，接着进行手动精调的调锥度、调宽操作。

(6) 使用结晶器锥度仪对结晶器窄面板的锥度状态进行测量，然后根据设定的锥度值与实际测量数值的差值，通过手动调节手轮进行微调，并使之达到设定的锥度位置状态。

(7) 结晶器左右两侧窄面板的锥度状态经手动调整到位后，需对结晶器上口的宽度尺寸进行复测和调整。

(8) 手动调锥、调宽操作全部结束后，可将手动调节手轮拔下、回收。

(9) 最后将处于松开状态的结晶器宽面板重新收紧，以夹住窄面板使其锁定。

3.3.7 Inspection of Mold Oscillation Device

(1) Check the lubrication system of the oscillation device to ensure normal operation.

(2) Release the electrical control interlock of oscillation and tension leveler, start the oscillation mechanism, and adjust the oscillation frequency to the highest working frequency matching the highest working speed.

(3) Observe and listen to the whole transmission process of the oscillation mechanism to en-

sure there is no abnormal sound.

(4) Check the oscillation frequency with a stopwatch or watch to ensure it is within the error range required by the process (±1time/min).

(5) Check the amplitude with a ruler to ensure it is within the error range (±0.5mm) required by the process.

(6) Observe the balance of the oscillation device. If there is any abnormality, the fitter shall be required to conduct further inspection.

(7) Adjust the oscillation frequency to the working frequency matching the average and minimum working casting speed, and then check the above items (3)~(6) respectively to ensure that they are normal.

3.3.7 结晶器振动装置的检查

(1) 检查振动装置的润滑系统，确保运行正常。

(2) 解除振动和拉矫机的电气控制联锁，开动振动机构，把振动频率调到与最高工作拉速相配的最高工作频率。

(3) 观察和倾听振动机构的整个传动过程，确保没有异声。

(4) 用秒表或手表，检查振频，确保在工艺要求误差范围内（±1 次/min）。

(5) 用直尺检查振幅，确保在工艺要求的误差范围内（±0.5mm）。

(6) 观察振动装置的平衡性，如有异常，应要求钳工做进一步的检查。

(7) 把振频调到与平均和最低工作拉速相匹配的工作频率，然后分别进行上述第(3)~(6) 项的检查，确保正常。

3.3.8 Maintenance of Oscillation Device

(1) At the end of casting, it is necessary to remove the protective slag, steel slag and other wastes around the mold oscillation device, mold roll which oscillates synchronously with the oscillation device, so as to ensure cleanness.

(2) After casting, check the lubrication device to ensure the system is normal.

(3) For lubrication points requiring manual oil filling, manual oil filling shall be conducted according to the interval time required by the process.

(4) According to the requirements of spot inspection, check and maintain the protective device attached to the oscillation device on time (to prevent the splashing of liquid steel) to ensure it is normal.

3.3.8 振动装置的维护

(1) 浇铸结束，必须清除结晶器、振动装置、与振动装置同步振动的结晶器辊等设备周围的保护渣、钢渣等垃圾，保证清洁。

(2) 浇铸结束，检查集中润滑装置，保证系统正常。

(3) 需人工加油的润滑点，按工艺要求的间隔时间做人工加油。

(4) 按点检要求，按时检查保养振动装置附属的防护装置（防止钢液飞溅），确保正常。

3. 3. 9 Inspection Method of Mold Oscillation Device

(1) Coin detection. The operation method of the coin detection method is to place the coin vertically on the oscillation device of the mold in the windless state, or on the four corners of the oscillation frame or on the horizontal plane of the inner and outer arcs of the mold. If the coin can vibrate with the oscillation device for a long time without moving or falling down, the oscillation state of the oscillation device is considered to be better.

(2) One bowl water test method. The operation method of a bowl of water detection method is to place a flat bottom bowl with more than half of the bowl of water on the inner arc side water tank or the outer arc side water tank of the mold, observe the fluctuation of the liquid level and the change of the ripples in the bowl of water, so as to determine the oscillation condition of the mold oscillation device. If the liquid level fluctuates obviously, the oscillation state of the oscillation device is relatively poor.

(3) Dial indicator test method. The operation of the dial indicator detection method can be divided into the detection of lateral displacement and shaking momentum and the detection of oscillation state in the vertical direction according to the detection content. Use the dial indicator to check the oscillation state in each direction and observe whether it meets the required deviation range.

3. 3. 9 结晶器振动装置检测方法

(1) 硬币检测法。硬币检测法的操作方法是在无风状态下将硬币垂直放置在结晶器振动装置上，或放在振动框架的 4 个角部位置或结晶器内、外弧水平面的位置上。如果硬币能较长时间随振动装置一起振动而不移动或倒下，则可认为该振动装置的振动状态比较好。

(2) 一碗水检测法。一碗水检测法的操作方法是将一只装有大半碗水的平底碗放置在结晶器的内弧侧水箱或外弧侧水箱上，观察这碗水中液面的波动及波纹的变化情况，来判定结晶器振动装置振动状况。如果其液面有明显波动，则说明该振动装置的振动状态比较差。

(3) 百分表检测法。百分表检测法的操作按其检测的内容可分侧向偏移与晃动量的检测及垂直方向的振动状态检测等。利用百分表检查各方向的振动状态，观察其是否符合要求的偏移范围。

Exercises:

(1) What is the function of the mold? How to ensure that the shell is not broken?

(2) Why does the mold oscillate during casting?

(3) Use the simulation software to complete the operation of the mold, check the cooling water, oscillation, mold powder feeding and monitor the change of the molten steel level.

思考与习题：

(1) 结晶器的作用是什么，如何保证坯壳不被拉断？

（2）结晶器在浇铸时为什么要进行振动？

（3）利用仿真软件完成结晶器相关操作，检查冷却水、振动、保护渣加入情况，监测钢液面变化。

Task 3.4　Operation of Secondary Cooling System
任务 3.4　二次冷却系统操作

Mission objectives:

任务目标：

（1）Master the structure and main parameters of secondary cooling system.

（1）掌握二次冷却系统的结构和主要参数。

（2）Master the secondary cooling mode and system

（2）掌握二次冷却方式和制度。

（3）Be able to check, operate and maintain the secondary cooling device.

（3）能够对二次冷却装置进行检查、操作和维护。

Task Preparation 任务准备

3.4.1　Cognition of Secondary Cooling Area

The secondary cooling area usually refers to the area below the mold to the area before the tension leveler. Although the billets coming out of the mold have been formed, the shell is generally only 10~30mm thick, so it needs to continue cooling in the secondary cooling area and complete solidification.

3.4.1　二次冷却区认知

二次冷却区通常是指结晶器以下到拉矫机以前的区域，从结晶器里出来的铸坯虽已成型，但坯壳厚度一般只有 10~30mm，需要在二冷区域继续冷却完全凝固。

The purpose of setting up the secondary cooling device is to force and evenly cool the billet to make the shell solidify rapidly, prevent the deformation of the shell from exceeding the limit, control the crack and steel leakage; at the same time, support and guide the billet and dummy bar; in the straight arc caster, the secondary cooling device must also bend the straight billet into an arc billet and enter the arc section; in the caster with dummy bar installed on it, It is also necessary to set up a driving roller in the secondary cooling zone to drive the dummy bar to realize billet drawing; for the multi radius arc caster, it also plays the role of sectional straightening of the arc billet.

设置二冷装置的目的，就是对铸坯通过强制而均匀的冷却，促使坯壳迅速凝固，预防坯壳变形超过极限，控制产生裂纹和发生漏钢；同时支撑和导向铸坯和引锭杆；在直弧形连铸机中，二冷装置还需要把直坯弯曲成弧形坯，进入弧形段；在上装引锭杆的连铸机中，还需在二冷区里设置驱动辊，以驱动引锭杆实现拉坯；对于多半径弧形连铸机，它又起到将弧形铸坯分段矫直的作用。

3.4.2 The Structure of Secondary Cooling Device

The pinch roll of the house type structure is all arranged on the support of the open memorial archway structure, and the whole secondary cooling area is composed of one or several sections of racks. Steel plates are used to form a closed room around the secondary cooling area, so it is called a room structure, as shown in Figure 3-22. The utility model has the advantages of simple structure, convenient observation equipment and casting billet. However, the fan capacity and floor area are large.

3.4.2 二次冷却装置结构

房式结构的夹辊全部布置在敞开的牌坊结构的支架上，整个二冷区由一段或若干段机架组成。在二冷区四周用钢板构成封闭的房室，故称为房式结构，如图 3-22 所示。其具有结构简单，观察设备和铸坯方便等一系列优点，但风机容量和占地面积较大。

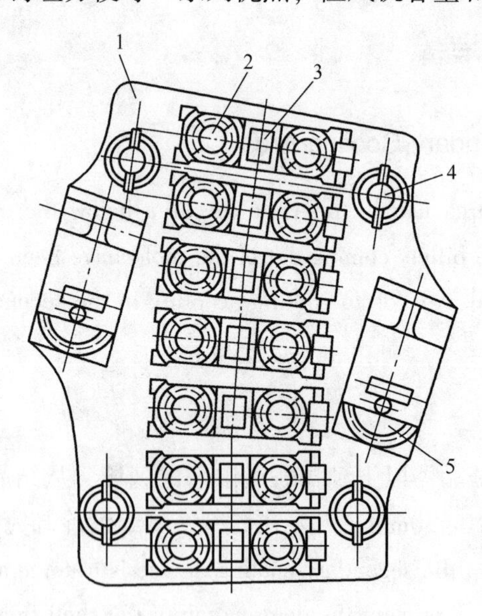

Figure 3-22 Room type structure

1—Archway frame; 2—Pinch roll; 3—Cushion block; 4—Pull rod; 5—Wheel (outgoing trolley)

图 3-22 房式结构图

1—牌坊架；2—夹辊；3—垫块 4—拉杆；5—车轮（开出式小车）

As the base of the secondary cooling device is under the action of high temperature and large drawing force for a long time, the secondary cooling support and guide device is installed on the foundation through a common base with strong rigidity (shown in Figure 3-23); 5 is the fixed support and 4 is the movable support, which is allowed to slide along the circular arc direction to avoid the wrong arc caused by the poor deformation resistance.

由于二次冷却装置底座长期处于高温和很大拉坯力的作用下，因此二冷区支导装置通过刚性很强的共同底座安装在基础上（见图 3-23）；图中 5 为固定支点，4 为活动支点，允许沿圆弧线方向滑动，以避免抗变形能力差导致的错弧。

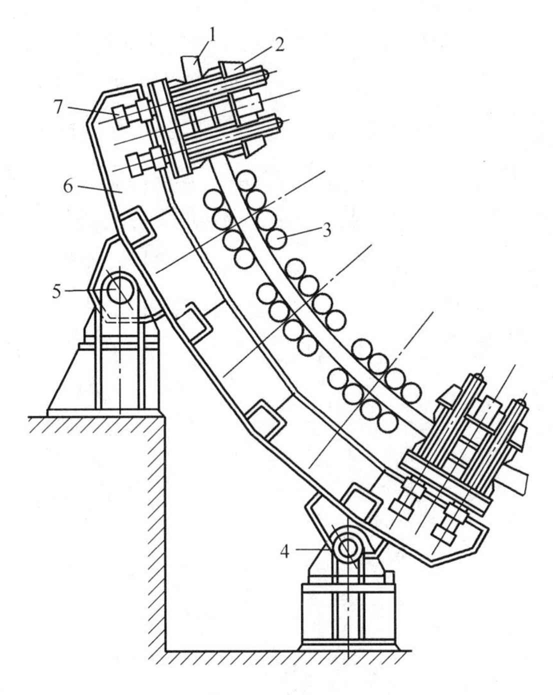

Figure 3-23 Base of secondary cooling support and guide device

1—Billet; 2—Sector section; 3—Pinch roll; 4—Movable fulcrum; 5—Fixed fulcrum; 6—Base; 7—Hydraulic cylinder

图 3-23 二冷区支导装置的底座

1—铸坯；2—扇形段；3—夹辊；4—活动支点；5—固定支点；6—底座；7—液压缸

3.4.2.1 The Secondary Cooling Device of Billet Caster

The section of small square billet is small, and the shell with enough thickness has been formed when it is out of the mold. Generally, the deformation will not occur. Therefore, the secondary cooling device of billet casters is very simple, as shown in Figure 3-24. Generally, only the upper part of the arc section is sprayed with water to cool the billet, and the lower part is not sprayed with water. There is little or no pinch roller in the whole arc section. The supporting guide device is used to feed the dummy bar. Therefore, there are only 4 pairs of pinch rolls, 5 pairs of side guide rolls, 2 guide plates and 14 spray rings in the secondary cooling zone of some billet casters. The distance between rollers can be adjusted by cushion block to adapt to the casting of different sections. For example, the Concast billet caster is equipped with a movable support plate at the outlet of the mold. In the stage of loading dummy bar and starting casting, the support plate plays a supporting role; when the billet passes through the support plate, the support plate can be moved back for a certain distance to prevent steel leakage damage and facilitate the handling of accidents. Another example is Rokop billet caster, which adopts rigid dummy bar, and the secondary cooling device can be more simplified.

3.4.2.1 小方坯连铸机二次冷却装置

小方坯铸坯断面小，在出结晶器时已形成足够厚度的坯壳，一般情况下，不会发生变形现象。因此，小方坯连铸机的二次冷却装置非常简单，如图 3-24 所示，通常只在弧形段的上半部喷水冷却铸坯，下半段不喷水。在整个弧形段少设或不设夹辊。其支撑导向装置是用来上引锭杆的。所以，有的小方坯连铸机的二次冷却区只有 4 对夹辊、5 对侧导辊、2 块导板和 14 个喷水环。用垫块调节辊距以适应不同断面的浇铸。如康卡斯特小方坯连铸机，在结晶器的出口处安装了一个能够移动的托板。在装引锭杆和开浇阶段，托板起支撑作用；当铸坯通过托板后，托板可以后移一段距离，以防止漏钢损坏，也便于处理事故。再如罗可普小方坯连铸机，采用刚性引锭杆，二次冷却装置可以简化。

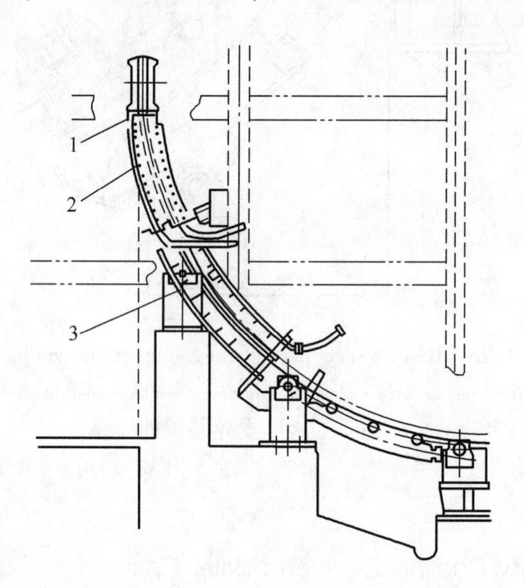

Figure 3-24 Secondary cooling device of billet caster

1—Foot roller section; 2—Movable spray section; 3—Fixed spray section

图 3-24 小方坯铸机二次冷却装置

1—足辊段；2—可移动喷淋段；3—固定喷淋段

3.4.2.2 Secondary Cooling Device of Bloom Caster

The bloom is thicker than billet, and the bloom may bulge after it comes out of the lower outlet of the mold. The secondary cooling device is divided into two parts. All sides of the upper part are supported by close row pinch roll and cooled by water spray; the billet strength of the lower part in secondary cooling zone is sufficient, and we don't need to set the pinch roll just like the lower part of the billet caster.

3.4.2.2 大方坯连铸机二次冷却装置

大方坯铸坯较厚，出结晶器下口后铸坯有可能发生鼓肚现象，其二次冷却装置分为两部分。上部四周均采用密排夹辊支撑，喷水冷却；二冷区的下部铸坯坯壳强度足够时，可像小方坯连铸机下部那样不设夹辊。

3.4.2.3 Secondary Cooling Device of Slab Caster

Slab caster is mainly composed of secondary cooling zero section and sector section.

3.4.2.3 板坯连铸机二次冷却装置

板坯连铸机主要由二冷零段和扇形段组成。

(1) Zero section of secondary cooling zone

The slab caster has a large section, thin shell at the bottom of the mold, and a long metallurgical length. Until the center of the slab in the straightening area is still in the liquid state, it is easy to have bulging deformation, which may cause steel leakage in serious cases. Therefore, the lower outlet of the mold is generally equipped with a dense row foot roller or a cooling grid. The roller group below the mold foot roller is called the secondary cooling zero section. Generally, 10~12 pairs of dense row pinch rollers can be used, either long pinch roller or multi section pinch roller can be used in this area, which plays a role in guiding and supporting the slab and quickly thickens the slab shell.

(1) 二冷区零段。板坯连铸机由于铸坯断面很大，出结晶器下口坯壳较薄，冶金长度较长，直到矫直区铸坯中心仍处于液态，容易发生鼓肚变形，严重时可能造成漏钢。所以结晶器下口一般安有密排足辊或冷却格栅，结晶器足辊以下的辊子组称为二冷零段，一般是用10~12对密排夹辊，可以用长夹辊，也可以用多节夹辊，起到对铸坯导向支撑的作用，快速增厚坯壳。

(2) Sector segment. According to the type of caster, the structure, number of segments, roll diameter and roll spacing of each sector segment after zero segment are quite different. The section is composed of nip roll and its bearing seat, upper and lower frame, roll gap adjustment device, nip device of nip stick, cooling water pipe, grease pipe, etc., as shown in Figure 3-25.

(2) 扇形段。从零段以后的各扇形段的结构、段数、夹辊的辊径和辊距，根据铸机的类型，所浇钢种和铸坯断面的不同有很大差别。扇形段由夹辊及其轴承座、上下框架、辊缝调节装置、夹辊的压下装置、冷却水配管、给油脂配管等部分组成，如图3-25所示。

The sector section can be equipped with power device, which plays the role of blank drawing and straightening. Generally, DC motor is used to drive through planetary gear reducer. The roller

Figure 3-25 Sector section

图 3-25 扇形段

gap adjustment device of sector section generally adopts hydraulic mechanism. The mold, secondary cooling zero section and each sector section must be aligned.

扇形段可以有动力装置，起拉坯和矫直作用。一般采用直流电机通过行星齿轮减速箱带动。扇形段的辊缝调节装置一般采用液压机构。结晶器、二冷零段、各扇形段必须对中。

3. 4. 3 Cognition of Secondary Cooling System

3. 4. 3. 1 Spray Water Intensity

The cooling intensity of secondary cooling area is generally expressed by specific water volume (spray water intensity). The specific water quantity is defined as the ratio of the cooling water consumed to the weight of the slab passing through the secondary cooling zone, in kg(water)/kg (steel) or L(water)/kg (steel). The specific water quantity is related to the type of caster, section size, steel grade and other factors. The selection of specific water quantity parameters is more complex and more factors are considered. Table 3-4 is the general relationship between steel grades and specific water volume.

3. 4. 3 二次冷却制度认知

3. 4. 3. 1 比水量

二次冷却区的冷却强度，一般用比水量来表示。比水量的定义是：所消耗的冷却水量与通过二冷区的铸坯重量的比值，单位为 kg(水)/kg(钢)或 L(水)/kg(钢)。比水量与铸机类型、断面尺寸、钢种等因素有关。比水量参数选择比较复杂，考虑因素较多。钢种与比水量大致关系可见表 3-4。

Table 3-4 Relationship between steel grades and specific water volume

表 3-4 钢种与比水量关系

Steel grades 钢种	Specific water volume 比水量/L · kg^{-1}
Common steel 普通钢	1. 0~1. 2
Medium and high carbon steel, alloy steel 中高碳钢、合金钢	0. 6~0. 8
Steel with high crack sensitivity (pipeline, low alloy steel) 裂纹敏感性强的钢（管线、低合金钢）	0. 4~0. 6
High speed steel 高速钢	0. 1~0. 3

3. 4. 3. 2 Parameters of Secondary Cooling

In order to obtain a good cooling effect, it is important to have a proper nozzle. When selecting the nozzle, the cold characteristics of the nozzle are mainly considered:

(1) Water flow characteristics. The water flow of the nozzle under different water supply pressure.

(2) Water flow density. The amount of water sprayed vertically to the unit area in unit time. With the increase of water flow density, heat transfer is accelerated.

(3) Water mist diameter. The smaller the water droplet diameter is, the better the atomization is, and the better the uniform cooling and heat transfer efficiency is.

(4) Water drop speed. The water drop speed is fast, easy to penetrate the vapor film, and the heat transfer efficiency is high.

(5) Spray area. Ensure a certain cone bottom area and uniform water distribution on the whole area to ensure uniform cooling.

3. 4. 3. 2 二次冷却参数

为获得良好的冷却效果，合适的喷嘴很重要。选择喷嘴时主要考虑喷嘴的冷态特性如下：

（1）水流量特性。不同供水压力下喷嘴的水流量。

（2）水流密度。单位时间喷嘴垂直喷射到单位面积上的水量。水流密度增加，传热加快。

（3）水雾直径。水滴直径越小，雾化越好，越有利于铸坯均匀冷却和传热效率的提高。

（4）水滴速度。水滴速度快，容易穿透汽膜，传热效率就高。

（5）喷射面积。保证一定的锥底面积并在整个面积上水量分布均匀，保证均匀冷却。

3. 4. 3. 3 Distribution of Cooling Water in the Direction of Slab Length

In the solidification process, the heat in the center of the slab is transferred to the surface of the slab through the shell, and the heat transfer slows down with the increasing thickness of the shell. The heat transferred from the shell to the surface is mainly carried away by the water droplets sprayed to the surface. Therefore, with the continuous increase of the shell thickness, the heat transferred to the surface is gradually reduced, and the natural cooling water volume should also be gradually reduced. However, in order to facilitate the water control, the secondary cooling is usually divided into several cooling zones, and then the total water is distributed to each cooling zone in a certain gradually decreasing proportion.

3.4.3.3 铸坯长度方向上冷却水的分配

在凝固过程中，铸坯中心的热量是通过坯壳传到铸坯表面的，而这种热量的传递随着坯壳厚度不断增加而变缓。而从坯壳传到表面的热量主要是由喷射到表面的水滴带走的，所以，随着坯壳厚度的不断增加，传到表面热量的逐渐减小，自然冷却水量也应随之逐渐减少。但是为了便于水量控制，通常将二次冷却区分成若干冷却区，随后将总水量按一定逐渐减小的比例分配给各个冷却区。

3.4.3.4 Water Distribution in Inner and Outer Arc of Casting Billet

Different from the vertical caster, the cooling conditions of the inner and outer arcs of the arc caster are quite different. When just out of the mold, the cooling section is close to the vertical arrangement, so the water distribution of inner and outer arcs cooling should be the same. As far away from the mold, for the inner arc, the water without vaporization will continue to flow down to cool down, while the spray water without vaporization in the outer arc will leave the billet immediately due to gravity. As the billet tends to level, the difference is more and more big. For this reason, the water volume of inner and outer arcs generally changes from 1 : 1 to 1 : 1.5.

3.4.3.4 铸坯内外弧的水分配

当刚出结晶器时，因冷却段接近于垂直布置，因此，内外弧冷却水量分配应该相同。随着远离结晶器，对于内弧来说，部分没有汽化的水会往下流继续起冷却作用，而外弧的喷淋水没有汽化部分则因重力作用而即刻离开铸坯。随着铸坯趋于水平，差别越来越大，内外弧的水量一般作 1 : 1 到 1 : 1.5 的比例变化。

3.4.3.5 Relation of Secondary Cooling Water and Casting Speed

The specific water quantity is considered by the quality of casting billet. The faster the casting speed is, the more the quality of casting billet per unit time is, and the greater the water supply per unit time is. On the contrary, the water quantity decreases.

3.4.3.5 二次冷却水与拉速的关系

比水量是以铸机内通过的铸坯重量来考虑的，拉速越快，单位时间通过铸坯重量越多，单位时间供水量也应越大。反之，水量则减小。

3.4.4 Cooling Mode and Equipment

3.4.4.1 Water spray cooling

The so-called water spray cooling is a cooling way that atomized by water pressure. The nozzle can be divided into solid cone, hollow cone, rectangle, flat shape, etc. according to the shape of the spray, as shown in Figure 3-26. Generally, solid conical nozzle is used for billet cooling, and hollow conical nozzle is also used. Rectangular or flat nozzle is used for slab cooling.

3.4.4 冷却方式及设备

3.4.4.1 水喷雾冷却

所谓水喷雾冷却就是靠水的压力使其雾化的一种冷却方式。喷嘴根据喷出水雾的形状可分为实心圆锥形、空心圆锥形、矩形、扁平形等，如图 3-26 所示。方坯冷却一般采用实心圆锥形喷嘴，也有采用空心圆锥形喷嘴。板坯冷却采用矩形或扁平形喷嘴。

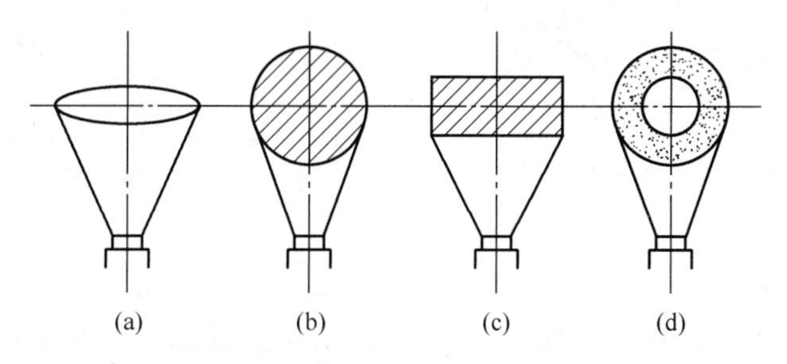

Figure 3-26 Several spray shape of atomizing nozzles

(a) Flat; (b) Cone (solid); (c) Rectangle; (d) Cone (hollow)

图 3-26 几种雾化喷嘴的喷雾形状

（a）扁平形；（b）圆锥形（实心）；（c）矩形；（d）圆锥形（空心）

There are two types of water spray arrangement in the secondary cooling zone of billet caster, ring tube type and single tube type, as shown in Figure 3-27. Because the single tube type is convenient for maintenance, and it is widely used.

小方坯连铸机二冷区喷水布置有环管式和单管式两种，如图 3-27 所示。由于单管式布置维修方便，所以采用此种布置较多。

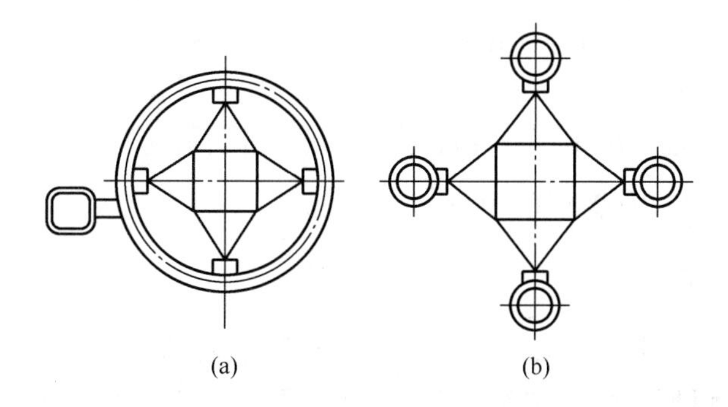

Figure 3-27 Layout of billet nozzle

(a) Ring tube type; (b) Single tube type

图 3-27 小方坯喷嘴布置

（a）环管式；（b）单管式

According to the arrangement of the number of nozzles, the water spray system in slab secondary cooling zone can be divided into single nozzle system (shown in Figure 3-28) and multi nozzle system (shown in Figure 3-29).

板坯二冷区喷水系统根据喷嘴数量的排列区分可分为单喷嘴系统（见图 3-28）和多喷嘴系统（见图 3-29）。

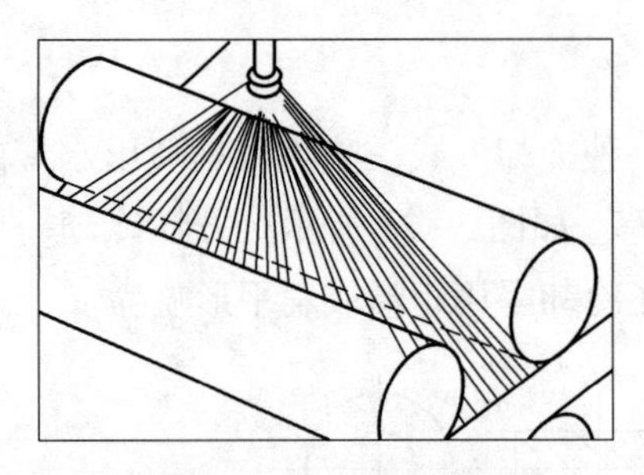

Figure 3-28 Single nozzle system

图 3-28 单喷嘴系统

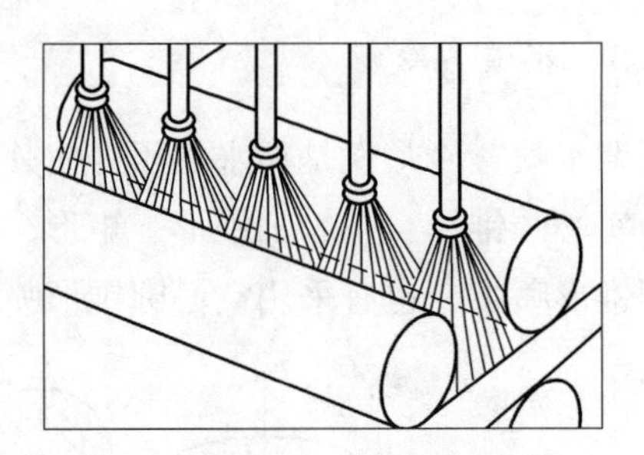

Figure 3-29 Multi nozzle system

图 3-29 多喷嘴系统

In the single nozzle system, only one large angle flat nozzle (sometimes two nozzles) is set in each roll gap to cover all cooling surfaces. The multi nozzle system is equipped with several small angle solid nozzles in each gap. It is arranged in a row, forming a spray surface to cover the cooling surface.

单喷嘴系统是每个辊缝间隙内只设一个大角度扁平喷嘴（有时也设两个），就把全部冷却面覆盖住。多喷嘴系统是每个辊缝间隙内设若干个较小角度实心喷嘴，排成一行，组成一个喷雾面把冷却面覆盖住。

3.4.4.2 Spray Cooling with Water and Air (air mist spray)

In terms of the number of nozzles, air-water cooling also has single nozzle and multiple nozzles. It is characterized by introducing compressed air into the nozzle and mixing with water to form a high-speed air mist at the nozzle outlet. This air mist contains a large number of water droplets with small particles, high speed and large kinetic energy, that is, a wide-angle jet stream with good atomization and high impact force, so as to achieve a high cooling effect and even uniformity, mostly used in slab and bloom caster, as shown in Figure 3-30 and Figure 3-31. Therefore, the cooling effect is greatly improved.

3.4.4.2 气-水喷雾冷却

气-水冷却就喷嘴数量而言也有单喷嘴和多喷嘴，它的特点是将压缩空气引入喷嘴，与水混合后使喷嘴出口形成高速“气雾”，这种“气雾”中含有大量颗粒小、速度快、动能大的水滴，即喷出雾化很好的、高冲击力的广角射流股，以达到对铸坯很高的冷却效果和均匀程度，多用在板坯及大方坯连铸机上，如图 3-30 和图 3-31 所示，因此冷却效果大大改善。

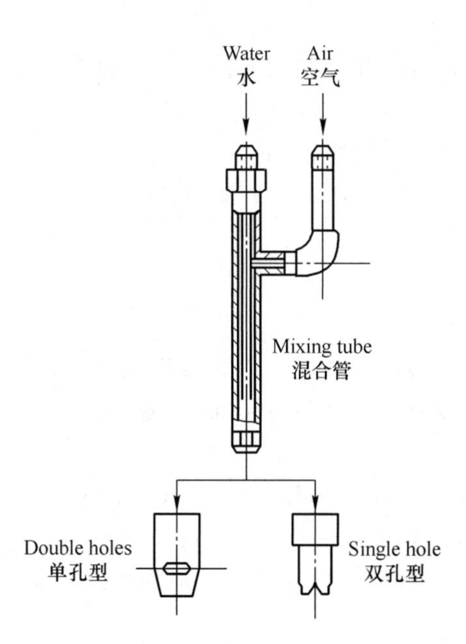

Figure 3-30 Structure of air mist nozzle

图 3-30 气-水喷嘴结构

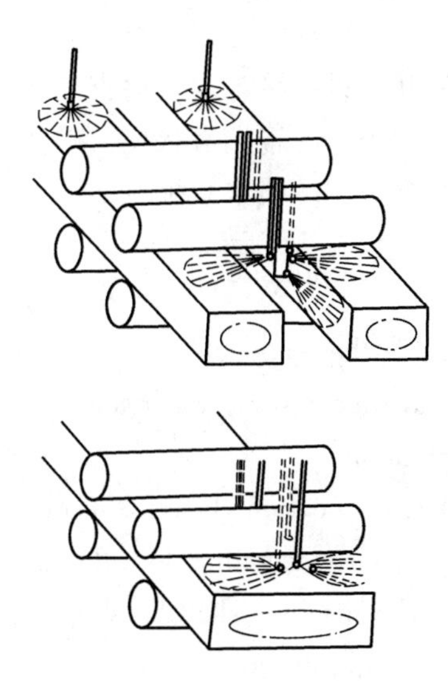

Figure 3-31 Air-water cooling spray system

图 3-31 气-水冷却喷雾系统

Task Implementation 任务实施

3.4.5 Inspection of Secondary Cooling Device

3.4.5.1 Operation Steps

(1) Use the counter arc template to check the secondary cooling radian.

(2) Check the opening and rotation of the secondary cooling pinch roll.

(3) If the roller clearance measuring instrument is used, the inspection contents of (1) ~ (2) can be completed by the operation of the roller clearance measuring instrument.

(4) Check the alignment between the mold and the secondary cooling side roll with a ruler or taper meter.

(5) Turn on the cooling water of the equipment and check the cooling water leakage of the secondary cooling equipment.

(6) Check the lubrication condition of the secondary cooling roller to ensure that centralized lubrication is in place or manual lubrication is in place.

(7) Check the secondary cooling water spray system.

3.4.5 二冷装置的检查

3.4.5.1 操作步骤

(1) 使用对弧样板检查二冷弧度。

(2) 检查二冷夹辊开口度和转动情况。

(3) 若使用辊缝测量仪，则可通过辊缝测量仪操作来完成（1）~(2) 的检查内容。

(4) 用直尺或锥度仪检查结晶器与二冷侧面辊的对中情况。

(5) 开启设备冷却水，检查二冷设备冷却的漏水情况。

(6) 检查二冷辊润滑状况，确保集中润滑到位，或手动润滑加油到位。

(7) 检查二冷喷水系统。

3. 4. 5. 2 Precautions

(1) For the sectional assembly of the secondary cooling section, the radian, opening, roll rotation, cold water, spray water and other conditions must be checked and meet the requirements before operation.

(2) On-line (on-board) secondary cooling opening, radian and roll rotation rate can be checked regularly according to different use conditions (according to the spot inspection system).

(3) The machine shall be shut down for maintenance when it is found that it does not meet the process requirements.

3. 4. 5. 2 注意事项

(1) 二冷段分段组装，在上机前对弧度、开口度、辊子转动、机冷水、喷淋水等情况必须进行检查并应符合要求。

(2) 在线的（机上）二冷开口度、弧度和辊子转动率可根据不同使用情况定期检查（按点检制度）。

(3) 凡检查发现不符合工艺要求时应停机检修。

3. 4. 6 Inspection of Secondary Cooling Nozzle

3. 4. 6. 1 Operation Steps

(1) Inspection before nozzle installation:

1) Theshape of old nozzle shall be intact.

2) The old nozzle shall be cleaned regularly to ensure the overall dimension and spray effect.

3) For new nozzles, calipers, feeler gauges or special measuring tools shall be used to spot check the external dimensions of some nozzles, especially the size of nozzle nozzles and spray angle.

4) Spot check some nozzles on the spray test bench to ensure the cold characteristics of nozzles (flow, water density distribution, water mist diameter, speed, spray area, etc.).

(2) Inspection after nozzle installation:

1) Check whether the nozzle is installed firmly and sealed.

2) Check whether the installation angle of the nozzle itself is correct, and ensure that the spray area of the nozzle does not fall on the secondary cooling roller.

3) Check whether the jet direction of nozzle is perpendicular to the surface of billet.

4) Check whether the size between nozzle and nozzle meets the process requirements.

5) Check that the distance between the nozzle and the billet is correct.

6) Check whether the nozzle model is consistent with the model required by the secondary cooling zone.

7) The above inspection can be carried out on-line (on-board) or off-line (on sector section commissioning platform).

(3) Check the secondary cooling water supply system to ensure the water level of cold and hot water pools. Spot check the water quality, start the water pump to ensure the normal operation of the water pump (check according to the spot inspection regulations).

(4) Open the water pump, adjust the secondary cooling control valves (simulate manual or automatic according to the pouring requirements), and ensure the normal pressure and flow.

(5) Under the condition of water supply, check the leakage of secondary cooling control room, on-board and spray pipeline system.

(6) Under the condition of water supply, check whether the drainage state of the caster is normal.

3.4.6 二冷喷嘴状态的检查

3.4.6.1 操作步骤

(1) 喷嘴安装前的检查：

1) 旧喷嘴要保证外形完整无损。

2) 旧喷嘴要定期清除结垢，保证外形尺寸和喷淋效果。

3) 新喷嘴要用卡尺、塞尺或专门量具等对部分喷嘴外形尺寸抽查，特别要注意喷嘴喷射口和喷射角大小的检查。

4) 在喷淋试验台上抽查部分喷嘴，确保喷嘴的冷态特性（流量、水密度分布、水雾直径、速度、喷射面积等）。

(2) 喷嘴安装后的检查：

1) 检查喷嘴是否安装牢固和密封。

2) 检查喷嘴本身安装的角度是否正确，确保喷嘴的喷射面积不落到二冷辊上。

3) 检查喷嘴射流方向是否与铸坯表面垂直。

4) 检查喷嘴与喷嘴之间的尺寸是否符合工艺要求。

5) 检查喷嘴与铸坯之间的距离是否正确。

6) 检查喷嘴型号是否与该二冷区要求的型号一致。

7) 上述检查可以在线（机上）检查，也可离线（在扇形段调试台上）检查。

(3) 检查二冷供水系统，保证冷、热水池水位。对水质进行抽查，开启水泵确保水泵正常运转等（按点检条例进行检查）。

(4) 开启水泵，调节二冷各项控制阀门（根据浇铸要求模拟手动或自动），确保压力和流量正常。

(5) 在通水的情况下，检查二冷控制室、机上、机旁喷淋管路系统的渗漏情况。

(6) 在通水的情况下，检查铸机排水状态是否正常。

3.4.6.2 Precautions

(1) Before installing the nozzle, the pipeline should be flushed to prevent the waste from blocking the nozzle.

(2) The nozzle installed offline must meet the process requirements before it is put on.

(3) If the spray state does not meet the requirements, casting is not allowed.

3.4.6.2 注意事项

(1) 在安装喷嘴前，应先对管路进行冲洗，以防垃圾堵塞喷嘴。

(2) 采用离线安装的喷嘴，上机前必须符合工艺要求。

(3) 喷淋状态不符合要求，不得进行浇铸。

Exercises:

(1) What is the function of secondary cooling? What are the requirements for water supply?

(2) What are the types and characteristics of secondary cooling zone nozzles?

(3) Used the simulation software to complete the operation of secondary cooling, observe the position of roll and the movement of billet, and monitor the status of cooling water in different areas.

思考与习题:

(1) 二次冷却的作用是什么，对供水有什么要求?

(2) 二次冷却区喷嘴有哪几种类型，各有什么特点?

(3) 利用仿真软件完成二次冷却相关操作，观察辊的位置和铸坯运动情况，监测不同区域的冷却水状态。

Task 3.5 Operation of Billet Withdrawing and Straightening Device

任务 3.5 拉坯矫直装置操作

Mission objectives:

任务目标:

(1) Master the structure and main parameters of the tension leveler system.

(1) 掌握拉矫装置系统的结构和主要参数。

(2) Master the working process of the tension leveler.

(2) 掌握拉矫机工作过程。

(3) Be able to inspect, operate and maintain the tension leveler.

(3) 能够对拉矫机进行检查、操作和维护。

Task Preparation 任务准备

3.5.1 Function of Tension Leveller

Among all kinds of continuous casting machines, there must be a billet withdrawing machine or a straightening machine. It is arranged at the tail of the guide device in the secondary cooling area, and bears the functions of billet withdrawing, straightening and dummy bar feeding.

3.5.1 拉矫装置的作用

在各种连铸机中，必须要有拉坯机或拉矫机。它布置在二次冷却区导向装置的尾部，承担拉坯、矫直和送引锭杆的作用。

The withdrawing and straightening machine consists of rollers or nippers. These rollers not only have the function of drawing billet, but also have the function of straightening billet. Therefore, the two processes of drawing billet and straightening of continuous casting billet are usually completed by one unit, so they are called the straightening machine of drawing billet, which is called the tension leveller for short.

拉坯矫直机由辊子或夹辊组成。这些辊子既有拉坯作用，也有矫直铸坯作用，所以连铸坯的拉坯和矫直这两个工序，通常由一个机组来完成，故称为拉坯矫直机，简称拉矫机。

Straightening is the process of making the billet arc-shaped, producing plastic deformation under the action of external torque, and becoming straight. The requirements for the straightening machine are: enough capacity to pull the billet, to overcome the resistance of each point of the billet; enough straightening force, to straighten the billet under the specified temperature; the speed of the billet can be adjusted.

矫直是使呈弧形铸坯，在外力矩作用下产生塑性变形，成为平直的过程。对拉矫机要求是：足够的拉坯能力，能克服铸坯各点阻力；有足够的矫直力，在规定的温度下能把铸坯矫直；拉坯速度可以调节。

3.5.2 Structure of the Tension Leveller

The form of tension leveller is usually nominal according to the number of rollers. Figure 3-32 is a five roll tension leveller, which is applied to billet caster. It consists of two identical stands and one lower roll. The upper pull roll 5 is driven by the air cylinder to swing up and down and driven by the motor to pull the billet. Two upper rollers and one middle lower roller form a straightening group to complete single point straightening. This five roller tension leveller is actually a two roller withdrawing and single point straightening machine.

3.5.2 拉坯矫直机的结构形式

拉矫机的形式通常按辊子多少来标称。图 3-32 是五辊拉矫机，它应用在小方坯连铸机上，它由两个相同的机架和一个下辊组成拉矫机组，上拉辊 5 由气缸带动能上下摆动并由电机驱动进行拉坯。而两个上辊和一个中间下辊这 3 个辊子组成一个矫直组，完成单点矫直，这种五辊拉矫机，实际上是两辊拉坯单点矫直的拉矫机。

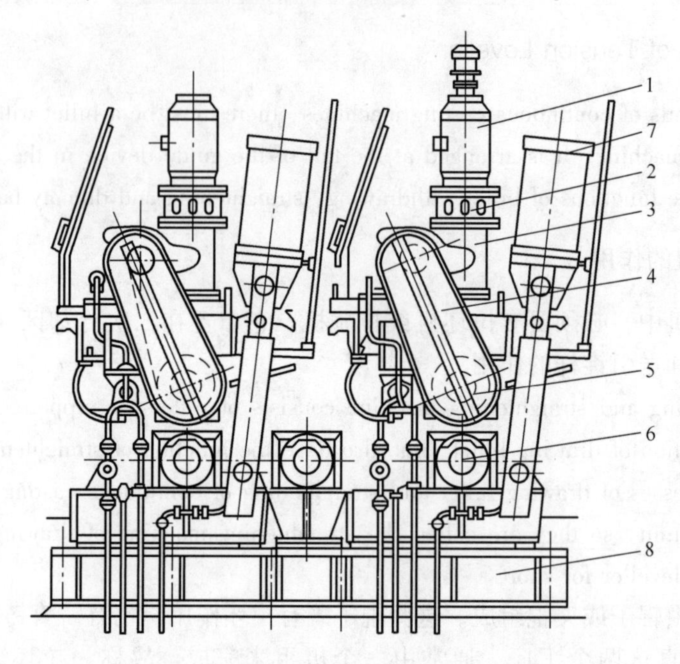

Fig 3-32　Five rollers tension leveller

1—Vertical DC motor；2—Brake；3—Reducer；4—Drive chain；5—Upper pull roller；6—Lower roller；7—Lower cylinder；8—Base

图 3-32　五辊拉矫机

1—立式直流电机；2—制动器；3—减速机；4—传动链；5—上拉辊；6—下辊；7—压下气缸；8—底座

The multi roll tension leveller is shown in Figure 3-33. This kind of tension leveller is mostly used in slab caster. It is a tension leveller with multi roll withdrawing and multi-point straightening.

多辊拉矫机如图 3-33 所示。这种拉矫机多使用在板坯连铸机上，它是多辊拉坯多点矫直的拉矫机机组。

3.5.3　Straightening Method of Continuous Casting Billet

The straightening of continuous casting billet can be divided into full solidification straightening and liquid core straightening according to the solidification state of billet during straightening, such as single point straightening, multi - point straightening and continuous straightening according to the arrangement of straightening roller.

3.5.3　连铸坯的矫直方式

连铸坯的矫直按矫直时铸坯凝固状态分有全凝固矫直和带液芯矫直，如按矫直辊布置方式分有单点矫直、多点矫直和连续矫直。

The thickness of billet is thin，such as small square billet，small rectangular billet，etc.，be-

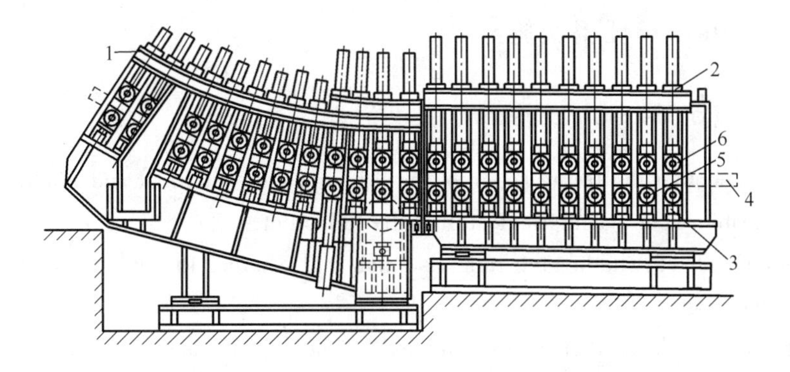

Figure 3-33　Multi roll tension leveller

1—Frame; 2—Upper roll pressing device; 3—Tension straightening roller and lifting device; 4—Billet; 5—Drive roller; 6—Driven roller

图 3-33　多辊拉矫机

1—机架；2—上辊压下装置；3—拉矫辊及升降装置；4—铸坯；5—驱动辊；6—从动辊

cause the thickness of billet is thin, solidification is fast, and the length of liquid core is short, it has all solidified when entering the straightening area. In this case, straightening is called full solidification straightening. Because the billet has been solidified completely, its strength is high and it can bear large strain, it can adopt single point straightening.

铸坯厚度较薄，如小方坯、小矩形坯等，由于铸坯厚度较薄，凝固较快，液芯长度较短，在进入矫直区时已全部凝固，在这种情况下矫直称为全凝固矫直。由于铸坯已全部凝固，强度较高，能承受较大的应变，可以采取单点矫直。

When the thickness of slab is large, such as slab, bloom, etc., the whole solidification time of slab is long, and the length of liquid core is also long. If solid-phase one point straightening is still used, its caster radius is large, so the straightening with liquid core is carried out. Because the strength of two-phase area of slab is very low, multi-point straightening (more than two points) is adopted to prevent internal crack caused by too large strain during single point straightening, that is, multi-point straightening with liquid core.

铸坯厚度较大时，如板坯、大方坯等，铸坯全部凝固时间较长，液芯长度也较长，如仍采用固相一点矫直，其铸机半径很大，因此采取仍有液芯的情况下进行矫直。由于铸坯两相区强度很低，为了防止单点矫直时应变过大而产生内裂，而采取多点矫直（两点以上称多点），即带液芯多点矫直。

Continuous straightening can also be adopted for straightening with liquid core. The so-called continuous straightening is the continuous straightening deformation of the slab in the straightening area, so its strain and strain rate are very low, which can greatly improve the stress state of the slab and improve the quality of the slab.

带液芯矫直还可采取连续矫直的方式，所谓连续矫直就是在矫直区内铸坯连续矫直变形，因此其应变和应变率都很低，可极大地改善铸坯受力状态，有利于提高铸坯质量。

Task Implementation 任务实施

3. 5. 4 Inspection of Tension Leveller

(1) Monitor whether the lifting action of the driving roller is normal and whether the brake of the reducer is flexible when casting and pulling the tail billet.

(2) Regularly check whether the roller is stuck and whether the crack and bending amount exceed the standard.

(3) Regularly check whether the opening and camber of the caster roller are within the specified range.

(4) Check the bearing regularly for abnormal noise and water leakage at the roller end.

(5) Regularly check whether there are residual steel and sundries on the upper and lower pinch roll and frame.

(6) Regularly check whether the pressure of the whole hydraulic system is normal, whether there is leakage in the hydraulic pipeline, whether the connecting bolt is loose or broken, and whether the connecting pin shaft of the clamping hydraulic cylinder is damaged.

(7) Regularly check whether the bearing dry oil lubrication is normal, whether there is leakage in the lubrication pipeline, and whether the oil is supplied on time. Add dry oil to universal joint shaft regularly.

(8) Regularly check whether the oil level and quality of the transmission gearbox are normal.

(9) Regularly check whether the cooling water and compressed air pipes have leakage, whether the nozzle is blocked, whether the rotary joint and the water connecting plate are easy to use and whether they are damaged.

(10) Regularly check whether the connecting bolts and locating pins are loose.

(11) Check the accuracy of the centering station regularly.

(12) After each maintenance, clean the dirt on each fitting surface and positioning reference surface, and apply oil to prevent rust.

3. 5. 4 拉矫机检查

(1) 开浇和拉尾坯时监视传动辊的升降动作是否正常、减速机的制动器开闭是否灵活。

(2) 定期检查辊子是否卡住，裂纹及弯曲量是否超标。

(3) 定期检查铸机辊子开口度、弧度是否在规定范围之内。

(4) 经常检查轴承是否有异常响声、辊端是否漏水。

(5) 定期检查上下夹辊及框架上是否有残钢、杂物。

(6) 定期检查整个液压系统压力是否正常，液压管路有无泄漏现象，连接螺栓是否松动断裂，夹持液压缸上下端连接销轴是否损坏。

(7) 定期检查轴承甘油润滑是否正常、润滑管路是否有泄漏，是否按时给油。定期给万向接轴人工加甘油。

（8）定期检查传动齿轮箱油位是否正常、油质是否良好。

（9）定期检查冷却水、压缩空气管路有无泄漏、喷嘴有无阻塞，旋转接头和水连接板是否好用、有无损坏现象。

（10）定期检查各连接螺栓和定位销有无松动。

（11）定期检查对中台的精度。

（12）每次检修后，清除干净各配合面和定位基准面上的污物，并涂油防锈。

Exercises:

(1) How does the tension leveller carry out the straightening process?

(2) Use the simulation software to adjust the casting speed and enter the withdrawing and straightening interface to observe the operation and deformation of the billet.

思考与习题：

（1）拉矫机是如何进行拉坯矫直过程的？

（2）利用仿真软件调整拉速，进入拉矫界面观察铸坯运行和变形情况。

Task 3.6 Operation of Dummy Bar Device
任务3.6 引锭装置操作

Mission objectives:

任务目标：

(1) Master the structure, function and main parameters of dummy bar device system.

（1）掌握引锭装置系统的结构、作用和主要参数。

(2) Master the working process of dummy bar.

（2）掌握引锭杆的工作过程。

(3) Be able to check and maintain the dummy bar device and complete the operation of dummy bar feeding.

（3）能够对引锭装置进行检查和维护，完成送引锭操作。

Task Preparation 任务准备

3.6.1 Cognition of Dummy Bar Device

Before the casting of continuous casting machine, the head of dummy bar blocks the lower outlet of the mold, temporarily forming the bottom of the mold, so that the molten steel does not leak out. When the liquid steel reaches a certain height, it starts to pull down the dummy bar through the pinch roller. At this time, the liquid steel has solidified at the head of the dummy bar, and the billet is gradually pulled out along with the dummy bar. After entering the tension leveller through the secondary cooling support device, the dummy bar completes the function of guiding the billet. At this time, the disengaging device separates the billet from the head of the dummy bar, and the tension leveller enters the normal working state of the billet and straightening state. The dummy bar shall be transported to the storage place for use in the next casting.

3.6.1 引锭装置认知

在连铸机开浇之前，引锭杆的头部堵住结晶器的下口，临时形成结晶器的底，使钢水不漏出。当钢水达到一定的高度时，通过拉辊开始向下拉动引锭杆，此时钢水已在引锭杆的头部凝固，铸坯随着引锭杆渐渐被拉出，经过二冷支导装置进入拉矫机后，引锭杆完成引导铸坯作用，此时脱引锭装置把铸坯和引锭杆头部脱离，拉矫机进入正常的拉坯和矫直工作状态。引锭杆运至存放处，留待下次浇铸时使用。

Before casting, put some scrap steel on the dummy bar head, and plug the gap with asbestos rope, so that the connection between the billet and the dummy bar head is firm and easy to separate dummy bar. Dummy bar device includes dummy bar (composed of dummy bar body and dummy bar head), dummy bar storage device and dummy bar removal device.

在浇铸前，引锭头上放些碎废钢，并用石棉绳塞好间隙，使得铸坯和引锭头既连接牢靠又利于脱锭。引锭装置包括引锭杆（由引锭杆本体和引锭头两部分组成）、引锭杆存放装置、脱引锭装置。

3.6.2 Structure and Classification of Dummy Bar

The dummy bar body can be divided into two types: flexible and rigid.

3.6.2 引锭杆的结构和分类

引锭杆本体可分为柔性和刚性两种。

3.6.2.1 Flexible Dummy Bar

Flexible dummy bar is composed of dummy bar head, transition piece and bar body. It is the earliest dummy bar, as shown in Figure 3-34. It is a chain connected by activities, so it is also called dummy block chain. The structure of the dummy bar is simple and the storage area is small. The arc caster basically adopts this structure.

3.6.2.1 柔性引锭杆

柔性引锭杆由引锭头、过渡件和杆身三部分组成，是最早的引锭杆，如图 3-34 所示。它是一根活动联结的链条，故又称为引锭链。这种引锭杆结构简单，存放占地小，弧形连铸机基本上都采用这种结构。

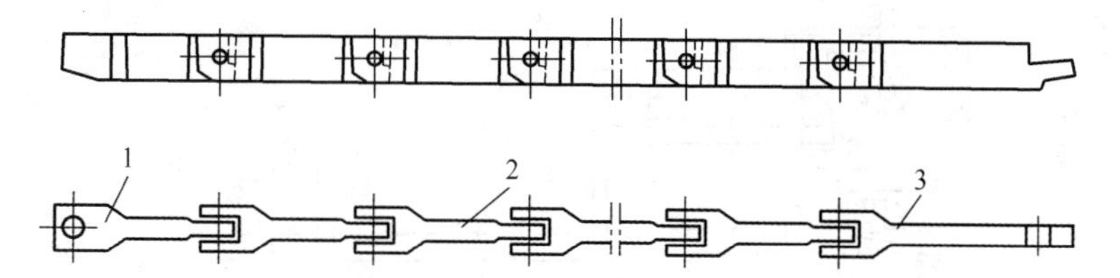

Fig 3-34 Flexible dummy bar for billet caster

1—Dummy bar head; 2—Dummy bar chain link; 3—Dummy bar tail

图 3-34 小方坯连铸机用的柔性引锭杆

1—引锭头；2—引锭杆链环；3—引锭杆尾

3.6.2.2 Rigid Dummy Bar

Rigid dummy bar is a rigid 90° arc bar. The bar body is a box structure welded by four steel plates. The outer arc radius of the arc plates on both sides is equal to the radius of the caster. The head of dummy bar is the dummy bar head of cast steel. The dummy bar head of starting casting and molten steel are condensed together, and the dummy bar tension is transferred pulling during casting. The dummy bar head is separated from the dummy bar after passing through the tension leveller, and then enters the head cutting box with the blank head after shearing.

3.6.2.2 刚性引锭杆

刚性引锭杆是一根刚性的 90°圆弧杆，杆身是由 4 块钢板焊成的箱形结构，两侧弧形板的外弧半径等于铸机半径。引锭杆的头部是铸钢的引锭头，开浇的引锭头与钢水凝结在一起，在开拉时传递引锭拉力，引锭头通过拉矫机后与引锭杆脱离，剪切后随坯头进入切头箱。

In recent years, this kind of dummy bar structure is often used in the arc billet caster. As shown in Figure 3-35, it is an arc-shaped steel plate structure with the same radius as the caster. The advantages are that the guiding device and dummy bar tracking device of the secondary cooling section are greatly simplified, the dummy bar is relatively stable, but the storage space of the rigid dummy bar is large.

近年来，弧形小方坯连铸机多采用此种引锭结构，如图 3-35 所示，它是一个与铸机相同半径的圆弧形钢板结构件。其优点在于大大简化了二冷段铸坯导向装置和引锭杆跟踪装置，引锭较平稳，但刚性引锭杆存放占空间很大。

3.6.2.3 Dummy Bar Head

The dummy head is mainly used to plug the lower outlet of the mold before casting, so that the molten steel will not leak down, and the molten steel will have enough time to solidify into the billet head in the mold. At the same time, the dummy head firmly connects the billet head with the dummy bar body, so that the billet can be pulled out of the mold continuously. According to the function of the dummy bar device, the dummy bar head should not only be firmly connected

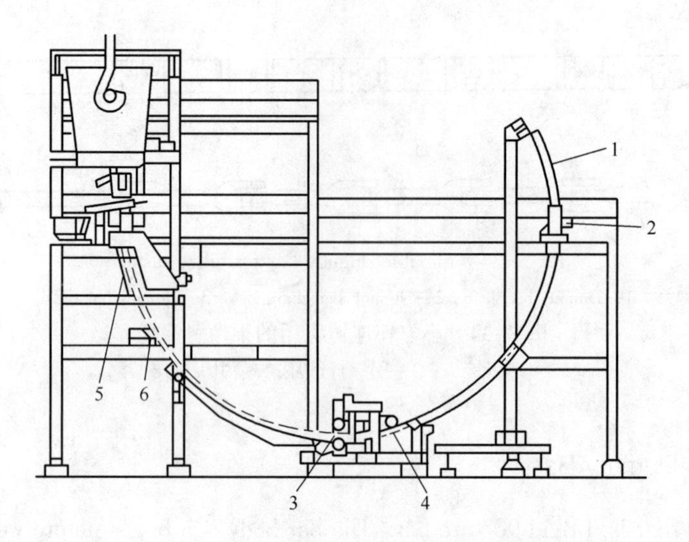

Figure 3-35 Schematic diagram of rigid dummy bar

1—Dummy bar; 2—Drive device; 3—Pull roller; 4—Straightening roller; 5—Secondary cooling zone; 6—Darrier roller

图 3-35 刚性引锭杆示意图

1—引锭杆；2—驱动装置；3—拉辊；4—矫直辊；5—二冷区；6—托坯辊

with the billet, but also be easy to be separated from the billet. There are two types of structure of dummy bar head: dovetail groove type and hook head type, as shown in Figure 3-36 and Figure 3-37.

3.6.2.3 引锭头

引锭头主要是在开浇前将结晶器下口堵住，使钢液不会漏下，并使浇入的钢液有足够的时间在结晶器内凝固成坯头，同时，引锭头牢固地将铸坯坯头与引锭杆本体连接起来，使铸坯能够被连续不断地从结晶器里拉出来。根据引锭装置的作用、引锭头既要与铸坯连接牢固，又要易于与铸坯脱开。引锭头的结构形式有燕尾槽式和钩头式两种，如图 3-36 和图 3-37 所示。

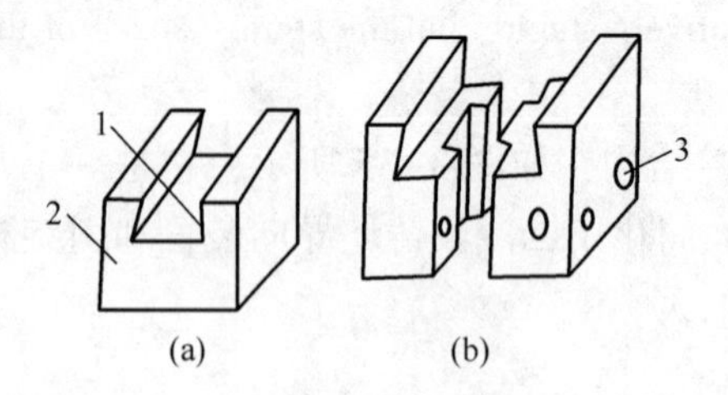

Figure 3-36 Schematic diagram of dovetail type dummy bar head

(a) Integral type; (b) Detachable type

1—Dovetail slot; 2—Dummy head; 3—Pin hole

图 3-36 燕尾槽式引锭头简图

(a) 整体式；(b) 可拆式

1—燕尾槽；2—引锭头；3—销孔

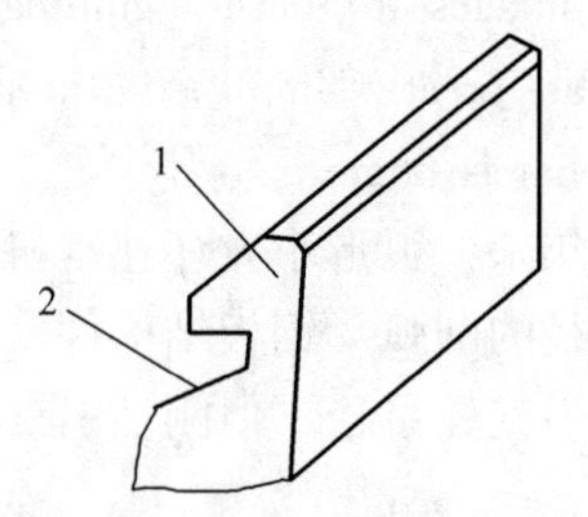

Figure 3-37 Schematic diagram of hook type dummy bar head

1—Dummy bar head; 2—Coupler head groove

图 3-37 钩头式引锭头简图

1—引锭头；2—钩头槽

3.6.2.4 Dummy Bar Storage Device

The function of the dummy bar storage device is to store the dummy bar in time after the dummy bar is separated from the billet, and to send the dummy bar into the mold by cooperating with the casting machine roller before the next casting. The storage device of dummy bar is related to the feeding mode of dummy bar. There are two ways of feeding dummy bar into mold, i. e. bottom feeding and top feeding. The storage device of dummy bar can also be divided into lower storage device and upper storage device.

3.6.2.4 引锭杆存放装置

引锭杆存放装置的作用是在引锭杆与铸坯脱离后，及时把引锭杆收存起来，并在下一次浇铸前，通过与铸机拉辊配合，把引锭杆送入结晶器内。引锭杆存放装置与引锭杆的装入方式有关，引锭杆装入结晶器的方式有两种，即下装式和上装式。引锭杆的存放装置也分为下装式存放装置和上装式存放装置。

3.6.2.5 Disconnecting Dummy Bar Device

The commonly used dummy head is mainly hook head type. The dummy bar head can be decoupled in cooperation with the tension leveller, as shown in Figure 3-38. After the dummy bar head passes through the tension roller, press the tail of the first dummy bar with the upper straightening roller, so that the dummy bar head and the billet are separated. In order to prevent the influence of thermal radiation, the part close to the billet should be cooled by water.

Some billet casters screw a bolt on the head of the dummy bar as the dummy bar head. After pulling out the billet, it is cut off together with the billet head, and the billet head and the bolt are discarded together. The structure is simple, the use is convenient and the cost is low.

3.6.2.5 脱引锭装置

常用的引锭头主要是钩头式。引锭头可与拉矫机配合实现脱钩，如图 3-38 所示，在引锭头通过拉辊后，用上矫直辊压一下第一节引锭杆的尾部，便使引锭头与铸坯脱开。脱锭装置为了防止热辐射的影响，对靠近铸坯的部分应通水冷却。

部分小方坯连铸机在引锭杆头部拧上一个螺栓作为引锭头，拉出铸坯后与坯头一起切断，将坯头和螺栓一起丢弃，结构简单，使用方便，成本低廉。

Task Implementation 任务实施

3.6.3 Inspection of Dummy Bar Device

(1) Before feeding dummy bar, it must be confirmed that the mold has been opened.

(2) Check the steel structure of the system regularly for deformation and damage, and take measures to deal with it in time.

(3) Regularly check whether the lubrication of all moving parts of the system is in good con-

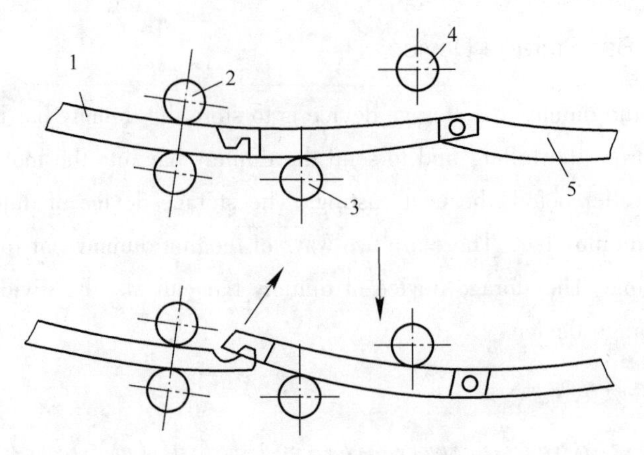

Figure 3-38　Diagram of disconnecting dummy bar head of tension leveller

1—Billet; 2—Pulling roll; 3—Lower straightening roll; 4—Upper straightening roll; 5—Long pitch dummy bar

图 3-38　拉矫机脱锭示意图

1—铸坯；2—拉辊；3—下矫直辊；4—上矫直辊；5—长节距引锭杆

dition, whether there is abnormal noise or jamming, and whether the wear exceeds the requirements.

(4) Check the hydraulic and lubricating pipelines regularly for leakage and normal pressure.

(5) Regularly check whether the fixed connecting bolts and nuts are loose and fasten them in time; regularly check whether the hydraulic and lubricating pipelines are leaking and whether the pressure is normal.

(6) Regularly check whether the steel wire rope has excessive wear and broken wires, and pay attention to the oil filling protection; if there is any damage, replace it firmly.

(7) Frequently check whether the brakes are easy to use and whether the brake pad is seriously worn.

(8) Frequently check whether the travel limit is accurate.

(9) Regularly check whether the drum bearing sound and temperature are normal, and timely take measures for maintenance.

3.6.3　引锭装置检查

(1) 送引锭之前，必须确认结晶器已经打开。

(2) 定期检查系统钢结构有无变形损坏情况，并及时采取措施处理。

(3) 定期检查系统各运动部位润滑情况是否良好，有无不正常响声或卡阻现象，磨损是否超规定。

(4) 定期检查液压、润滑管路有无泄漏，压力是否正常。

(5) 定期检查固定联结螺栓、螺母有无松动现象并及时紧固；定期检查液压、润滑管路有无泄漏，压力是否正常。

(6) 定期检查钢丝绳是否有磨损超标和断丝，并注意加油保护；如有损坏超标要坚决予以更换。

（7）经常检查各制动器是否好用，闸片磨损是否严重。

（8）经常检查行程极限是否准确。

（9）定期检查卷筒轴承声音、温度是否正常，并及时采取措施维修。

3.6.4 Dummy Bar Feeding Operation

(1) Check the external dimension of dummy bar head to ensure that the dimension is correct and that the dummy bar head is clean and free of residual steel slag.

(2) Install the dummy bar head on the dummy bar correctly.

(3) Start the hydraulic system of the tension leveller and raise the roller of the tension leveller.

(4) Start the dummy bar storage device, place the dummy bar on the conveying roller table, and use the roller table to send the dummy bar into the tension leveller.

(5) Press and pull down the straightening roller, start the straightening machine in the direction of dummy bar feeding, and feed the dummy bar at the specified speed.

(6) Pay close attention to the movement of dummy bar to find any slight obstruction.

(7) Stop pulling the dummy bar when its head is close to the lower outlet of the mold.

(8) Put the low-pressure lamp in the mold, and the operator shall observe the position of dummy bar head.

(9) The operator is ready to move the crowbar of the dummy bar head.

(10) The operator commands to start the tension leveller in the direction of dummy bar feeding, and feed the dummy bar head into the mold at the specified low speed.

(11) The dummy bar head shall be sent to the position to be cast beyond the process specification, and the tension leveller shall be stopped.

(12) The tension leveller switches to the direction of drawing billet, starts the tension leveller, and pulls the dummy bar head to the position to be cast at a slow speed (generally at 1/3 of the height of the mold).

(13) At the end of dummy bar feeding, wait for the operation of plug ingot head.

(14) If there is an automatic dummy bar feeding device, when the dummy bar enters the tension leveller and all the indicator or computer display equipment conditions are normal, the automatic dummy bar feeding program can be used. The dummy head is still manually positioned in the mold.

3.6.4 送引锭操作

（1）检查引锭头外形尺寸，确保尺寸正确并确保引锭头清洁无残钢渣。

（2）将引锭头正确装在引锭杆上。

（3）开动拉矫机液压系统，升起拉矫机辊。

（4）开动引锭杆存放装置，将引锭杆放置在输送辊道上，用辊道将引锭杆送入拉矫机内。

（5）压下拉矫辊，以送锭方向开动拉矫机，按规定速度送引锭杆。

（6）密切注意引锭杆的运动，以发现任何微小的受阻现象。

（7）当引锭头接近结晶器下口时，停止拉引锭杆。

（8）在结晶器内放入低压照明灯，操作工观察引锭头位置。

（9）操作工准备好拨动引锭头的撬棒。

（10）操作工指挥以送引锭方向开动拉矫机，按规定的低速将引锭头送入结晶器内。

（11）将引锭头送到超过工艺规定的待浇位置，拉矫机停车。

（12）拉矫机换向到拉坯方向，起动拉矫机，以慢速将引锭头拉到待浇位置（一般在结晶器高度的1/3处）。

（13）送引锭结束，等待塞引锭头操作。

（14）若有自动送锭装置，当引锭杆进入拉矫机内时，所有指示灯或计算机显示设备条件全部正常时，可采用自动送引锭程序。引锭头在结晶器内定位仍采用手动操作。

3.6.5 Plug Dummy Bar Head Operation

3.6.5.1 Operation Steps

(1) Before checking and adjusting the mold taper, the dummy bar head shall be fed into the mold according to the specified procedure.

(2) Prepare the tools and materials required for the plug of dummy bar head, asbestos rope or V-shaped block, steel plate, scrap iron, silica gel, refined oil, etc.

(3) The gap between the dummy bar head and the inner wall of the mold shall be carefully filled and sealed with 10~15mm diameter asbestos rope or V-shaped block. The asbestos rope must be filled, filled and slightly higher than the upper surface of the dummy head, and a layer of the same material shall be paved on both sides of the hook groove of the dummy bar head.

(4) A proper amount of iron chips shall be evenly spread on the dummy bar head, which shall be paved with a thickness of about 20~30mm, and steel plates shall be placed as required.

(5) Place the cooling scrap steel where the steel flow is easy to reach, and keep a distance of 10mm from the copper plate of the mold.

(6) A layer of silica gel should be evenly applied to four corners of the combined mold.

(7) Before casting, a layer of refined oil shall be evenly applied around the inner wall of the mold.

After the above work is completed, all remaining materials and tools shall be taken away and the required protective slag shall be prepared.

3.6.5 塞引锭头操作

3.6.5.1 操作步骤

（1）在进行结晶器锥度检查、调整前应按规定程序将引锭头送入结晶器。

（2）准备钢筋棍、石棉绳或V形块、钢板、铁屑以及硅胶、精制油等塞引锭头所需的工具和材料。

（3）在引锭头和结晶器内壁缝隙之间用直径为10~15mm的石棉绳或V形块对引锭头与结晶器的间隙进行仔细填充、密封。石棉绳必须填满、填实并略高于引锭头上表面，并在引锭头的钩槽两侧铺上一层相同材料。

（4）在引锭头上均匀撒放适量铁屑，铺平，厚度约为20~30mm，并按要求放置钢板。

（5）放入冷却废钢，位于钢流易冲到之处，并注意需与结晶器铜板保持有10mm的间距。

（6）对组合式结晶器，需在4个角部均匀涂抹一层硅胶。

（7）开浇前在结晶器内壁四周均匀涂擦一层精制油。

上述工作做完后，应将所有剩余的材料、工具取走，并准备好所需的保护渣。

3.6.5.2 Precautions

(1) Before the dummy bar head is plugged, it must be dry and clean. Otherwise, it shall be blown with compressed air or wiped dry with dry cloth.

(2) The materials used must be dry, clean and rust free.

(3) The reinforcing bar of plugging dummy bar head shall not impact the inner wall of the mold to prevent the inner wall from being damaged.

(4) Place the protection plate on the surface of the copper plate to avoid scratching the copper plate when dummy bar feeding.

(5) After plugging, water and foreign matters shall be prevented from falling into the mold.

3.6.5.2 注意事项

（1）塞引锭头前，必须保证干燥和干净，否则用压缩空气吹扫或用干布擦干。

（2）所用材料必须干燥、清洁、无锈。

（3）塞引锭头的钢筋棍不能冲击到结晶器内壁，以防止内壁损坏。

（4）将保护板放于铜板表面，以免送引锭时划伤铜板。

（5）塞好后，应防止水和异物落入结晶器内。

Exercises:

(1) What is the function of dummy bar? How many structures are there?

(2) What is the function of putting iron scraps, steel blocks, springs and other items into the mold when plugging the dummy bar head?

(3) Use simulation software to complete the operation of dummy bar feeding and installation, and observe the operation of dummy bar.

思考与习题：

（1）引锭杆的作用是什么，有几种结构形式？

（2）塞引锭头时放入铁屑、钢块、弹簧等物品的作用是什么？

（3）利用仿真软件完成送引锭和引锭安装操作，观察引锭杆运行情况。

Task 3.7 Operation of Cutting Device
任务3.7 切割装置操作

Mission objectives:

任务目标:

(1) Master the structure and function of the billet cutting device system.

(1) 掌握铸坯切割装置系统的结构和作用。

(2) Master the working process of billet cutting.

(2) 掌握铸坯切割的工作过程。

(3) Be able to check, operate and maintain the cutting system.

(3) 能够对切割系统进行检查、操作和维护。

Task Preparation 任务准备

3.7.1 Cognition of Billet Cutting Operation

After the primary cooling device of water-cooled mold and the secondary cooling device cooling, the molten steel has basically solidified. The straightened billet needs to be cut to meet the sizing requirements of the next process, facilitate transportation and storage, and ensure the smooth progress of the continuous casting process. At present, there are two kinds of cutting devices commonly used in continuous casting machine: flame cutting machine and mechanical cutting machine. The cutting device in the caster shall have the following two functional requirements:

(1) The cutting device can cut the straightened billets into billets with fixed length or several times of fixed length according to the requirements of users or the next process;

(2) The cutting device moves synchronously with the billet pulled out from the tension leveller, and completes cutting in the synchronous movement with the billet. The cutting action should have a certain speed to prevent bending and other defects when the billet is cut.

3.7.1 铸坯切割操作认知

钢液经过水冷结晶器的一次冷却装置和二次冷却装置两次冷却之后，已基本上凝固。已经矫直完成的铸坯需要进行切割，以满足下道工序的定尺要求，方便运输和存放，保证连铸工艺过程的顺利进行。目前连铸机上常用的切割装置主要有火焰切割机、机械剪切两种。连铸机中的切割装置应具有以下两点功能要求：

（1）切割装置能够比较准确地把矫直后的铸坯，按照用户或下步工序的要求切割成定尺或数倍定尺长的铸坯；

（2）与从拉矫机拉出的铸坯同步运动，并在与铸坯同步运动中完成切割，切割动作应具有一定的速度，以防止铸坯切割时出现弯曲和其他缺陷。

3.7.2 Flame Cutting Device

3.7.2.1 Working Process of Flame Cutting

The flame cutting machine uses the flame that preheat oxygen and combustible gas (acetylene, propane, natural gas, coke oven gas, hydrogen, etc.) mix and burn to melt the metal at the cutting seam, and blow off the molten metal with high-pressure oxygen until the billet is cut off. The advantages of the flame cutting device are: light weight equipment, easy processing and manufacturing; good cutting quality, and not limited by the billet temperature and section size; the external dimension of the equipment is small, which is particularly suitable for multi strand caster. The disadvantages are long cutting time, wide cutting seam and large material loss. The smoke and slag generated during cutting pollute the environment and require heavy slag cleaning work. The metal loss is large, which is about 1%~1.5% of the weight of the billet. The cutting speed is slow. Iron oxide, waste gas and heat are produced during the cutting. The slag transport equipment and dust removal facilities are needed. When the cutting length is short and fixed, the secondary cutting is needed. A large amount of oxygen and gas are consumed.

3.7.2 火焰切割装置

3.7.2.1 火焰切割的工作过程

火焰切割机利用预热氧气和可燃气体（乙炔、丙烷、天然气、焦炉煤气、氢气等）混合燃烧的火焰，将切缝处的金属熔化，同时用高压氧气把熔化的金属吹掉，直至把铸坯切断。火焰切割装置的优点是：设备轻，加工制造容易；切缝质量好，且不受铸坯温度和断面大小的限制；设备的外形尺寸较小，对多流连铸机尤为适合。缺点是切割时间长、切缝宽、材料损失大，切割时产生的烟雾和熔渣污染环境，需要繁重的清渣工作。金属损失大，约为铸坯重的1%~1.5%，切割速度较慢，在切割时产生氧化铁、废气和热量，需要运渣设备和除尘设施；当切割短定尺时需要增加二次切割；消耗大量的氧和燃气。

3.7.2.2 Equipment Structure of Flame Cutting Machine

Flame cutting machine is composed of cutting mechanism, synchronous mechanism, return mechanism, electricity, water, gas, oxygen and other pipelines, as shown in Figure 3-39.

3.7.2.2 火焰切割机的设备结构

火焰切割机由切割机构、同步机构、返回机构、电、水、燃气、氧气等管线组成，如图3-39所示。

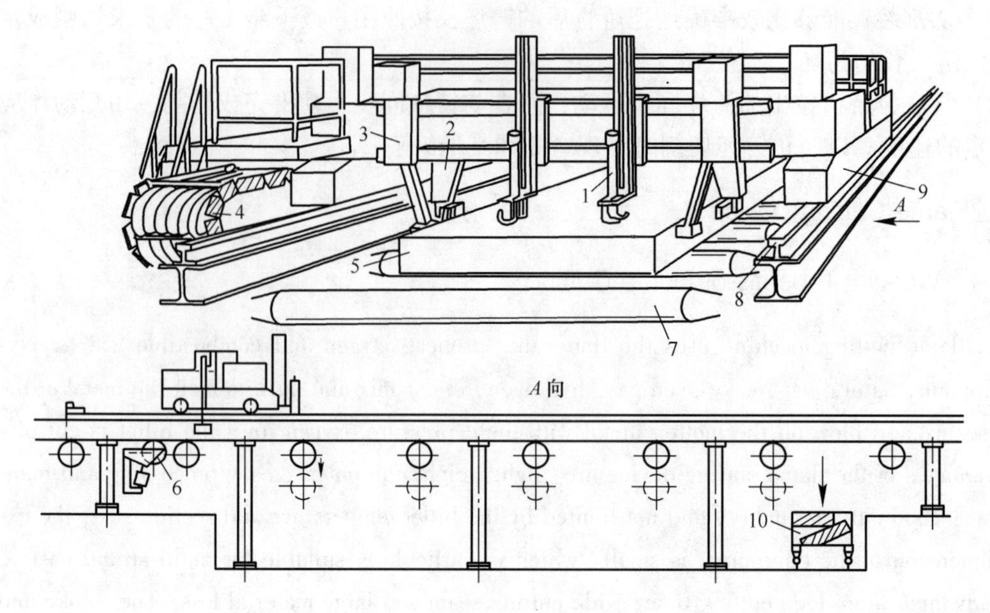

Figure 3-39　Flame cutting device

1—Cutting gun; 2—Synchronous mechanism; 3—Edge detector; 4—Hose reel; 5—Billet; 6—Fixed length mechanism; 7—Roller table; 8—Track; 9—Cutting trolley; 10—Cutting collection vehicle

图 3-39　火焰切割装置

1—切割枪；2—同步机构；3—边部检测器；4—软管盘；5—铸坯；6—定尺机构；7—辊道；8—轨道；9—切割小车；10—切头收集车

(1) Cutting mechanism. The cutting mechanism is the key part of the flame cutting device, which is mainly composed of the cutting gun and the driving mechanism.

The cutting gun consists of a gun body and a cutting nozzle. When cutting, the billet is preheated to the melting point, and then the molten metal is blown away by a high-speed oxygen flow to form a cutting seam. The cutting nozzle is made of copper alloy and cooled by water.

Internal mixing means that the gas and preheating oxygen are mixed in the cutting nozzle and then ejected for combustion, while external mixing means that the gas and preheating oxygen are mixed for combustion outside the nozzle. The external mixing cutting nozzle of the external mixing cutting gun has three rows of holes, which respectively supply cutting oxygen, gas and preheating oxygen. As shown in Figure 3-40, it will not cause tempering, and the flame is long. The cutting nozzle can be cut 50～100mm from the surface of the billet, and the hot cleaning efficiency is high.

(1) 切割机构。切割机构是火焰切割装置的关键部分，主要由切割枪和传动机构两部分组成。

切割枪由枪体和切割嘴组成，切割时先把铸坯预热到熔点，再用高速氧气流把熔化的金属吹去，形成切缝。切割嘴分为内混合和外混合两种形式，利用铜合金制造并通水冷却。

内混合是燃气和预热氧在割嘴内混合后再喷出燃烧，外混合则是燃气和预热氧在喷嘴外混合燃烧。外混式切割枪的外混合割嘴有 3 排孔，分别供应切割氧、燃气及预热氧，如图 3-40 所示，故不会引起回火，且火焰长，割嘴距离铸坯表面 50～100mm 即可以切割，热清理效率高。

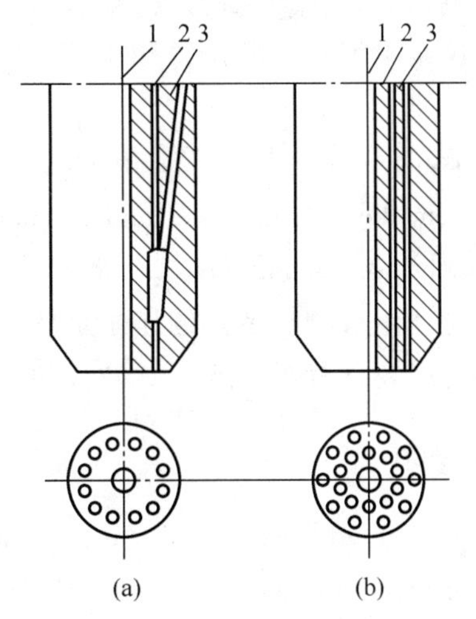

Figure 3-40　Pattern of cutting nozzle

(a) Internal mixing type; (b) External mixing type

1—Cutting oxygen; 2—Preheating oxygen; 3—Gas

图 3-40　切割嘴的形式

(a) 内混式; (b) 外混式

1—切割氧; 2—预热氧; 3—燃气

Generally, when the width of billet is less than 600mm, single gun is used for cutting; when the width of slab is more than 600mm, double gun is used for cutting, but two cutting guns are required to move in the same straight line to prevent uneven cutting seam. During cutting, the cutting gun shall be able to move horizontally and up and down.

通常，当铸坯宽度小于 600mm 时，用单枪切割；宽度大于 600mm 的铸坯，用双枪切割，但要求两支切割枪在同一条直线上移动，以防切缝不齐。切割时，割枪应能横向运动和升降运动。

(2) Synchronization mechanism. The synchronous mechanism ensures that the cutting gun keeps synchronous with the billet in the process of cutting the billet to ensure that the cutting seam is neat and that the cutting trolley can cut the billet without relative movement of the billet.

(2) 同步机构。同步机构保证割枪在切割铸坯过程中与铸坯保持同步以保证割缝整齐，保证切割小车在铸坯无相对运动的条件下切割铸坯。

(3) Return mechanism. The return mechanism of the cutting trolley generally adopts the common trolley running mechanism, which is equipped with an automatic transmission device, so as to automatically decelerate when approaching the original position.

(3) 返回机构。切割小车的返回机构一般采用普通小车运行机构，配备有自动变速装置，以便在接近原位时自动减速。

（4）Edge detector. The edge detector is installed on the walking carriage of the cutting gun. Its main function is to guide the cutting gun to a certain position on the side edge of the slab, so that the cutting gun can be accurately stopped at the preheating position regardless of the position.

（4）边部检测器。边部检测器安装在切割枪走行拖板上，其主要作用是把切割枪引导到铸坯侧边缘一定的位置上，这样不管铸坯的位置和宽度如何，都能保证切割枪准确地停止在预热位置上。

（5）Measuring device for casting billet sizing. The function of the casting billet length measuring device is to get the signal from the moving billet and to control the shearing machine automatically. There are three commonly used sizing devices: mechanical, pulse and photoelectric.

（5）铸坯定尺测量装置。铸坯定尺测量装置的作用是从行进中的铸坯取得信号，准确地控制剪机自动剪切。常用的定尺装置有机械式、脉冲式和光电式三种。

3.7.3 Mechanical Cutting Device

Mechanical shear equipment is called mechanical shear or shearing machine for short. Because shearing is carried out in the process of movement, the shearing machine used in the continuous casting machine is also called flying shear. Mechanical shear is adopted, the equipment is large, but its shear speed is fast, the shear time is only 2~4s, and the sizing accuracy is high. Especially when producing billets with short sizing, it is widely used in billet caster because of its no metal loss and convenient operation.

3.7.3 机械切割装置

机械剪切设备简称机械剪或剪切机，由于剪切是在运动过程中进行的，所以连铸机上用的剪机又称为飞剪。采用机械剪切，设备较大，但其剪切速度快，剪切时间只需2~4s，定尺精度高，特别是生产定尺较短的铸坯时，因其无金属损耗且操作方便，在小方坯连铸机上应用较为广泛。

3.7.3.1 Classification of Mechanical Shear

According to the power source, mechanical shear can be divided into electric shear and hydraulic shear. Both mechanical shear and hydraulic shear use the upper and lower parallel blades to do relative movement to complete the cutting of the billet in operation, but the way of driving the blades up and down is different.

3.7.3.1 机械剪切的分类

机械剪按动力源可分为电动剪切和液压剪切两类，机械飞剪和液压飞剪都是用上下平行的刀片做相对运动来完成对运行中铸坯的剪切，只是驱动刀片上下运动的方式不同。

3.7.3.2 Principle of Mechanical Cutting

The shear mechanism of the mechanical flying shear is driven by the crank connecting rod mechanism, as shown in Figure 3-41. The upper and lower tool tables are driven by the eccentric shaft and move in the vertical direction in the guide groove. Eccentric shaft is driven by motor through pulley and open gear. When the eccentric axis is at 0°, the scissors are open; when it is rotated 180°, the scissors are sheared; when it is rotated 360°, the upper and lower tool tables are returned to their original positions and complete a shear. In the process of shearing, the device can only synchronize with the slab by the pull rod swings an angle, so the length of the pull rod should be considered from the swing angle, but should not be too large.

3.7.3.2 机械切割原理

机械飞剪的剪切机构是由曲柄连杆机构带动，如图 3-41 所示。上、下刀台由偏心轴带动，在导槽内沿垂直方向运动。偏心轴由电机通过皮带轮及开式齿轮传动。当偏心轴处于0°，剪刀张开；当其转动 180°，剪刀进行剪切；当转动 360°时，使上、下刀台回到原位，完成一次剪切。在剪切过程中拉杆摆动一个角度才能与铸坯同步，因而拉杆长度应从摆动角度需要来考虑，但不宜过大。

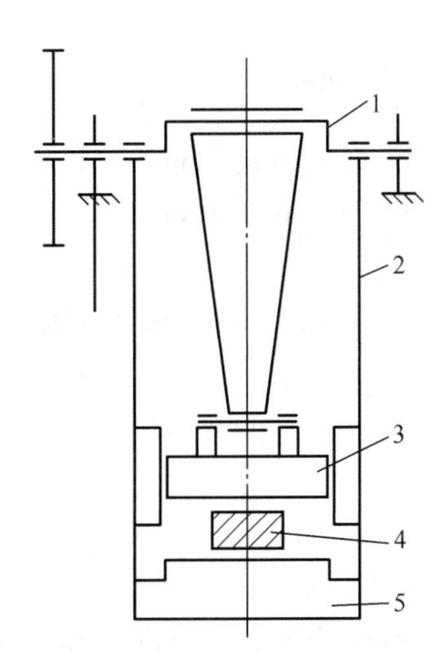

Figure 3-41 Principle of mechanical flying shear work

1—Eccentric shaft; 2—Pull rod; 3—Upper tool table; 4—Casting billet; 5—Lower tool table

图 3-41 机械飞剪工作原理图

1—偏心轴；2—拉杆；3—上刀台；4—铸坯；5—下刀台

The hydraulic shear device is installed together with the main hydraulic cylinder to drive the upper or lower tool table in order to complete the cutting task through the hydraulic plug. The shearing machine completes the task of cutting the billet by moving the cutting table up, that is, the down-cut type, which is widely used.

液压剪切装置是刀台与主液压缸安装在一起，通过液压塞柱来驱动上刀台或下刀台完成剪切任务。剪切机通过下刀台上移完成剪切铸坯任务，即下切式，此方式应用较为广泛。

3.7.3.3 Requirements of Mechanical Cutting Temperature

The steel shear resistance changes with the temperature. At high temperature (over 900℃), the shear resistance is not very large, but when it cools to below 750℃, the shear resistance increases greatly. Therefore, the minimum shear temperature of the general shear is 750℃ (or 800℃), which requires the continuous casting process to ensure that the surface temperature of the billet when it reaches the shear position is not lower than the minimum shear temperature.

3.7.3.3 机械切割温度要求

钢的剪切阻力随着温度的变化而变化，在高温状态下（大于900℃）时其剪切阻力不是很大，但当其冷却到750℃以下时，剪切阻力就大幅度增加。因此，一般剪机的最低剪切温度为750℃（或800℃），要求连铸工艺保证铸坯到达剪切位置时其表面温度不低于最低剪切温度。

Task Implementation 任务实施

3.7.4 Inspection of Flame Cutting Device

(1) During manual operation, the preheating flame shall be turned on first, and then the cutting oxygen; when it is turned off, the oxygen shall be turned off first, and then the preheating flame.

(2) Apply dry oil to worm, sliding plate, guide rail, track, screw rod, rack and gear regularly.

(3) Check the cooling water regularly for leakage and check the cooling condition of the cutting gun.

(4) Check whether the transmission structure is blocked and whether the joints are flexible.

(5) Always check whether the position of edge detector is accurate.

(6) Regularly clean the filter screen of energy pipeline and various valves.

(7) If the flame is scattered and not concentrated, check whether the cutting nozzle of the cutting gun is blocked; if the cutting seam is too large, replace the cutting nozzle; if the cutting seam is not on the same line, adjust the position of the two guns.

(8) Frequently check whether there is leakage in the energy pipeline and whether the pressure of the pressure reducing valve is normal. Regularly check whether the rubber pipes are aged.

(9) Check the compressed air pipeline for leakage and the cylinder for leakage.

3.7.4 火焰切割装置的检查

(1) 在手动操作时，应先开预热火焰，后开切割氧气；关闭时，应先关切割氧气，后关预热火焰。

(2) 由人工定期在蜗杆、滑板、导轨、轨道、丝杆、齿条、齿轮上涂抹甘油。

(3) 经常检查各冷却水管线有无渗漏现象，检查切割枪的冷却情况。

(4) 检查各传动结构有无卡阻现象，各关节部位是否灵活。

(5) 经常检查边部检测器位置是否准确。

(6) 定期清理能源管道及各类阀门的过滤网。

(7) 如果火焰分散、不集中，应检查切割枪的割嘴是否有堵塞；如果割缝过大，应更换割嘴；如果割缝不在一条线上，应调整两枪的位置。

(8) 经常检查能源管道是否有泄漏现象、减压阀压力是否正常。定期检查橡胶管道是否有老化现象。

(9) 检查压缩空气管路有无泄漏现象，气缸有无内外泄漏。

Precautions:

(1) Strictly follow the gas operation regulations, and the ignition sequence of the cutting gun shall not be reversed, so as to avoid harm.

(2) Adjust the flame length and cutting roller table equipment, do not damage the roller table.

注意事项：

(1) 严格按煤气操作规程进行，割枪点火顺序切勿颠倒，以免造成危害。

(2) 调整好火焰长度及切割辊道设备，切勿损坏辊道。

3.7.5 Use of Flame Cutting Device

(1) When the continuous casting slab moves to the cutting position, the signal will be generated through the sizing device; when the continuous casting slab reaches the sizing length, the cutting trolley will rely on the synchronous mechanism to synchronize the continuous casting slab and the cutting device, and the cutting gun will quickly approach the continuous casting slab.

(2) When the cutting gun is close to the edge of the continuous casting slab, the cutting gun is controlled by the limit switch, which is quickly converted to the starting cutting speed. After the cutting gun is running for a certain time, it is converted to the cutting speed for cutting.

(3) After cutting, the cutting gun quickly returns to the terminal; when the synchronous mechanism is opened, the cutting car quickly returns to the original place, ready to be cut again, at the same time, the slab casting roller is connected, and the continuous casting billet is output.

3.7.5 火焰切割装置的使用

(1) 连铸坯运行至切割位置时，通过定尺装置发生信号；当连铸坯达到定尺长度时，切割小车靠同步机构使连铸坯与切割装置同步，同时切割枪快速向连铸坯靠拢。

(2) 切割枪接近连铸坯边缘时，由限位开关控制切割枪，快速转换为始切速度，待切割枪运行一定时间后转换为切割速度进行切割。

(3) 切割完后，切割枪快速返回终端；同步机构打开，切割小车快速返回原处，准备接受再次切割，同时出坯辊道接通，输出连铸坯。

3.7.6 Use of Mechanical Cutting Device

(1) When the continuous casting billet is running to the shear, the signal is sent out through

the sizing device; when the continuous casting billet reaches the sizing length, the upper and lower tool tables of the shear hold the continuous casting billet, so that the shear and the continuous casting billet move synchronously.

(2) Through the transmission system, the upper and lower tool tables are driven to move, so that the upper and lower tool tables are closed and the continuous casting billets are cut.

(3) After the continuous casting slab is cut, the upper and lower tool tables are restored to the original open position by the reset mechanism, waiting for the next cutting, at the same time, the billet casting roller is connected, and the continuous casting billet is output.

3.7.6 机械切割装置使用

(1) 连铸坯运行至剪切机时，通过定尺装置发出信号；当连铸坯达到定尺长度时，剪切机的上下刀台咬住连铸坯，使剪切机与连铸坯同步运动。

(2) 通过传动系统带动上下刀台移动，使上下刀台合拢，剪断连铸坯。

(3) 剪断连铸坯后，由复位机构使上下刀台回复到原来的张开位置，等待下一步剪切，同时出坯辊道接通，输出连铸坯。

Exercises:

(1) What are the methods and characteristics of billet cutting?

(2) How to turn on combustion supporting air and acetylene gas during flame cutting?

(3) Using the simulation software to complete the cutting operation of the billet, observe the cut length of the billet, the position change of the cutting gun and the cutting section.

思考与习题：

(1) 铸坯的切割方式有哪几种，各有什么特点？

(2) 火焰切割时如何开启助燃空气和乙炔气体？

(3) 利用仿真软件完成铸坯切割操作，观察铸坯切割长度，割枪位置变化和切割断面情况。

Project 4 Continuous Casting Process Control and Operation
项目4 连铸生产工艺控制及操作

Task 4.1 Molten Steel Preparation
任务4.1 钢液准备

Mission objectives:

任务目标:

(1) Understand the influence of molten steel state on continuous casting process.

(1) 了解钢水状态对连铸过程的影响。

(2) Be able to judge the casting state of molten steel.

(2) 能够判断钢水的浇铸状态。

(3) Be able to explain the relationship between the quality of molten steel and accidents in production.

(3) 能够阐述钢水质量与生产中事故的关系。

Task Preparation 任务准备

Continuous casting has strict requirements on the quality of molten steel. It not only needs to ensure stable and appropriate molten steel temperature and degree of deoxidization to meet the casting ability, but also to minimize the sulfur and phosphorus impurities and gas content in the steel, so as to ensure the smooth operation of continuous casting, the improvement of billet quality and the timely supply of qualified molten steel, which is the basis and premise of improving continuous casting production. It is mainly manifested in the temperature, composition and purity of molten steel.

连铸对钢水质量有着严格的要求，它既要保证稳定适宜的钢水温度和脱氧程度，以满足可浇性；又要最大限度地降低钢中S、P杂质及气体含量，以确保连铸的顺行和铸坯质量的提高，保证合格钢水的及时供应，是提高连铸生产的基础和前提。主要表现在钢水的温度、成分、纯净度等方面。

4.1.1 Supply of Molten Steel

The characteristics of continuous casting production require the timely supply of molten steel. If the supply time of molten steel is advanced, the temperature drop of molten steel will be large, and the freezing flow will stop casting; on the contrary, if the supply time is delayed, the liquid level of the tundish will fluctuate too much, slag will be entrained, and the inclusion will increase, which will affect the quality of the billet. In serious cases, the slag will be dropped, steel will leak, and the flow will stop casting. Therefore, it is necessary to control the production rhythm of steelmaking, refining, continuous casting and other processes to ensure the stable supply of molten steel.

4.1.1 钢水的供应

连铸生产的特点要求按时供应钢水。若钢水供应时间提前会造成钢水温降大，冻流停浇；反之，供应时间滞后，可能造成中间包液面波动过大，出现卷渣，使夹杂物增多影响铸坯质量，严重时会造成下渣漏钢、断流停浇事故。因此，要控制好炼钢、精炼、连铸等工序的生产节奏，保证钢水的稳定供应。

4.1.2 Preparation of Molten Steel Temperature

4.1.2.1 Requirements for Temperature Control of Continuous Casting Molten Steel

The casting temperature of molten steel is an important process parameter of casting, and the proper casting temperature is the basis of smooth continuous casting. If the temperature of molten steel is too low, it is easy to cause the tundish nozzle to freeze and force the casting to be interrupted; if the temperature of molten steel is too high, it is easy to cause the tundish nozzle out of control, which will make the shell thin and uneven in thickness, resulting in steel leakage.

4.1.2 钢水的温度准备

4.1.2.1 连铸钢水温度控制要求

钢水的浇铸温度是浇铸的重要工艺参数，合适的浇铸温度是顺利连铸的基础，钢水温度过低，容易引起中间包水口冻结，迫使浇铸中断；钢水温度过高，容易引起钢水包水口失控，会使坯壳变薄和厚度不均，造成漏钢。

Proper casting temperature is the basis of obtaining good quality of billet. The low casting temperature will make the molten steel in the mold form a cold shell, worsen the surface quality of the billet, make the non-metallic inclusions in the steel difficult to float up and remove, reduce the purity of the steel; the high casting temperature will aggravate the reoxidation of the molten steel, intensify the erosion of the refractory for the ladle lining, increase the non-metallic inclusions in the steel, make the billet bulge, internal crack, center porosity and segregation, etc.

The temperature requirements of continuous casting molten steel can be summarized as high temperature, uniformity and stability.

合适的浇铸温度是获得良好铸坯质量的基础。浇铸温度偏低，使结晶器内钢液形成冷壳，恶化铸坯的表面质量，且使钢中的非金属夹杂物难以上浮排除，降低钢的纯净度；浇铸温度偏高，会加剧钢水的二次氧化，加剧对钢水包衬耐火材料的侵蚀，增加钢中非金属夹杂物，导致铸坯鼓肚、内裂、中心疏松和偏析等缺陷的产生。

连铸钢水的温度要求可总结为高温、均匀和稳定。

4.1.2.2 Determination of Casting Temperature of Continuous Casting Steel

The casting temperature of continuous casting molten steel generally refers to the temperature of molten steel in the tundish. The casting temperature of molten steel is equal to the liquidus temperature of the steel type plus the appropriate superheat of molten steel in the tundish.

$$T_C = T_L + \Delta T$$

Where T_C——Casting temperature, ℃;

T_L——Liquidus temperature of steel, ℃;

ΔT——Superheat of tundish steel, ℃。

4.1.2.2 连铸钢水浇铸温度的确定

连铸钢水的浇铸温度，一般是指中间包内的钢水温度，钢水的浇铸温度等于该钢种的液相线温度加上中间包钢水合适的过热度。

$$T_C = T_L + \Delta T$$

式中 T_C——浇铸温度，℃；

T_L——钢的液相线温度，℃；

ΔT——中间包钢水的过热度，℃。

(1) Calculation of liquidus temperature of steel. The liquidus temperature of molten steel is the basis to determine the casting temperature, which depends on the nature and content of the elements contained in the molten steel.

$$T_L = 1536 - \{88[C\%] + 8[Si\%] + [Mn\%] + 30[P\%] + 25[S\%] + 5[Cu\%] + 4[Ni\%] + 2[Mo\%] + 1.5[Cr\%] + 2[V\%] + 7\}\ ℃$$

(1) 钢的液相线温度的计算。钢水的液相线温度是确定浇铸温度的基础，它取决于钢水中所含元素的性质和含量。

$$T_L = 1536 - \{88[C\%] + 8[Si\%] + [Mn\%] + 30[P\%] + 25[S\%] + 5[Cu\%] + 4[Ni\%] + 2[Mo\%] + 1.5[Cr\%] + 2[V\%] + 7\}\ ℃$$

(2) Determination of superheat. The superheat of continuous casting steel will affect the yield of continuous casting machine and the quality of billet. The high superheat will reduce the casting speed, increase the risk of leakage, and cause serious central segregation, but it is conducive to the floating of inclusions. On the contrary, the low superheat will increase the casting speed, reduce the probability of leakage, reduce the central segregation, and make it difficult for inclusions to float up.

(2) 过热度的确定。连铸钢水过热度会影响连铸机产量和铸坯质量，高的过热度，会使拉速降低，增加拉漏的危险性，中心偏析严重；但有利于夹杂物的上浮。相反，低的过热度会使拉速提高，拉漏的几率减小，中心偏析减轻；夹杂物上浮困难。

Generally speaking, if the carbon content of the steel is low and the section of the billet is small, the superheat degree of the molten steel is higher. On the contrary, if the content of C, Si and Mn in the steel is high and the section of the billet is large, the superheat degree of the molten steel is lower. The values of superheat for different steel grades are shown in Table 4-1.

一般来说，钢种含碳量低，铸坯断面小，则钢水的过热度取高些，相反，钢种中 C、Si、Mn 含量高，铸坯断面大，则钢水的过热度取低些。不同钢种的过热度取值见表 4-1。

Table 4-1 Values of superheat for different steel grades

表 4-1 各类钢种连铸时的参考钢水过热度值 (℃)

Casting steel grade 浇铸钢种	Slab, bloom 板坯、大方坯	Billet 小方坯
High carbon steel 高碳钢, High manganese steel 高锰钢	10	15~20
Alloy structural steel 合金结构钢	5~10	15~20
Al-killed steel 铝镇静钢, Low alloy steel 低合金钢	15~20	25~30
Stainless steel 不锈钢	15~20	20~30
Silicon steel 硅钢	10	15~20

4.1.2.3 Adjustment of Molten Steel Temperature

(1) Uniform temperature treatment. The temperature distribution of the molten steel from the steelmaking furnace to the ladle is uneven. Due to the heat absorption of the ladle lining and the heat dissipation of the ladle surface, the temperature of the molten steel around the ladle lining is low, while the temperature of the central area of the ladle is high. If the molten steel is injected into the tundish, the temperature of the molten steel in the tundish will be lowered too much due to the heat absorption of the tundish lining and the low temperature of the molten steel at the bottom of the ladle too large and close to liquidus temperature, resulting in freezing of nozzle and interruption of casting.

4.1.2.3 钢水温度的调节

(1) 均匀温度处理。从炼钢炉出到钢包的钢水，在钢包内钢水温度分布是不均匀的，由于包衬吸热和钢包表面的散热，在包衬周围钢水温度较低，而钢包中心区域温度较高，如这样把钢水注入中间包，由于中间包衬的吸热再加上钢包底部钢水温度较低，就会造成中间包钢水温度降低过大而接近液相线温度，导致水口冻结，浇铸中断。

With the help of stirring with blowing inert gas, the temperature of low temperature molten steel near the bottom of ladle will move to the surface and homogenize the temperature in ladle. This is the purpose of argon blowing in ladle. Nitrogen can also be used for stirring, but it should be noted that nitrogen will react with some elements in steel to generate nitride, which will affect the quality of steel.

借助吹惰性气体搅拌来均匀温度，使包底附近的低温钢水向表层移动，并且使包内温度均匀化，这就是钢包吹氩的目的，也可使用氮气进行搅拌，但要注意氮会与钢中的一些元素反应生成氮化物，影响钢的质量。

（2）Cooling treatment of scrap steel. At the same time of blowing and stirring, light clean scrap steel is added to the molten steel to cool down, and the cooling effect is relatively stable. When 1% scrap is added, the temperature of molten steel in ladle can be reduced by about 14℃, i. e. 0. 7kg scrap is needed for every 1℃.

（2）加废钢降温处理。在吹气搅拌的同时向钢水中加入轻型的洁净的废钢降温，冷却效果比较稳定。加入1%的废钢，可使包内钢水温度降低约14℃，即每降温1℃，需加废钢0. 7kg。

（3）Heating adjustment. At present, arc heating and chemical heating are mainly used to heat the molten steel in ladle.

1）The electric arc heating method, also known as ladle furnace（LF furnace）method, which uses submerged arc heating, has high thermal efficiency, good desulfurization and diffusion deoxidization effect, low cost, and more used.

2）In chemical heating method, oxygen is blown by spray gun to oxidize the aluminum and other elements added into the molten steel, release a large amount of heat, make the molten steel rise rapidly, and the oxidation efficiency is high.

（3）升温调节。在钢包内对钢水进行加热处理，目前加热主要采用电弧加热法和化学加热法。

1）电弧加热法，也称钢包炉（LF炉）法，此种加热方法使用埋弧加热，热效率高，脱硫和扩散脱氧的效果好，费用低，应用较多。

2）化学加热法，用喷枪吹氧使加入钢水中的铝等元素氧化，释放大量的热量，使钢水快速升温，氧化效率高。

4. 1. 3 Composition Control of Molten Steel

The composition of molten steel should first meet the requirements of the steel specification, but the molten steel meeting the requirements may not be completely suitable for continuous casting. Therefore, the composition of molten steel must be strictly controlled according to the special requirements of continuous casting process and billet quality. It mainly includes component stability, crack resistance sensitivity, casting ability of molten steel and so on.

4. 1. 3 钢水的成分控制

钢水成分首先应满足钢种规格的要求，但符合要求的钢水不一定完全适宜连铸。因此，必须根据连铸工艺和铸坯质量的特殊要求，对钢水的成分进行严格的控制。主要包括成分稳定性、抗裂纹敏感性、钢水的可浇性等内容。

Carbon is the most basic element in steel and the one that has the greatest influence on the structure and properties of steel. It must be precisely controlled. When multiple furnaces are continuously cast, the difference of carbon content in molten steel between each ladle is required to be 0. 02%. As shown in Figure 4-1, the influence of carbon content on the longitudinal crack of billet is shown. Pay attention to avoid the sensitive range in production.

碳是钢中最基本的也是对钢的组织性能影响最大的元素，必须对它进行精确控制，多炉连浇时，各包次之间钢水的含碳量的差别要求 0.02%，如图 4-1 所示为碳含量对铸坯纵裂纹的影响，生产中注意避开敏感范围。

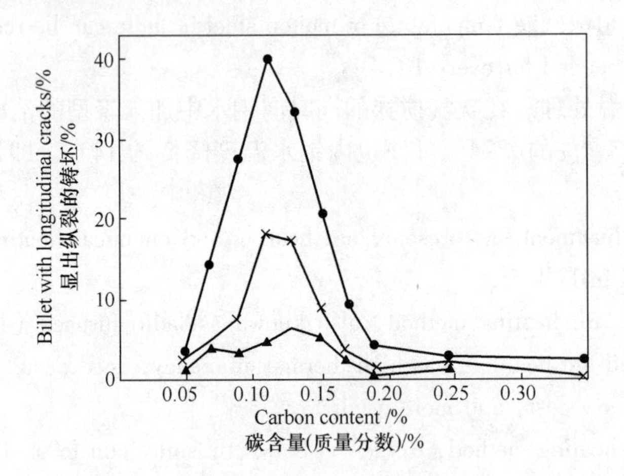

Figure 4-1　Effect of carbon content on longitudinal crack of billet

图 4-1　碳含量对铸坯纵裂纹的影响

Si and Mn are controlled in a narrow range, which affects the casting ability of molten steel. In order to ensure the mutual stability of Si and Mn in the continuous casting steel, it is necessary to fine tune the composition of the steel after refining to meet the requirements of steel grades. The fluctuation of general composition requires that $w[Si]=\pm0.05\%$, $w[Mn]=\pm0.10\%$, and then increase the $w[Mn]/w[Si]$ ratio to improve the fluidity of molten steel. The content of silicon and manganese must be controlled strictly, and the requirements can only be met through the fine adjustment of the composition of refining outside the furnace.

Si、Mn 的控制在较窄的范围内，影响着钢液的可浇性。保证连浇钢水的 Si、Mn 含量相互稳定，必须经过炉外精炼的成分微调，达到钢种的要求。一般成分波动要求 $w[Si]=\pm0.05\%$、$w[Mn]=\pm0.10\%$，其次提高 $w[Mn]/w[Si]$ 比以改善钢水流动性。严格控制硅、锰等含量，必须经过炉外精炼的成分微调才能达到要求。

Phosphorus and sulfur content shall be controlled at the lower limit. The amount of sulfur content in continuous casting liquid steel directly affects whether the continuous casting process can be carried out normally and the quality of billet, which is mainly reflected in the influence on the sensitivity of steel hot crack. With the increase of sulfur content, the furnace defect rate also increases. In the process of crystallization, the segregation tendency of phosphorus is larger, which makes the grain boundary embrittlement of steel and the cracking tendency of steel increase. At present, the development trend of continuous casting technology is to reduce the content of sulfur and phosphorus in steel as much as possible through hot metal pretreatment and various refining technologies.

P、S 含量应在下限控制。连铸钢液硫含量的多少直接影响到连铸工艺能否正常进行和铸坯质量的好坏，主要表现在对钢热裂纹敏感性的影响，随着含硫量的提高，炉次缺陷率也增加。磷元素在结晶过程中偏析倾向较大，使钢的晶界脆化，从而使钢的裂纹倾向增大。目前，连铸技术发展的趋势是通过铁水预处理以及各种精炼技术，尽量降低钢中硫和磷的含量。

The residual elements are brought into the steel by raw materials rather than intentionally added, which can not be removed in the refining process of molten steel and remain in the steel such as: Cu, Sn, Sb, Pb, etc. The total amount is 0.20%.

残余元素是钢中原料带入而不是有意加入的，在钢水的精炼过程中又不能去除，残留在钢中的元素，如 Cu、Sn、Sb、Pb 等。其总量（质量分数）为 0.20%。

Other elements, such as chromium, vanadium, titanium and rare earth, which can form refractory compounds with oxygen in steel, will increase the viscosity of molten steel.

其他与钢中氧生成难熔化合物的元素，如铬、钒、钛、稀土等，将增加钢液的黏度。

4.1.4 Control of Molten Steel Purity

The purity of molten steel refers to the content, shape and distribution of nitrogen, hydrogen, oxygen and non-metallic inclusions in the steel. With high purity, it can improve the properties of steel, reduce the defects of billet, improve the internal quality, the mechanical and service properties of the final product.

4.1.4 钢水纯净度控制

钢水纯净度是指钢中气体氮、氢、氧和非金属夹杂物的含量、形态和分布。纯净度高，可改善钢的性能，减少铸坯的缺陷，提高内部质量，提高最终产品的力学性能和使用性能。

4.1.4.1 Generation of Inclusions in Molten Steel

Oxygen in molten steel is controlled by [C] in the molten pool and (FeO) in the slag. Before tapping, the oxygen content in molten steel is higher than that in equilibrium, and increases with the decrease of carbon content. When the carbon content in molten steel is below 0.10%, the oxygen content in molten steel will increase sharply with the decrease of carbon content. The excess oxygen is removed by deoxidation, and the deoxidized product is not removed completely, and remains in the steel as a non-metallic inclusion. The higher the oxygen content is, the higher the inclusion content is.

The existence of inclusions will affect the casting ability of the molten steel, and it is difficult to ensure the continuous casting process and worsen the quality of the billet. Inclusions in steel can be divided into endogenetic inclusions and foreign inclusions. The endogenetic inclusions are mainly deoxidized products. The foreign inclusions include the reoxidation products of molten steel in the casting process, which come from the eroded refractories, the involved ladle slag, tundish slag and mold powder, etc. In order to ensure the quality of the final billet, the content of non-metallic inclusions in the steel should be minimized.

4.1.4.1 钢水夹杂物的产生

钢水的氧受熔池中［C］和炉渣中（FeO）的控制。出钢前钢水中氧含量高于平衡氧含量，且随着碳含量的降低而增多，当钢水中含碳量在0.10%以下时，随碳含量的降低钢中氧含量猛增。通过脱氧去除过剩氧，生成的脱氧产物没有排除干净，残留于钢中成为非金属夹杂物。终点钢水中氧含量越高，夹杂物含量也越高。

夹杂物的存在会影响钢水的可浇性，难以保证连铸顺行，危害钢坯质量。钢中夹杂物可分为内生夹杂物和外来夹杂物。内生夹杂物主要是脱氧产物；外来夹杂物包括在浇铸过程中钢水的二次氧化产物，来源于被冲蚀的耐火材料以及卷入的钢包渣、中间包渣和结晶器浮渣等。为了确保最终铸坯质量，要尽量降低钢中非金属夹杂物的含量。

4.1.4.2 Control of Molten Steel Deoxidation

(1) Deoxidation of silicon and manganese. Manganese and silicon are the two most widely used deoxidizers. The mechanical strength of steel and the fluidity of molten steel can be improved by properly increasing the manganese content according to the requirements of steel grades. The deoxidizing ability of silicon is stronger than that of manganese, but the precipitation of deoxidizing product (SiO_2) will increase the viscosity of molten steel and worsen the casting ability of molten steel. Therefore, the silicon content should be controlled according to the requirements of steel grades. When only using Si and Mn to deoxidize the steel, $w[Mn]/w[Si]>3.0$, the particle size of deoxidized product is large, and it is liquid silicate, which can improve the fluidity of molten steel and is conducive to the floatation of inclusions. Therefore, the ratio of [Mn]/[Si] should be increased within the allowable range of steel composition.

4.1.4.2 钢液脱氧控制

(1) 硅和锰的脱氧。锰和硅是使用最为广泛的两种脱氧剂。按照钢种要求适当提高锰含量（中上限控制）能提高钢的机械强度并改善钢水的流动性。硅的脱氧能力比锰强，但其脱氧产物（SiO_2）的析出会增加钢液的黏度而恶化钢液的可浇性，因此应按钢种要求控制硅含量（中下限控制）。当仅用Si、Mn脱氧的钢，$w[Mn]/w[Si]>3.0$时，脱氧产物颗粒尺寸较大，且是液态的硅酸盐，可以改善钢液的流动性，有利于夹杂物上浮。因此，应在钢种成分允许的范围内适当增加［Mn］/［Si］比。

(2) Deoxidation of aluminum. Aluminum is a strong deoxidizer. Generally, aluminum is used as the final deoxidizer in steel, often with ferromanganese and ferrosilicon. Al_2O_3 inclusions formed after aluminum oxidation are clustered in steel, which has the characteristics of dendrite. Although the dendrite with large particles and complex shape contains steel, it increases the density of inclusions, is difficult to float up and remove, and it is easy to block the nozzle.

In addition, it can be seen from the equilibrium calculation that with the same amount of [Si], increasing the $w[Mn]/w[Si]$ ratio can improve the deoxidization ability of aluminum. It must be pointed out that calcium has a strong deoxidizing ability. Usually, silicon calcium powder is injected or silicon calcium cored wire is added to control the aluminum content in the steel and reduce the Al_2O_3 inclusion content in the molten steel, so as to avoid the nozzle clogging.

（2）铝的脱氧。铝是强脱氧剂，一般钢种都是把铝作为终脱氧剂，经常与锰铁、硅铁一起使用。铝氧化后形成 Al_2O_3夹杂在钢中呈群簇状，具有树枝形的特点，虽然颗粒很大，外形复杂的树枝状内部含有钢，加大了夹杂物的密度，难以上浮排除，还容易堵塞水口。

此外，通过平衡计算可知，在［Si］含量相同的情况下，提高 $w[Mn]/w[Si]$ 比，铝的脱氧能力提高。必须指出，钙具有很强的脱氧能力，通常喷吹硅钙粉或加入硅钙包芯线来控制钢中铝含量，减少钢水中 Al_2O_3夹杂物含量，以避免水口结瘤和堵塞。

4.1.4.3 Method of Purifying Molten Steel

At present, it is mainly used to purify the molten steel by refining outside the furnace, and transfer some or all of the refining tasks completed in the conventional steelmaking furnace, such as removing gas, inclusions, uniform composition and temperature, to the ladle or other containers. Therefore, the off furnace refining is also called secondary refining or ladle metallurgy. Its main role is to provide purified molten steel for continuous casting.

4.1.4.3 净化钢液方法

目前主要利用炉外精炼来净化钢液，将常规炼钢炉中完成的精炼任务，如去除气体、夹杂物、均匀成分和温度，部分或全部转移到钢包或其他容器中进行。因此，炉外精炼也称炉外精炼或钢包冶金，其主要作用就是向连铸提供净化的钢液。

Task Implementation 任务实施

4.1.5 Control of Molten Steel Temperature and Fluidity

（1）Master the steel grades, temperature of molten steel and tapping.

（2）Pay attention to the characteristics of steel flow, brightness, color and fluidity of liquid steel during tapping. The higher the temperature is, the brighter the liquid steel is, and the whiter the color is; the better the fluidity and the higher the temperature are when the steel composition is the same.

（3）After starting casting, the temperature of molten steel can be further determined according to the conditions of steel injection, nozzle clogging (cold steel) and so on.

（4）To master the influence of different steel composition on liquid steel fluidity, the general rule is as follows:

1）The lower the carbon content, the worse the fluidity;

2）When the steel contains titanium, vanadium, copper, aluminum and rare earth elements, the fluidity of liquid steel will be poor;

3）When the content of inclusions in molten steel is high, the fluidity is poor.

（5）Pay attention to the ladle condition and turnover, and pay attention to the factors that cause the abnormal fluctuation of molten steel temperature.

4.1.5 钢液温度与流动性的控制

（1）掌握所冶炼的钢种及钢液升温、出钢测温情况。

（2）出钢过程中注意观察钢流的特征、钢液的亮度、颜色、流动性等。温度越高，钢液越亮，颜色越白；钢种成分相同时，流动性越好，温度就越高。

（3）开浇后，根据铸流、水口结瘤（结冷钢）等情况进一步判断钢液温度。

（4）掌握不同的钢种成分对钢液流动性的影响，其一般规律为：

1）碳含量越低，则流动性越差；

2）当钢中含钛、钒、铜、铝及稀土元素等成分时，将使钢液流动性变差；

3）当钢液中夹杂物含量高时，流动性差。

（5）关注钢包状况和周转情况，注意引起钢液温度异常波动的因素。

4.1.6 Measures to Reduce the Temperature Drop of Molten Steel

In the process of production, the temperature of molten steel gradually decreases with time after entering the ladle, mainly including the temperature drop during tapping, the temperature drop during transportation to the refining station outside the ladle after tapping, the temperature drop during ladle treatment, the temperature drop after ladle treatment and before ladle casting, and the temperature drop during ladle injection into the tundish.

The temperature drop of tapping process and tundish is the largest, and the fluctuation is relatively large, so the key to reduce the temperature drop of process is to reduce the temperature drop of tapping process and tundish. The following measures can be taken to reduce the temperature drop of molten steel:

(1) Red ladle tapping. If the temperature of ladle lining is lower than 800℃, it shall be baked with gas.

(2) Shorten waiting time before ladle tapping.

(3) Ladle liquid level is added with covering agent.

(4) Use the heat insulation layer and ladle cover in the ladle.

(5) Speed up ladle turnover.

(6) In order to reduce the heat loss of tundish, the tundish must be preheated before casting, and the tundish must be covered with carbonized rice husk.

4.1.6 减少钢水温度降低的措施

生产过程中，钢水进入钢包后，随着时间的推移，其温度逐渐降低，主要为出钢时的温降、出钢后到钢水炉外精炼站运输过程中的温降、钢水在钢包处理过程中的温降、钢包从处理后到开浇前的温降、钢水从钢包注入中间包内的温降。

其中出钢温降和中间包温降最大，且波动比较大，因此降低过程温降关键是要降低出钢过程温降和中间包温降。减少钢水温降可采用如下措施：

（1）红包出钢，包衬温度低于800℃应用煤气烘烤。

（2）缩短钢包出钢前的等待时间。

（3）钢包液面加覆盖剂。

（4）钢包加绝热层和钢包加盖。

（5）加速钢包周转。

（6）减少中间包的热损失，开浇前中间包必须预热，中间包加盖或加炭化稻壳覆盖。

Exercises:

(1) What are the requirements of continuous casting process for molten steel?

(2) What should be paid attention to when casting molten steel with high aluminum content?

(3) Use the simulation software to check the temperature, composition and other parameters of the molten steel, and observe the temperature change during the casting.

思考与习题:

（1）连铸工艺对钢水有什么要求？

（2）浇铸铝含量较高的钢水需注意哪些问题？

（3）利用仿真软件查看钢水的温度、成分等参数，观察钢水浇铸时的温度变化。

Task 4.2 Start Casting Operation
任务 4.2 开浇操作

Mission objectives:

任务目标:

(1) Master the operation key points of starting casting.

（1）掌握连铸开浇的操作要点。

(2) Be able to check the equipment before casting according to the post.

（2）能够根据岗位在开浇前对设备进行检查。

(3) Be able to complete ladle starting casting, tundish liquid steel control, tundish starting casting, emergence time control and casting process speed control as required.

（3）能够按要求完成钢包开浇、中间包钢液控制、中间包开浇、出苗时间控制和开浇过程的拉速控制等工作。

Task Preparation 任务准备

The start casting operation refers to the operation that the molten steel reaches the casting platform until the steel liquid is injected into the mold, and the casting speed turns to normal. Start casting operation is an important operation in continuous casting operation, which has practical significance for stabilizing continuous casting operation, improving productivity and reducing accidents.

连铸机开浇操作是指钢液到达浇铸平台直至钢液注入结晶器，拉坯速度转入正常这一段时间内的操作。开浇操作是连铸操作中比较重要的操作，对于稳定连铸操作，提高生产率，减少事故的发生具有现实意义。

Task Implementation 任务实施

4. 2. 1 Inspection and Preparation before Casting

Good inspection and preparation before casting can ensure normal operation of equipment and product quality, so as to reduce accidents and improve production capacity and quality of continuous casting billet.

4. 2. 1 浇铸前的检查与准备

浇铸前的检查与准备工作做得好，就能保证设备运转正常和产品的质量，从而减少事故，提高生产能力和连铸坯的质量。

4. 2. 1. 1 Inspection and Preparation of the Director of Continuous Caster and Casting Worker

(1) Equipment inspection. The equipment inspected by the director of continuous caster and the casting worker before casting includes: ladle turret or other ladle support equipment, manipulator for ladle injection protection, tundish, tundish car, mold, mold oscillation device and secondary cooling device.

4. 2. 1. 1 机长和浇钢工的检查与准备

(1) 设备检查。机长和浇钢工在浇铸前检查的设备有钢包回转台或其他钢包支撑设备、钢包铸流保护的机械手、中间包、中间包车、结晶器、结晶器振动装置和二次冷却装置。

1) Ladle turret or other ladle support equipment. For ladle supporting equipment, it is required to rotate left and right before casting to check whether the rotation is normal, whether the stop position is accurate, whether the limit switch and indicator light are easy to use, and whether the relevant electrical and mechanical systems are normal.

1) 钢包回转台。对于采用回转台式的钢包支撑设备，浇铸前应左旋和右旋两圈，检查旋转是否正常，停位是否准确，限位开关和指示灯是否好用，有关电气和机械系统是否正常。

2) Manipulator for ladle injection protection. Manipulator is a common form of stream shrouding from ladle to tundish. Before casting, it is required to check whether the swivel arm and control lever are flexible, whether the supporting ring and fork head are free of residual steel, and whether the trolley is running normally, whether there are residual steel and foreign matters on the

trolley track, and whether they are placed stably and properly; prepare the counterweight, argon blowing hose and quick connector.

2）钢包铸流保护的机械手。机械手是钢包到中间包铸流保护浇铸的一种常见形式。浇铸前应检查旋臂及操纵杆使用是否灵活，检查托圈、叉头，要求无残钢、残渣，转动良好；检查小车运行是否正常，小车轨道上有无残钢异物，放置是否平稳且到位适中；准备好平衡重锤、吹氩软管及快速接头。

3）Tundish. Check whether the shell is deformed and cracked, and whether there is steel sticking, to ensure that the inside of the tundish is clean and undamaged. When the tundish adopts the stopper to control the injection flow, it is required that the mechanical operation is flexible, the stopper size meets the requirements, the stopper head and the nozzle are closed tightly, and the stopper rod is placed accurately. When the tundish adopts slide brick to control the flow, the control system shall be flexible. When the tundish is opened, the upper and lower slide plates shall be concentric. When the tundish is closed, the lower slide plate can seal the upper slide plate. When the submerged nozzle is used for casting, check whether the internal and external surfaces of the submerged nozzle are clean, whether there are cracks and missing angles, whether they are tight and firm, whether the size and shape meet the requirements, whether the protruding part is vertical to the bottom of the tundish and whether the side holes are installed correctly. For tundish with slag retaining dam, the shape and installation position of the slag retaining dam shall be accurate and fixed at the same time. Finally, baking is carried out according to the masonry condition of tundish. Generally, only the nozzle is baked in the cold tundish, and the baking temperature shall be higher than 800℃; the lining in the hot tundish shall be baked, and the baking temperature shall be higher than 1100℃.

3）中间包。检查其外壳是否变形开裂、有无粘钢，确保包内清洁无损。当中间包采用塞棒式控制注流时，要求机械操作灵活，塞棒尺寸符合要求，塞头与水口关闭严密，塞棒落位准确。当中间包采用滑板砖控制注流时，要求控制系统灵活，开启时上下滑板注流口同心，关闭时下滑板能封住上滑板注流口。采用浸入式水口浇铸时，使用前检查浸入式水口内外表面是否干净，有无裂纹缺角，是否上紧、上牢固，尺寸和形状是否符合要求，伸出部分是否和中间包底垂直及侧孔是否装正。采用挡渣墙的中间包，挡渣墙的形状及安装位置应准确，同时安装牢固。最后，根据中间包的砌筑情况进行烘烤。一般冷中间包只烘烤水口，其烘烤温度应大于800℃；热中间包烘烤包衬，其烘烤温度应大于1100℃。

4）Tundish car. Check whether the lifting and traverse of the tundish are normal, whether the splash plate on the tundish is intact, and whether there are obstacles on the track.

4）中间包车。检查中间包车升降、横移是否正常，中间包车上的挡溅板是否完好，轨道上有无障碍物。

5）Mold. Check the cover plate at the upper inlet of the mold and its matching with the mold. The cover plate shall be matched in size, placed flat, free of residual steel and residue, and flush with the mold entrance. The gap between the cover plate and the mold interface shall be blocked with asbestos rope and sealed with fire-resistant mud. Check the surface of the copper plate on the inner wall of the mold. It is required that the surface shall be smooth and free of residual steel, residue and dirt, and the surface damage shall be less than 1mm. If there are residual steel, residue and dirt, they must be removed completely. The surface of the copper plate shall be slightly scratched and polished with sandpaper. If the surface damage is greater than 1mm, the mold shall be replaced before casting. Check the water inlet and outlet pipes and joints of the mold, and there shall be no water leakage, bending or blocking. Test the cooling water pressure and water temperature of the mold. Generally, the cooling water pressure is about 0.6MPa, the water inlet temperature is less than or equal to 40℃, and there is no water leakage. The water cut-off alarm of the mold works normally. Regularly check the mold size and taper.

5）结晶器。检查结晶器上口的盖板及与结晶器配合情况，要求盖板大小配套，放置平整，无残钢、残渣，与结晶器口平齐，其盖板与结晶器接口处间隙用石棉绳堵好并用耐火泥料堵严、抹平。检查结晶器内壁铜板表面，要求表面平整光滑，无残钢、残渣、污垢，表面损伤（刮痕、伤痕）小于1mm；如果有残钢、残渣、污垢，必须除尽，铜板表面轻微划伤用砂纸打磨，表面损伤大于1mm时，则应更换结晶器后再浇铸。检查结晶器的进出水管及接头，不应有漏水、弯折或堵塞现象。测试结晶器冷却水压和水温，一般冷却水压为0.6MPa左右，进水温度小于或等于40℃，且无漏水、渗水现象，结晶器断水报警器工作正常。定期检测结晶器尺寸和倒锥度。

6）Mold oscillation device. There shall be no shaking or jamming phenomenon in the mold oscillation, and the oscillation frequency and amplitude shall meet the process requirements.

6）结晶器振动装置。振动时不应有抖动或卡住现象，振动频率和振幅符合工艺要求。

7）Secondary cooling device. Check the arc alignment of the mold and the secondary cooling device. The arc alignment error is required to be less than 0.5mm. Check the opening of the secondary cooling pinch roll to make it meet the process requirements. When the hydraulic control clamp roll is used, the hydraulic pressure is normal and the clamp roll is adjusted normally. Check the secondary cooling roll to make sure that there is no bending, deformation, crack, adhesion and flexible rotation. Check the secondary cooling water supply system and the nozzle. All of them are free of blockage, the joint is firm, the water volume is adjustable within the specified range, and the shape and atomization of cooling water sprayed by the nozzle meet the requirements. When the cooling grid is used, blow out the residue in the grid hole, and observe whether the grid grinding plate is broken or burned.

7）二次冷却装置。检查结晶器与二次冷却装置的对弧情况，要求对弧误差不大于0.5mm；检查二冷夹辊的开口度，使之满足工艺要求；采用液压调节夹辊时，液压压力正常，夹辊调节正常；检查二冷辊子，要求无弯曲变形、裂纹，无黏附物，转动灵活；检查二冷水供给系统，要求喷嘴均无堵塞，接头牢固，水量在规定范围内可调，喷嘴喷出冷却水形状及雾化情况满足要求。当采用冷却格栅时，吹扫格栅孔内的残渣，观察格栅磨板有无断裂和烧伤情况，若有则及时处理。

（2）Preparation of tools, instruments and raw materials:

1）Prepare slag tank and overflow tank without water or moisture, which can hold residual steel and residue of ladle and tundish.

2）If the protective tube is used to protect the ladle flow, the protective tube shall be baked in the baking furnace.

3) Prepare a certain amount of tundish covering slag, mold fluxes (mold powder) or lubricating oil, and the variety and quality meet the requirements of steel grade and operation.

4) When the cold tundish is used for casting, a certain amount of heating agent is added around the nozzle to prevent the nozzle from being blocked due to the low temperature.

5) Prepare steel casting and accident handling tools. For example, tundish stopper handle, slag rake, slag pusher, sampling spoon, sampling mold, temperature measuring gun, aluminum bar, oxygen pipe, oxygen acetylene cutting gun, etc.

(2) 工器具和原材料的准备:

1) 准备好无水或无潮湿物的渣罐和溢流槽，能盛接钢包和中间包的残钢、残渣。

2) 若采用保护管保护钢包铸流，则应将保护管在烘烤炉内烘烤好。

3) 准备一定数量的中间包覆盖渣、结晶器保护渣或润滑油，其品种、质量符合钢种和工艺操作要求。

4) 采用冷中间包浇铸时，水口周围加一定数量的发热剂，以防开浇时温度低堵水口。

5) 准备好浇钢及事故处理工具。如中间包塞棒压把、捞渣耙、推渣棒、取样勺、取样模、测温枪、铝条、氧气管、氧-乙炔割枪等。

(3) Feeding and blocking dummy bar. After receiving the dummy bar feeding instruction, the casting worker shall keep in touch with the dummy bar driver through the operation board on the caster platform, pay attention to observe the rising of the dummy bar, and prevent the equipment from being damaged due to the deviation of the dummy bar head. When the dummy bar head is sent to about 500mm away from the lower outlet of the mold, the steel casting worker visually moves the dummy bar head to the specified distance from the upper inlet of the mold to stop, and then uses dry and clean asbestos rope or paper rope to embed the gap between the dummy bar head and the copper wall of the mold, and evenly spreads clean, dry, oil-free and sundry free steel filings of 20 ~ 30mm thickness on the dummy bar head. Finally, place the cooling square steel block.

(3) 送堵引锭。接到送引锭指令后，浇钢工通过连铸机平台上的操作板与引锭工保持联系，注意观察引锭杆的上升情况，防止引锭头跳偏而损坏设备。当引锭头送到距结晶器下口 500mm 左右时，浇钢工目视引锭头进行点动送引锭操作，将引锭头送至距结晶器上口规定的距离停止，然后用干燥、清洁的石棉绳或纸绳嵌紧引锭头和结晶器铜壁之间的间隙，并在引锭头上均匀铺撒 20~30mm 厚的干净、干燥、无油、无杂物钢屑，最后放置冷却方钢块。

4.2.1.2 Inspection and Preparation of Operators in the Main Control Room

Carefully check various instruments, fault display, interphone in the main control room to ensure normal operation; cooperate with the casting workers to check the oscillation frequency of the mold, test the mold cooling water and secondary cooling water; start the fan, and inform the director of caster to check whether the fan has abnormal sound; prepare all recording papers. When the casting plan is issued, the relevant posts shall be informed immediately.

4.2.1.2 主控室操作工的检查与准备

对主控室内各种仪表、故障显示、对讲机认真检查，确保操作正常；配合浇钢工做好结晶器振动频率的检查、试结晶器冷却水和试二冷水的工作；启动风机，并通知机长检查风机有无异常响动；准备好所有记录纸张。当浇铸计划下达后，立即向各有关岗位发出通知。

4.2.1.3 Inspection and Preparation of Dummy Bar Workers

(1) Equipment inspection:

1) Dummy bar head and dummy bar body. Before casting, the sundries and cold steel on the dummy bar head and dummy bar body shall be cleaned. The shape and specification of dummy bar head and dummy bar body must meet the requirements without damage and deformation, and the chain link of dummy bar body is well connected.

2) Dummy bar head roaster in good condition.

3) Tension leveler. The upper and lower rollers operate normally, and the distance between the upper and lower rollers is consistent with the section thickness of the mold.

4) Operation console. Check whether all operation elements and lights on the console are normal, and whether the control equipment acts normally.

5) Dummy bar removal equipment in normal operation.

6) Disconnecting dummy bar device in normal operation.

4.2.1.3 引锭工的检查与准备工作

(1) 设备检查：

1) 引锭头和引锭杆本体。浇钢前对引锭头和引锭杆的杂物、冷钢要清理干净；引锭头和引锭杆的形状和规格必须满足要求，不可损伤和变形，引锭杆本体链节联结良好。

2) 引锭头烘烤器。处于良好状态。

3) 拉矫机。上、下辊运转正常，上、下辊距与结晶器断面厚度相符。

4) 操作台。检查操作台上的各种部件、灯光显示是否正常，控制设备动作是否正常。

5) 引锭杆移出设备。运行正常。

6) 脱锭装置。运行正常。

(2) Dummy bar feeding operation:

1) Before feeding the dummy head, it must be heated to about 200℃, so as not to get wet when passing through the secondary cooling section, thus causing explosion during casting.

2) Confirm the contact with the steel casting worker. When the steel casting worker selects the button on the platform operation board to 'send dummy bar', the dummy bar worker can start the operation.

3) Start the button of 'feeding dummy bar' on the operation platform, and the mode of automatic feeding dummy bar will start to work. The procedure is: feed the dummy head into the tension leveler, start the tension leveler and quickly press down and compress the dummy bar. When the dummy head runs to the specified distance below the mold, the operation of feeding the dummy

bar will be stopped automatically.

（2）送引锭操作：

1）在送引锭头之前必须将其加热至200℃左右，以免通过二冷段时被弄湿，从而导致浇铸时爆炸。

2）确认与浇钢工联系。当浇钢工将平台操作板上的按钮选择到“送引锭”时，引锭工方可开始操作。

3）启动操作台上“送引锭”按钮，则自动送引锭方式开始工作。其程序是：将引锭头送入拉矫机，拉矫机启动并快速压下压紧引锭杆，当引锭头运行到距结晶器下方规定距离时，自动停止送引锭操作。

4.2.1.4 Inspection of Cutter

Flame cutting inspection operation:

(1) Check whether all kinds of light display and buttons on the console are normal.

(2) Check whether the operation and return mechanism of the cutting trolley are normal.

(3) Adjust and check the working data of cutting nozzle according to the thickness of casting billet and steel type, and check whether each joint leaks air.

(4) Connect the closed-circuit water of cutting gun and cooling water of cutting machine, and adjust and observe the normal working parameters.

(5) Turn on oxygen and combustible gas and ignite, and check the flame length.

(6) The length of the pre selected cutting head and the length of the pre selected continuous casting billet cut to a certain length according to the steel requirements.

(7) Check whether the spare accident cutting gun is easy to use.

4.2.1.4 切割工的检查

火焰切割检查步骤：

（1）检查操作台上各种灯光显示及按钮是否正常。

（2）检查切割小车的运行和返回机构是否正常。

（3）根据所浇铸坯厚度及钢种，调整、检查切割嘴的工作数据，检查各接头是否漏气。

（4）接通切割枪闭路水和切割机冷却水，并调整观察至正常工作参数。

（5）接通氧气和可燃气体并点火，检查火焰的长度。

（6）预选切头长度和按钢种要求预选连铸坯定尺切割长度。

（7）检查备用的事故切割枪是否好用。

4.2.2 Key Points of Starting Casting Operation

The starting casting operation shall be fast and stable. The so-called 'fast' means that the ladle and tundish should be placed in place quickly, and the ladle should be opened and cast quickly, so as to reduce the temperature loss of molten steel; 'stable' means that the tundish should be opened and cast stably, and the tension leveler should be started stably, so as to prevent the mold from leaking during the casting.

4.2.2 开浇操作要点

开浇操作要做到快和稳。所谓“快”是钢包、中间包就位要快，钢包开浇要快，减少钢液温度损失；“稳”是中间包开浇要稳，拉矫机启动要平稳，防止开浇时结晶器拉漏。

(1) Temperature measurement of casting platform. The measurement of the molten steel temperature in the ladle of the casting platform is to ensure that the caster does not cast ‘too high’ and ‘too low’ molten steel, to provide operating parameters for the control of the tundish opening time and emergence time, and to provide basis for the continuous casting rejection. During the temperature measurement of the casting platform, it is necessary to operate it carefully. The depth of the temperature gun inserted into the molten steel of the ladle shall be kept at 300~400mm, and the temperature measurement place shall be in the middle of the ladle to prevent the inaccurate temperature measurement from affecting the normal operation of the casting operation.

(2) Ladle starting casting. When using sliding nozzle to control ladle flow, there are two situations: one is natural drainage, that is, when opening the sliding gate, the liquid steel will flow out of the nozzle automatically. Second, it can't be drained naturally. Artificial drainage is used, that is, the nozzle is burned with oxygen. Three points should be paid attention to during ladle casting:

1) In order to shorten the drainage time, the artificial drainage should be prepared when the natural drainage is carried out; once the natural drainage is not successful, the artificial drainage should be carried out immediately.

2) After opening the slide gate, if it is found that the slide gate can't be closed or opened, handle it in time.

3) After the ladle is casting, it is cast with a full open sliding gate.

(1) 浇铸平台测温。测量浇铸平台钢包内钢液温度是保证连铸机不浇铸“过高”和“过低”的钢液，为控制中间包开浇时间、出苗时间等提供操作参数，为连铸拒浇提供依据。在浇铸平台测温时，要认真操作，测温枪插入钢包钢液的深度要保持在300~400mm，测温地方要在钢包中部，以防测温不准确而影响开浇操作的正常进行。

(2) 钢包开浇。采用滑动水口控制钢包铸流时，其开浇有两种情况：一是自然引流，即打开滑动水口时，钢液自动从水口中流出。二是不能自然引流，采用人工引流，即用氧气将水口烧开。钢包开浇过程要注意3点：

1) 为了缩短引流时间，在进行自然引流时，要做好人工引流的准备；一旦自然引流不成功，立即进行人工引流。

2) 打开水口后要检查滑动水口关闭情况，如果发现水口关不死或打不开要及时处理。

3) 钢包开浇后，采用全开滑动水口浇铸。

(3) Tundish liquid steel control. The liquid level control of tundish steel is to make the liquid steel in tundish rise as fast as possible to reach the liquid steel height of tundish. This is beneficial to reduce the loss of molten steel temperature in tundish.

For abnormal conditions, such as low temperature of molten steel measured by casting platform, long drainage time of ladle and long waiting time of molten steel caused by equipment accident, tundish start casting is carried out when the liquid level of tundish steel is low.

（3）中间包钢液控制。中间包钢液面控制是钢包开浇后，尽量使中间包钢液快速升高，达到中间包开浇的钢液高度，有利于减少中间包钢液温度的损失。

对于不正常的情况，如浇铸平台测量钢液温度较低、钢包引流时间长和设备事故使钢液等待时间过长时，中间包钢液面较低时就进行中间包开浇。

（4）Tundish starting casting. Tundish starting casting refers to the operation of opening the tundish nozzle to inject molten steel into the mold. The main purpose of tundish starting casting is to make the molten steel injection into the mold smoothly, to ensure the emergence time and the smooth progress of billet drawing. Under normal conditions, the tundish opening operation shall be as stable as possible, that is, the liquid steel shall be injected into the dummy head groove stably, and then it shall rise slowly in the mold according to the emergence time. Start casting of tundish shall not be too strong. If the casting is too strong, the liquid steel is easy to plug and melting the dummy bar head, which makes it difficult to start the casting and pull out the dummy bar; at the same time, it is easy to cause the hanging steel and the emergence time of the mold can't be guaranteed.

On the basis of mastering the stable start-up of tundish, if the tundish is controlled by slide gate or stopper, the operation of 'test slide' or 'test stopper' shall be carried out to check whether the control system is flexible, so as to prevent the poor control of injection flow after start-up from affecting the operation.

For abnormal conditions, in order to ensure the success of casting, the control of tundish nozzle is more important. In the case of low temperature of molten steel or poor baking of the nozzle, the nozzle can be opened wider when tundish is opened to increase the impact force of the molten steel at the nozzle to prevent the molten steel from forming cold steel. At the same time, the emergence time can be shortened and the billet casting speed can be increased. If the temperature of molten steel is high, the nozzle should be controlled smaller and the emergence time should be increased.

（4）中间包开浇。中间包开浇是指打开中间包水口，使钢液注入结晶器这段过程的操作。中间包开浇的主要目的是使钢液平稳注入结晶器，保证出苗时间和拉坯顺利进行。在正常情况下，中间包开浇操作要尽量做到平稳，即钢液平稳注入引锭头沟槽，在结晶器内根据出苗时间的长短慢慢上升。中间包开浇不能过猛。开浇过猛，钢液易冲堵引锭头材料和冲熔引锭头，造成开浇拉漏和脱引锭困难，同时易造成结晶器挂钢和出苗时间不能保证。

在掌握中间包开浇平稳的基础上，对于采用滑动水口或塞棒控制中间包铸流的，还要进行“试滑”或“试棒”操作，检查控制系统是否灵活，防止开浇后注流控制不好而影响操作。

对于不正常情况，要想保证开浇成功，其中间包水口的控制就更重要。对于钢液温度低或水口烘烤不良的情况，中间包开浇时，可将水口开得大些，增加水口处钢液的冲击力，防止钢液结冷钢，同时出苗时间可缩短，拉坯速度增快。如果钢液温度较高，中间包开浇时，水口要控制小些，增加出苗时间。

（5）Control of emergence time. The emergence time refers to the period from the injection of molten steel into the mold to the start of withdrawal billet by the straightener. The emergence time is the necessary time for continuous casting to solidify into the shell with enough thickness in the mold. The solidification thickness of the shell is related to the temperature of the liquid steel, the steel grade and the section of the continuous casting billet. When the section of continuous casting billet is large or the temperature of molten steel is high, the upper limit of emergence time is used; when the section of continuous casting billet is small or the temperature of molten steel is low, the lower limit of emergence time is used. The emergence time is generally between 30s and 90s.

(5) 出苗时间的控制。出苗时间是指钢液注入结晶器到拉矫机开始拉坯的这段时间。出苗时间是保证连铸在结晶器内凝固成足够厚度的坯壳所必需的时间。坯壳的凝固厚度与钢液温度、钢种、连铸坯断面等因素有关。当浇铸的连铸坯断面较大或浇注的钢液温度较高时，采用出苗时间的上限控制；当浇铸的连铸坯断面较小或浇铸的钢液温度较低时，采用出苗时间的下限控制。出苗时间一般在 30~90s 之间。

(6) The casting speed control of staring casting process. The starting casting speed of continuous casting machine can be selected in advance or not from zero. Generally, it is better to adopt the operation method of not selecting the speed in advance for large section billet. The control principle of the casting speed is: the speed should be stable, and the speed up process should be slow.

When the temperature of molten steel is abnormal, such as the temperature of molten steel is low, the casting speed should be increased as much as possible on the premise of no leakage.

(6) 开浇过程的拉坯速度控制。连铸机的起步拉坯速度可采用预先选定某一拉速度或不预先选定从零开始拉速两种方法。一般对于大断面铸坯采用不预先选速的操作方法较好。开浇拉坯速度的控制原则是：升速要平稳，升速过程要慢。

当钢液温度不正常时，如钢液温度较低，应在保证不拉漏的前提下，尽量提高拉坯速度。

4.2.3 Operation of Starting Casting

(1) Each post shall check their own preparations and various instruments to confirm that the preparations are ready.

(2) After the liquid steel reaches the casting platform, the steel casting worker carries out the temperature measurement operation. When the temperature measurement meets the requirements, command the crane to place the ladle stably on the ladle turret or other supporting equipment.

(3) Drive the prepared tundish and tundish car to the casting position, and align them. If the cold tundish is used for casting, immediately put the drainage sand (or calcium silicate powder) into the tundish nozzle.

(4) Make use of ladle turret or other supporting equipment to keep ladle in casting position.

(5) Connect cooling water to the mold.

(6) Ladle starting casting. If the ladle flow protection casting is adopted, after the ladle opening casting is normal, close the ladle nozzle, quickly cover the prepared protection pipe into the nozzle, ensure that the protection pipe and the ladle nozzle are on the same central line, manually press the control lever (or use the manipulator) to install the long nozzle, open the ladle

nozzle, after confirming the normal flow, hang the balance weight, immediately connect the argon blowing pipe and hit it , then open the argon blowing valve.

(7) When the liquid level of the tundish reaches 1/2 of the height, the covering slag can be added to the tundish. When using ladle casting protection, covering slag can be added after the molten steel emerged protective tube, the quantity of adding depends on the specific situation. It is required to cover the steel level of the tundish evenly, with a thickness of 10~30mm.

(8) After the liquid level of tundish steel meets the requirements of staring casting, open the stopper or slide gate, and start casting of tundish. Once the tundish is casting, the operator in the main control room shall report the time to the director of caster in 5s to confirm the time of emergence; the casting worker shall press the steel flow in the side hole of the nozzle with the slag rakes to prevent the mold from hanging steel.

(9) When the molten steel level of the mold submerges the side hole of the submerged nozzle, the mold flux is pushed into the mold rapidly, and the quantity of the flux is based on the principle of completely covering the molten steel surface.

(10) When the emergence time is up and the liquid level of the mold steel is 100mm from the upper inlet (depending on the billet specification), start the tension leveler to start the billet withdrawing. Pay attention to whether the mold oscillates, start the extraction steam fan at the same time, and gradually open the secondary cooling water of each section from top to bottom. Once the casting starts, the operator in the main control room reports the time to the director of caster again in 10s, so that the director can control the casting speed during the casting process.

(11) After billet withdrawing, the dummy bar operator shall closely monitor the operation of the dummy bar, and deal with any abnormality in time.

4.2.3 开浇过程操作

（1）各岗位最后检查各自的准备工作和各种仪表情况，确认准备工作做好。

（2）钢液到达浇铸平台后，浇钢工进行测温操作。当测温符合要求时，指挥吊车将钢包稳定地放置在钢包回转台或其他支撑设备上。

（3）将准备好的中间包及中间包车开到浇铸位置，对中落位。如采用冷中间包浇铸时，立即将引流砂（或硅钙粉）放入中间包水口里。

（4）利用钢包回转台或其他支撑设备，使钢包处于浇铸位置。

（5）结晶器接通冷却水。

（6）钢包开浇。如采用钢包保护浇铸，在钢包开浇正常后，关闭钢包水口，迅速将预先准备好的保护管套入水口，确保保护管与钢包水口在一条中心线上，压住操纵杆（或采用机械手）安装长水口、打开钢包水口，确认铸流正常后，立即接上吹氩管并打开吹氩阀门。

（7）当中间包液面达到 1/2 高度时，向中间包加入覆盖渣。当采用钢包铸流保护浇铸时，钢液面淹没保护管下口就可加入覆盖渣，加入数量视具体情况而定，一般要求均匀覆盖中间包钢液面，厚度为 10~30mm。

（8）在中间包钢液面达到开浇要求后，打开塞棒或滑动水口，中间包开浇。一旦中间包开浇，主控室操作工以 5s 为单位向机长报出时间，以确认出苗时间；浇钢工用捞渣耙压住水口侧孔的钢流，严防结晶器挂钢，或接通结晶器润滑油。

（9）当结晶器钢液面淹没浸入式水口侧孔时，迅速向结晶器内推入保护渣，其加入数量以完全覆盖钢液表面为原则。

（10）当到了出苗时间，结晶器钢液面距上口100mm（据铸坯规格而定）时，启动拉矫机开始拉坯。注意结晶器是否振动，同时启动抽蒸汽风机，按从上到下顺序逐步打开各段的二次冷却水。一旦开始拉坯，主控室操作工以10s为单位重新向机长报告时间，以便机长对开浇过程的拉坯速度进行控制。

（11）拉坯后，引锭工要严密监视引锭杆的运行情况，发现异常及时处理。

Exercises:

(1) What are the requirements of continuous casting process for molten steel?

(2) What should be paid attention to when casting molten steel with high aluminum content?

(3) Use the simulation software to check the temperature, composition and other parameters of the molten steel, and observe the temperature change during the casting.

思考与习题：

（1）连铸开浇前如何对设备进行检查？

（2）浇铸铝含量较高的钢水需注意哪些问题？

（3）利用仿真软件查看钢水的温度、成分等参数，观察钢水浇铸时的温度变化。

Task 4.3　Normal Casting Operation

任务4.3　正常浇铸操作

Mission objectives:

任务目标：

(1) Master the operation key points of normal continuous casting.

（1）掌握连铸正常浇铸的操作要点。

(2) Be able to control the casting speed, liquid level and cooling system according to the process requirements, and complete the protection casting, dummy bar stripping and cutting operations.

（2）能够根据工艺要求控制拉速、液面、冷却制度，完成保护浇铸、脱锭和切割操作。

(3) Complete the sequence casting task by replacing the ladle and tundish.

（3）通过更换钢包、中间包完成多炉连浇任务。

Task Preparation 任务准备

Normal casting operation refers to the operation from the finish of starting casting and billet casting speed of the continuous caster to the end of the last ladle molten steel casting. The normal casting operation mainly includes the control of the casting speed, the protective casting, the liquid level control, the cooling system control, the dummy bar stripping operation and the cutting operation. In order to realize sequence casting, ladle and tundish replacement should be carried out.

正常浇铸操作是指连铸机开浇、拉坯速度转入正常以后，到本浇次最后一炉钢包钢液浇完为止这段时间的操作。正常浇铸操作主要是拉坯速度控制、保护浇铸、液面控制、冷却制度的控制、脱锭操作和切割操作。为实现多炉连浇，还要进行钢包和中间包更换操作。

4. 3. 1 Control of Casting Speed

The casting speed usually refers to the length of billet pulled out by the caster per first-class unit time (m/min). Casting speed is an important control parameter in normal casting operation. The temperature of molten steel in tundish is the key to control and adjust the casting speed. Casting speed directly affects the solidification speed of molten steel, internal quality and billet quality, and also determines the production capacity and safety of caster.

4. 3. 1 拉坯速度的控制

拉坯速度通常指连铸机每一流单位时间拉出的铸坯长度（m/min）。拉坯速度是正常浇铸操作中的重要控制参数，中间包内钢液温度是控制和调节拉坯速度的关键，其大小直接影响到钢液的凝固速度、内部质量、钢坯质量，同时也决定了铸机生产能力和安全性。

When multi flow casting is adopted, due to the influence of the injection position of molten steel, the velocity of each flow port in tundish is different, so the speed of each flow is different. The velocity of the nozzle far away from the injection point of molten steel is smaller than that near the injection point. In the actual production, the orifice diameter of the nozzle can be properly increased for the nozzle with the far steel liquid inflow point, and the nozzle with the near steel liquid injection point can be properly reduced or maintained the normal orifice diameter to maintain the relative balance and stability of the casting speed of each flow.

当采用多流浇铸时，在浇铸过程中，由于钢液注入位置的影响，中间包各流水口的流速有一定的差别，因而各流的拉坯速度不同，距钢液注入点较远的水口流速比近点的水口流速要小。在实际生产中，对钢液流入点较远的水口，可适当加大水口孔径；而钢液注入点较近的水口，可适当缩小或维持正常水口孔径，以保持各流的拉坯速度相对均衡和稳定。

4. 3. 2 Liquid Level Control

4. 3. 2. 1 Level in Tundish Control

The stability of the liquid level in tundish has great influence on the quality of continuous casting billet and breakout accident. During normal casting, the liquid level should be controlled at 400~600mm or 50~100mm from the overflow port of the tundish. The control method is mainly to control the flow from the ladle to the tundish. For example, if the ladle nozzle cannot be closed, the liquid level in the tundish will rise.

4.3.2 液面控制

4.3.2.1 中间包液面控制

中间包液面的稳定，对连铸坯质量及漏钢事故影响较大。正常浇铸时，液面应控制在400~600mm或距中间包溢流口50~100mm，其控制方法主要是对钢包到中间包的注流进行控制。比如钢包水口关不住时，中间包液面会上升。

4.3.2.2 Level in Mold Control

The mold level control is the key link to realize the automation of continuous casting equipment, and it is also the premise to produce flawless billet and ensure the integration of billet hot delivery and continuous casting and rolling.

The level in mold control is the key link to realize the automation of continuous casting equipment, and it is also the premise to produce flawless billet and ensure the integration of billet hot delivery and continuous casting and rolling. Generally, the liquid level of steel in the mold shall be controlled at 100mm from the top of the mold or 70mm from the slag surface to the top of the mold, and the liquid level fluctuation shall be within ±10mm. Some enterprises also control the liquid level within 85%±3% according to the percentage.

The control of the steel level in the mold is realized by the operator's observation or the automatic control system. The Co-60 and Cs-137 ray methods can be used in the automatic control system. The γ-ray can detect the height of the steel liquid level. The linkage device continuously controls the opening of the tundish nozzle and the casting speed of the continuous caster to maintain the set value of the height of the steel liquid level, as shown in Figure 4-2.

4.3.2.2 结晶器液面控制

结晶器液面控制是连铸设备实现自动化的关键性环节，也是生产无缺陷铸坯，保证铸坯热送和连铸连轧一体化的前提。一般结晶器内钢液面应平稳地控制在距离结晶器上口100mm处或渣面距结晶器上口70mm左右，液面波动在±10mm以内，也有企业按照百分比将液面控制在85%±3%范围内。

结晶器内钢液面高度的控制是靠操作人员观察，或自动控制系统来实现的。其中自动控制系统可采用Co-60、Cs-137射线法。射线法是通过射线探测到钢液面高度，由联动装置连续控制中间包水口的开口度及连铸机的拉坯速度，以保持钢液面高度的设定值，如图4-2所示。

4.3.3 Control of Cooling System

In the normal casting process, the control of cooling system includes two aspects: the control

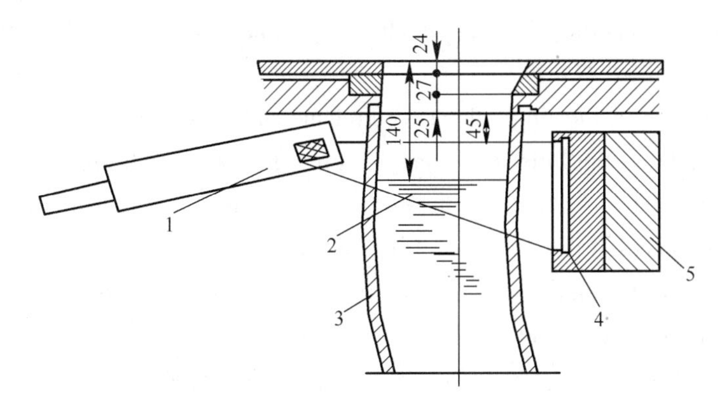

Figure 4-2　Schematic diagram of automatic control of mold liquid level

1—Scintillation counter；2—Ideal control height of liquid level；3-Copper tube of mold；4—Radiation source；5—Lead cylinder

图 4-2　结晶器液面自动控制原理图

1—闪烁计数器；2—液面理想控制高度；3—结晶器铜管；4—放射源；5—铅筒

of mold cooling system (primary cooling) and the control of secondary cooling section cooling system. The former determines the thickness of the initial solidified shell and some surface defects of the continuous casting billet, while the latter determines the internal structure and defects of the continuous casting billet.

4.3.3　冷却制度的控制

在正常浇铸过程中，冷却制度的控制包括两方面：结晶器冷却制度的控制（一次冷却）和二次冷却段冷却制度的控制。前者决定结晶器中初生凝固坯壳的形成厚度和连铸坯的一些表面缺陷；后者决定连铸坯的内部组织和内部缺陷。

The function of the mold is to ensure that the shell has enough thickness at the outlet of the mold, to bear the static pressure of the molten steel, to prevent leakage, and to make the shell cool evenly in the mold at the same time, to prevent the occurrence of surface defects.

结晶器的作用是保证坯壳在结晶器出口处有足够的厚度，以承受钢水的静压力，防止拉漏，同时又要使坯壳在结晶器内冷却均匀，防止表面缺陷的发生。

At present, the control methods of secondary cooling water include instrument method and automatic control.

(1) Instrument control method. The instrument is used to control the quantity of secondary cooling water in the early stage of continuous casting. The secondary cooling zone is divided into several sections, each section is equipped with electromagnetic flowmeter, and the water supply quantity of each section is adjusted by the regulator according to the process requirements (such as casting speed, steel grade, billet section). In production, when the process parameters change, the set value of the regulator shall be changed manually and timely, and the water supply of each section shall be changed accordingly. The instrument control system of secondary cooling water is mainly used for the caster with single type and size of billet.

(2) Automatic control method. There are three kinds of automatic control methods for sec-

ondary cooling water: proportional control method, parameter control method and target surface temperature control method.

The basic principle of the proportional control method is to measure the temperature at the secondary cooling outlet of the billet (the temperature before the billet enters the tension leveler) by the thermometer installed in front of the leveler, and send the value to PLC or computer for comparison with the process value, and feed the comparison value back to the water control system of the last section, so as to compensate and adjust the water quantity in this section, so as to make the surface temperature of the billet reach the set value.

目前二次冷却水的控制方法有仪表法和自动控制两大类别。

（1）仪表控制法。它是将二冷区分成若干段，每段装设电磁流量计，根据工艺要求（如拉速、钢种、铸坯断面）每一段的给水量，通过调节器按比例调节。生产中，当工艺参数发生变化时，由人工及时改变调节器的设定值，相应地改变各段的给水量。二冷水仪表控制系统，这种控制方式多用于铸坯品种和尺寸单一的连铸机。

（2）自动控制法。二次冷却水量自动控制有比例控制法、参数控制法和目标表面温度控制法三种，实际生产经常采用比例控制法来控制冷却水量的大小。

比例控制法的基本原理是通过拉坯矫直机前装的测温计来测量铸坯二冷出口温度（铸坯进入拉矫机前的温度），并将此值送入 PLC 或计算机，与工艺值相比较，并将该比较值反馈到最后一段的水量控制系统，用以补偿调节该段的水量，从而使铸坯表面温度达到设定值。

4.3.4 Operation of Dummy Bar Stripping

After the dummy bar leads the billet through the tension leveler, the dummy bar task is completed. At this time, it is necessary to separate the dummy bar and the billet, send the dummy bar to the storage place, and the billet will continue to move into the next cutting process, which is called dummy bar stripping operation (disconnecting the dummy bar).

4.3.4 脱锭操作

引锭杆牵着铸坯通过拉矫机后，便完成了引坯任务，此时需要把引锭杆与铸坯分开，将引锭杆送入存放处，铸坯继续行进将进入下道切割工序，这一操作称为脱锭操作。

4.3.5 Cutting Operation

The cutting operation is that the continuous casting billet of the tension leveler is cut into fixed length or multiple length according to the requirements. Refer to operation of cutting device in Task 7 of Project 3 for relevant operation steps.

4.3.5 切割操作

切割操作是指出拉矫机经脱锭后的连铸坯，按照要求将连铸坯切成定尺或倍尺长度的操作。相关操作步骤参考项目三中任务七中切割装置操作。

4.3.6 Sequence Casting Operation

4.3.6.1 Cognition of Sequence Casting Operation

Multi ladle continuous casting technology (sequence casting) includes continuous casting of the same steel type, continuous casting of different steel type and continuous casting of different sections (namely section width adjustment technology). Sequence casting is an important measure to improve the operation rate of continuous casting machine, increase the output of continuous casting billet and continuous casting ratio, and reduce the metal loss, so that the advantages of continuous casting machine can be fully exerted. The operation of sequence casting mainly includes: the operation of ladle replacement; the operation of rapid tundish replacement; the operation of different steel continuous casting.

4.3.6 多炉连浇操作

4.3.6.1 多炉连浇操作认知

多炉连浇技术包括同钢种连浇、异钢种连浇、不同断面连浇（即断面调宽技术）等。多炉连浇是提高连铸机作业率、提高连铸坯产量及连铸比、降低金属损失的重要措施，使连铸机的优点得到了充分的发挥。多炉连浇操作主要有：更换钢包的操作；快速更换中间包的操作；异钢种连浇的操作。

4.3.6.2 Operation of Sequence Casting of Different Steel Grades

The main problem of sequence casting of different steel grades is how to minimize the 'intermediate mixing zone'. The so-called 'intermediate mixing zone' is that the chemical composition of the continuous casting billet is between two steel grades. At present, the solution is to insert a 'solid bridge' into the mold when changing the tundish as a liquid steel separator.

The solid bridge device can be divided into two types, continuous casting billet connector and liquid steel separator. As shown in Figure 4-3, the solid bridge separator can be immersed in the molten steel of the mold and form a bridge under the liquid steel level. During each operation, the separator can be inserted quickly by machinery.

4.3.6.2 异钢种连浇的操作

不同钢种连浇的主要问题是如何使“中间混合区”最小。所谓“中间混合区”就是这段连铸坯的化学成分界于两个钢种之间。目前解决的方法是：更换中间包时往结晶器中插入一个“固体桥”，作为钢液分隔装置。

固体桥装置大体上可分为两类，即连铸坯连接件和钢液分隔器。如图 4-3 所示，固体桥钢液分隔器可以浸没在结晶器的钢液中，并在钢液液面下形成一个桥，在每次操作时，这种分隔器可以用机械迅速地插入。

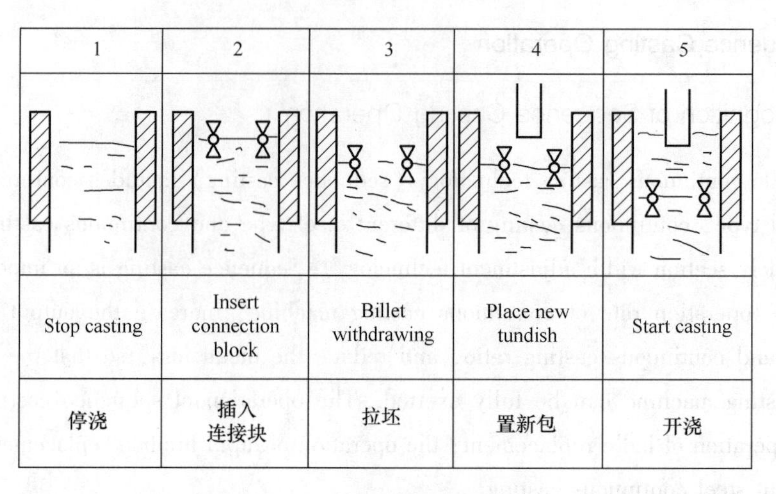

Figure 4-3 Diagram of connector use

图 4-3 连接件的使用示意图

Task Implementation 任务实施

4. 3. 7 Control Method of Casting Speed

In continuous casting production, stable casting speed is an important technological condition to ensure billet quality and smooth continuous casting operation. The casting speed should be suitable for the casting speed of molten steel from tundish to mold. Therefore, in addition to the diameter of the tundish nozzle should be suitable for the flow of molten steel, the molten steel in the tundish should also be kept at a suitable height (i. e. stable molten steel volume).

According to billet section, steel grade, tundish capacity, casting temperature and other factors, each factory keeps the tundish molten steel height stable to obtain stable casting speed. For this reason, modern continuous casting machines have automatic control devices for the molten steel height of the surface of the tundish. When the casting parameters fluctuate, the opening degree of the sliding nozzle of the ladle can be automatically controlled, so that the molten steel volume of the tundish steel is always same, creating conditions for stable casting speed.

4. 3. 7 拉速的控制方法

连铸生产中，稳定的拉速是保证铸坯质量和顺利进行连铸操作的重要工艺条件。拉速应和中间包向结晶器中浇铸钢水的速度相适应。为此除了中间包水口直径应和钢水的流量相适应以外，中间包内的钢水还应保持合适的高度（即钢水量稳定）。

各工厂根据铸坯断面、钢种、中间包容量、浇铸温度等因素使中间包钢水高度保持稳定，以获得稳定的拉速。为此，现代连铸机都有自动控制中间包钢水面高度的装置，当浇铸参数发生波动时，可自动控制钢包滑动水口的开启程度，使中间包钢水量始终不变，为稳定拉速创造条件。

4. 3. 8 Control of Cooling Water in Mold

According to different sections, different casters and different steel grades, the specific requirements of mold cooling water are determined. During the casting process, monitor the instrument display (or through the central control room display) at any time to ensure the cooling conditions of the mold. The specific requirements are as follows:

(1) Water pressure control. The pressure is controlled at about 0. 4～0. 6MPa. The proper water pressure can ensure the flow velocity of water in the water seam to be 6～12m/s, and prevent the intermittent boiling in the water seam of the mold and affect its heat transfer.

(2) Water temperature control. The inlet water temperature shall be less than or equal to 40℃, and the difference between the inlet and outlet water temperature shall not exceed 10℃. If the water temperature is too high, it is easy to produce scale and reduce heat transfer.

(3) Water quantity control. The control of water quantity can be determined according to the sectional area of the water seam and the flow rate of water in the water seam. For example, the flow of 140mm×140mm square billet is 72～146m^3/h (per flow).

(4) Water quality control. Soft water is used.

4. 3. 8 结晶器冷却水控制

根据不同断面、不同铸机、不同的钢种，确定结晶器冷却水特定要求。浇铸过程中要随时监视仪表显示（或通过中央控制室显示屏）以保证结晶器冷却条件。其特定要求如下：

（1）水压控制。压力控制在0. 4～0. 6MPa左右。合适的水压，可保证水缝内水的流速在6～12m/s，防止结晶器水缝中产生间断沸腾和影响其传热。

（2）水温控制。进水温度应小于等于40℃，进出水温差不应超过10℃。水温过高，容易产生水垢，减弱传热。

（3）水量控制。水量的控制可根据结晶器水缝的断面积和水缝内水的流速来确定。如140mm×140mm方坯的流量为72～146m^3/h（每流）。

（4）水质控制。水质控制为软水。

4. 3. 9 Control of Cooling Water in Secondary Cooling Section

(1) Water temperature control. The maximum water supply temperature is 40℃.

(2) Water pressure control. About 0. 7～0. 9MPa.

(3) Water quantity control. Due to different factors such as steel grade, section size of continuous casting billet and casting speed, the water quantity is also different, generally 0. 5～1. 5L/kg.

(4) Water quality control. Soft water is used.

4. 3. 9 二次冷却段冷却水控制

（1）水温控制。最高供水温度为40℃。

（2）水压控制。0.7~0.9MPa 左右。

（3）水量控制。因钢种、铸坯断面尺寸和拉坯速度等因素，水量也不同，一般为0.5~1.5L/kg。

（4）水质控制。软水。

Attention should be paid to the following problems during operation：

（1）Adjust and control the water quantity in the secondary cooling area in strict accordance with the process regulations.

（2）Spray cooling should make the longitudinal and transverse temperature distribution of the billet surface as uniform as possible to prevent the sudden change of temperature.

（3）In the secondary cooling area，the deformation（bulge，rhomboidity，etc.）caused by uneven cooling must be prevented.

（4）Operators of secondary cooling system shall avoid the damage caused by splashing of molten steel and explosion in water caused by steel leakage of caster.

操作中应注意如下问题：

（1）严格按照工艺规定进行二次冷却区的水量调节、控制。

（2）喷水冷却要使铸坯表面纵向和横向温度的分布尽可能均匀，防止温度的突变。

（3）在二次冷却区内必须防止铸坯因冷却不均匀造成的变形（鼓肚、菱变等）。

（4）二冷操作工要避免铸机漏钢等事故造成的钢液飞溅、遇水爆炸等引起的伤害。

4.3.10 Operation of Dummy Bar Stripping

（1）Reduce the casting speed to below the specified value before stripping.

（2）Pay attention to the operation of the continuous casting billet，and start to disconnect the dummy bar in time when it reaches the stripping position.

（3）When removing dummy bar，first select automatic dummy bar stripping，and carefully monitor the preparation for manual dummy bar stripping at any time. In case of failure to remove the dummy bar automatically，the manual operation shall be used immediately. If it still cannot be removed，stop casting in time. The dummy bar stripping of billets is usually completed after the continuous casting billets are cut off.

（4）After the dummy bar is removed，start the dummy bar removal device and store the dummy bar.

4.3.10 脱锭操作

（1）脱锭前将拉坯速度降到规定以下。

（2）注意连铸坯的运行情况，达到脱锭位置时，及时开始脱锭。

（3）脱锭时，首先选择自动脱锭，并认真监视随时做好手动脱锭的准备。在自动脱不掉时，立即改手动补救脱锭。仍然脱不掉时，要及时停浇。小方坯的脱锭往往在切断连铸坯后完成。

（4）引锭杆脱锭后，启动引锭杆移出装置，将引锭杆存放起来。

4.3.11 Operation of Replacing Ladle

Before finishing casting the molten steel in the ladle, pay close attention to the ladle flow. Once slag is found, close the nozzle immediately. If protective casting is adopted, the nozzle shall be closed 2~3min in advance according to the casting time of the ladle, and the ladle steel shall be continuously cast after the protective casing (tube) is removed.

Before replacing the ladle, try to control the liquid level of the tundish as high as possible to ensure that the billet casting speed is not reduced when replacing the ladle. The starting casting operation of the new ladle is same as the previous one. If the drainage takes a long time, the casting speed can be changed in a small range (about 20%). It is not allowed to stop the machine, and strictly prevent the slag from falling in the tundish.

4.3.11 更换钢包操作

钢包内钢液浇铸完毕前，严密注视钢包铸流情况，一旦发现下渣，立刻关闭水口。如采用保护浇铸时，应根据钢包的浇铸时间提前 2~3min 关闭水口，卸下保护管后，再继续将此炉钢浇完。

更换钢包前，尽量将中间包液面控制高一些，确保换钢包时不降低拉坯速度。新钢包开浇操作与前述开浇操作相同。如引流所需时间长，拉坯速度可在较小范围（约 20%）内变化，不得停机并严防中间包下渣。

4.3.12 Operation of Quick Replacement of Tundish

It is easy to realize multi ladles continuous casting in one tundish. This process is mainly to replace the ladle. If you want to prolong the continuous casting time of multiple ladles and increase the number of continuous casting ladles, you can interrupt the casting of tundish in a short time, change the tundish quickly, and then continue the casting. The specific operations are as follows:

(1) After the ladle stops casting, it should leave the casting position of the tundish, and at this time, the speed of billet drawing should be properly reduced.

(2) When 2/3 of molten steel in tundish is left, the new tundish which has been baked in advance will be driven by tundish car to the side of mold for use.

(3) When 1/2 of the molten steel in the tundish is left, the operation of protective slag exchange in the mold shall be carried out.

(4) When the liquid level of the tundish drops to 150mm (pay attention not to let the slag enter the mold), stop pouring immediately, quickly move the tundish and remove the debris of the nozzle. For medium and small section continuous casting billets, immediately insert the connector into the molten steel in the mold. One half of the connector is immersed in the molten steel, and the other half is left outside as the connector.

(5) When the liquid level of steel in the mold drops to the position of dummy bar head, the billet drawing shall be stopped and the secondary cooling water volume shall be reduced appropriately.

(6) Place the new tundish in the right position. If the connection body is used, the tundish nozzle shall not collide with the connection piece when lowering.

(7) When the new tundish is in place and confirmed to be in good condition, the ladle shall begin casting, and then follow the operation standard of starting casting.

(8) The tundish replacement time should not be too long. Generally, it is required to interrupt the casting time no more than 2min to prevent the steel leakage accident caused by the poor welding of the joint mark.

(9) The cutter shall leave the joint mark 200~300mm at the end of the continuous casting billet.

(10) In principle, it is not allowed to change the tundish at the same time when changing the ladle, and it is not allowed to change the tundish at the beginning or near the end of casting.

4.3.12 快速更换中间包操作

一个中间包实现多炉连浇比较容易，这个过程主要是更换钢包的操作。如果想延长多炉连浇的时间和提高连浇炉数，可在短时间内中断中间包浇铸，快速更换中间包，然后继续浇铸。具体操作如下：

(1) 在钢包停浇后，应离开中间包浇铸位置，此时适当降低拉坯速度。

(2) 当中间包钢液剩余2/3时，将预先烘烤好的新中间包，由中间包车开到结晶器旁待用。

(3) 当中间包钢液剩余1/2时，进行结晶器换渣操作。

(4) 当中间包液面降到150mm时（注意不让渣子进入结晶器），立即停止浇铸，快速开走中间包，捞出水口碎片。对于中小断面连铸坯，立即向结晶器内的钢液插入连接件。连接件的一半浸入钢液中，另一半留在外面，作为接头之用。

(5) 当结晶器内钢液面降至引锭头位置时，停止拉坯，并适当减少二次冷却水量。

(6) 新中间包就位、落位。若使用连接体，中间包水口下降时不得与连接件相碰。

(7) 新中间包落位时确认良好无误时，钢包到位开浇，然后按开浇操作标准执行。

(8) 更换中间包时间不能太长，一般要求中断浇铸时间不超过2min，以防接痕焊接不良引起漏钢事故。

(9) 切割工应将接痕留在连铸坯尾部200~300mm处。

(10) 换中间包时原则上不能同时换钢包，且不能在钢包钢液开浇初期或临近浇铸终了时更换中间包。

4.3.13 Operation of Sequence Casting of Different Steel Grades

(1) After ladle finished casting, close the nozzle and leave the casting position of tundish.

(2) When 1/2 of molten steel is left in the tundish, the old slag in the mold shall be removed and replaced with new slag. The new slag shall be evenly covered with a thickness of about 100mm.

(3) When the liquid level of the tundish drops to 150mm, stop casting immediately, quickly move away the tundish, remove the debris from the nozzle, control the liquid level at 100~150mm

from the upper inlet of the mold, and stop the billet drawing.

(4) The 'solid bridge' is quickly inserted into the molten steel in the mold. If necessary, evenly sprinkle 20mm cooling chips on the liquid surface to make the liquid surface solidify.

(5) The new tundish is put in place and dropped rapidly, and the liquid level in the mold is lowered to the position of dummy bar head.

(6) The ladle shall be casting in place, and then the operation standard of starting casting shall be followed.

(7) The cutter shall control the size well. The mixture area is 400mm in front of the joint mark line and 300mm behind the joint mark line. During cutting, the above 700mm shall be left at the end of the billet and not on the high-quality continuous casting billet.

4.3.13 异钢种连浇操作

(1) 钢包浇完后，关闭水口，离开中间包浇铸位置。

(2) 当中间包钢液剩余 1/2 时，捞出结晶器内旧渣，换上新渣，新渣均匀覆盖厚度 100mm 左右。

(3) 当中间包液面降到 150mm 时，立即停止浇铸，快速开走中间包，捞出水口碎片，液面控制在离结晶器上口 100~150mm，停止拉坯。

(4) 迅速将“固体桥”插入结晶器内的钢液中。必要时可在液面上均匀撒上 20mm 冷却屑，使液面凝固。

(5) 新中间包迅速到位、落下，同时把结晶器内的液面高度降到引锭头位置停放。

(6) 钢包到位开浇，然后按开浇操作标准执行。

(7) 切割工应控制好尺寸，接痕线前 400mm 处和后面 300mm 处为混合区，切割时将上述 700mm 留在坯尾，且不留在优质连铸坯上。

4.3.14 Sampling of Finished Products

The purpose of sampling is to analyze the chemical composition of the steel, in addition, the deoxidization status of the steel can be estimated preliminarily, which is convenient for the sequence of the next process.

The sampling must be representative and can fully represent the chemical composition of the molten steel. During sampling, the sample mold must be dry, clean, oil-free and rust free, otherwise the chemical composition of the steel sample will be affected, and the sample taken will not represent the molten steel. In addition, sampling time must also be correctly selected and representative. Generally, the samples are taken 10~15min after the ladle is opened for casting.

4.3.14 成品取样操作

取样是为了分析钢的化学成分，另外可以初步估计钢的脱氧状况，便于后道工序的顺利进行。

取样必须要有代表性，能充分表示所浇钢液的化学成分。取样时，样模内必须干燥、清洁、无油、无锈，否则要影响钢样的化学成分，所取的样也就不能代表所浇钢液。另外，取样时刻也必须正确选择和有代表性。一般在钢包开浇 10~15min 后才取样。

4. 3. 14. 1 Operation of Sampling with Ladle

(1) Tool preparation:

1) Sampling ladle with a certain length and clean without cold steel and residue.

2) A clean sampling mold without rust and sundries.

(2) Reduce the ladle flow and hold the sample ladle.

(3) Align the sample ladle with the ladle flow, and take back the sample ladle after the sample ladle is filled with liquid steel.

(4) Pour the molten steel in the sample ladle into the prepared sample mold.

(5) Knock the sample ladle and remove the cold steel inside the sample ladle for the next use.

(6) The sample taken out after the solidification of the molten steel in the sample ladle is the finished sample taken.

(7) Write the steel sample number on the steel sample with white paint or other equipment.

4. 3. 14. 1 样瓢取样操作

(1) 工具准备：

1) 有一定长度，且完好清洁无冷钢及残渣的样瓢。

2) 清洁无锈、内无杂物的样模一只。

(2) 关小钢包钢流，手握样瓢。

(3) 将样瓢对准钢包钢流，待样瓢内钢液注满后取回样瓢。

(4) 将样瓢内的钢液倒入事先准备好的样模内。

(5) 敲击样瓢，将样瓢内冷钢清除以便下次再用。

(6) 待样模内钢液凝固后取出的试样即为所取的成品样。

(7) 在钢样上用白漆或其他器材写上钢样编号。

4. 3. 14. 2 Operation of Sampling with Sampler

(1) Insert the sampling paper tube firmly on the sampling rod.

(2) Insert the sampling rod vertically under the liquid level of the tundish, so that the paper tube is more than 1/2 of the liquid level.

(3) The sampling paper tube must be kept in the molten steel for 5~10s.

(4) Pull out the sampling rod, and then remove the sampling paper tube from the sampling rod.

(5) Knock out the steel sample from the sampling paper tube.

(6) Write the steel sample number on the steel sample with white paint or other equipment.

4. 3. 14. 2 取样器取样操作

(1) 将取样纸管牢固插在取样棒上。

(2) 将取样棒垂直插入中间包液面下，使纸管埋入液面 1/2 以上。

(3) 取样纸管在钢液中必须保持 5~10s。

(4) 拔出取样棒，然后将取样纸管从取样棒上取下。
(5) 从取样纸管中敲打出钢样。
(6) 在钢样上用白漆或其他器材写上钢样编号。

Exercises:

(1) Describe the operation content of normal casting process.

(2) Use the simulation software to finish normal casting operation, and ensure casting speed and molten steel level in tundish and mold are stable.

(3) Replace ladle and tundish by simulation software to complete the sequence casting operation of three ladles.

思考与习题:

(1) 阐述正常浇铸过程的操作内容。
(2) 利用仿真软件进行正常浇铸操作，保证拉速稳定以及中间包和结晶器的页面稳定。
(3) 利用仿真软件更换钢包和中间包，完成3炉钢水的连浇操作。

Task 4.4 Stop Casting Operation
任务4.4 停浇操作

Mission objectives:

任务目标:

(1) Master the key points of stopping casting operation.

(1) 掌握连铸停止浇铸的操作要点。

(2) Be able to operate the equipment to complete the stop casting task, and carry out inspection and cleaning.

(2) 能够操作设备完成停浇任务，并进行检查和清理。

(3) Be able to complete ladle finishing casting, speed reduction, capping, tail billet output and other operations according to the process requirements.

(3) 能够根据工艺要求完成钢包浇完、降速、封顶、尾坯输出等操作。

Task Preparation 任务准备

4.4.1 Cognition of Stop Casting

Stop casting operation refers to the operation of ladle molten steel casting finish, tundish molten steel casting finish, continuous casting billet sending out of the continuous casting machine and inspection and cleaning after casting. The main content of stopping casting operation is ladle casting finish operation, speed reduction operation, capping operation, tail billet output operation and cleaning operation and inspection after casting.

4.4.1 停浇操作认知

停浇操作是指钢包钢液浇完，中间包钢液浇完，连铸坯送出连铸机及浇铸完检查和清理的操作。停浇操作的主要内容是钢包浇完操作、降速操作、封顶操作、尾坯输出操作及浇铸完的清理和检查。

4.4.2 Key Points of Stop Casting Operation

(1) Try to avoid ladle slag entering the tundish.

(2) The casting speed should be reduced with the decrease of the liquid level of tundish, so as to prevent the tundish slag from entering the mold, and increase the solidification of continuous casting billet, so as to prepare for the capping operation.

(3) In order to prevent the final molten steel in the tundish from entering the mold with slag, the method of retaining a certain amount of molten steel in the tundish should be adopted.

(4) When the molten steel in tundish is almost poured out, the mold powder should be removed to prepare for capping.

(5) In order to make the tail billet completely solidified, the casting speed is usually controlled at the 'creeping' speed. In order to make the inclusions of the billet in the mold float up as much as possible before solidification, the method of gently stirring liquid steel is often used.

(6) The tail billet must be completely solidified before it comes out of the mold. The tail billet with incomplete solidification cannot be pulled out of the mold.

(7) In order to ensure that the continuous casting billet can be straightened within the specified temperature range, the speed of 'tail-billet output' must be adopted.

4.4.2 停浇操作要点

(1) 要尽量避免钢包渣进入中间包。

(2) 拉坯速度应随着中间包液面的降低而降低，以防中间包渣进入结晶器，同时增加连铸坯的凝固，为封顶操作做好准备。

(3) 为防止最后的中间包钢液带渣进入结晶器，应采用保留一定中间包钢液的方法。

(4) 在中间包钢液快浇完时，要捞净结晶器内保护渣，为封顶做好准备。

(5) 为了使尾端连铸坯完全凝固，连铸机的拉坯速度往往控制在“蠕动”或“爬动”速度。为了使结晶器内铸坯在凝固前夹杂物尽量上浮，往往采用轻轻搅拌液态钢液的方法。

(6) 尾坯在出结晶器之前一定要完全凝固，凝固不完全的尾坯不能拉出结晶器。

(7) 尾坯输出时，必须采用“尾坯输出”速度，以保证连铸坯在规定温度范围内矫直。

4.4.3 Cognition of Capping Operation

The purpose of capping operation is to solidify and crust the molten steel at the end of the tail billet, so as to avoid the leakage of the non solidified molten steel when pulling out the mold and causing accidents. After solidification of the tail billet, it is ready to pull out the tail billet.

4.4.3 封顶操作认知

封顶操作的目的在于将尾坯末端的钢液凝固结壳，避免在拉出结晶器时未凝固的钢水漏出，造成事故。铸坯尾部凝固后准备进行尾坯拉出操作。

Task Implementation 任务实施

4.4.4 Operation Steps of Stop Casting

(1) After receiving the notice from the main control room or the director of caster to stop casting, turn the selection switch on the casting operation box to 'manual'.

(2) When there is a certain amount of molten steel left in the tundish, according to the director's instruction, the flow shall be blocked one by one according to the fixed length. After the billet is manually monitored to reach the third roller of the output roller table, the tundish steel casting worker of the flow shall be informed to stop casting.

(3) Pay attention to the distance between the tail billet and the tension leveler. When the tail billet is 300~500mm away from the tension leveler, the speed control knob on the steel casting operation box will return to zero. Press the lifting button of the frame of the tension leveler on the main control panel to avoid the frame falling down.

(4) After the final molten flow casting stop, drive the tundish car to the bottom of the accident slag pan, then swing the groove, and pull out the plug cone after cooling.

(5) When the residual steel in the tundish cools down, command the crane to lift the tundish away, then remove the residual steel in the swing groove, repair the refractory materials, and place the new tundish.

(6) Remove the sticking steel, residue and sundries around the mold and in the working area of the cover plate.

(7) Prepare for the next casting.

4.4.4 停浇操作步骤

(1) 接到主控室或机长停浇的通知后，将浇钢操作箱上选择开关打到“手动”。

(2) 当中间包剩一定量钢水时，根据机长指令，按定尺依次堵流，人工监视铸坯到达输出辊道第 3 个辊后，通知该流中间包浇钢工停浇。

(3) 密切观察尾坯距拉矫机的距离，当铸坯尾端距拉矫机 300~500mm 时，浇钢操作箱上的调速旋钮回零。在主控盘上按动拉矫机机架抬起按钮，避免机架砸下。

（4）在停浇最后一流后，将中间包车开到事故渣盘下，然后摆开摆槽，冷却后拔下堵眼锥。

（5）当中间包中残钢冷却后指挥吊车将中间包吊走，然后清除摆槽内的残钢，并修补耐火材料，坐新中间包。

（6）清除结晶器周围和盖板工作区域内的粘钢、残渣及杂物。

（7）做下一次浇铸前的准备。

4.4.5 Operation Steps of Ladle Casting Finish

(1) According to the casting time of molten steel in ladle or the casting length of continuous casting billet or the weight display of molten steel in ladle, the remaining molten steel in ladle can be correctly judged. If the protective pipe is used to protect the casting, the protective pipe shall be removed 2~3min before the ladle molten steel is finished casting.

(2) Watch the flow injection, close the ladle nozzle as soon as the slag is discharged.

(3) After closing the ladle nozzle, if there is any nodulation, use steel pipe to poke it out, and remove the relevant control mechanism of the nozzle.

(4) Send the ladle away.

4.4.5 钢包浇完操作步骤

（1）根据钢包内钢液的浇铸时间或连铸坯的浇铸长度或钢包内钢液的重量显示来正确判断钢包内剩余钢液量。如果采用保护管保护浇铸时，应在钢包钢液浇完前 2~3min 拆下保护管。

（2）目视注流，一见下渣立即关闭水口。

（3）关闭水口后，如果有水口结瘤，用钢管将其捅掉，并拆下水口有关控制机构。

（4）将钢包送走。

4.4.6 Operation Steps of Speed Reduction

(1) After the ladle finished casting, generally stop adding new protective slag to the mold, and prepare the slag raking, slag bailing, and start to reduce the speed of billet drawing.

(2) When the liquid level of steel in tundish drops to about 1/2, decrease the casting speed to about 50% of the normal speed. At this time, the mold slag is fished. During the slag bailing operation, the slag rakes shall be used to scoop up all the powder slag and molten slag along the surface of the molten steel, and the preparation for closing the nozzle shall be made at the same time.

(3) Pay attention to the liquid level of tundish steel. When the liquid level of the tundish steel drops to a level that may cause the tundish slag to flow into the mold, the tundish nozzle shall be closed immediately, and the billet casting speed shall be reduced to ‘creep’.

(4) Raise the tundish car, drive the car away from the mold and stop at the place where the slag is discharged from the tundish.

(5) Once again, remove the residual protective slag and nozzle debris in the mold, and prepare for capping operation.

4.4.6 降速操作步骤

（1）在钢包浇完后，一般停止向结晶器内添加新的保护渣，并准备好捞渣耙，捞渣，同时开始降低拉坯速度。

（2）当中间包内钢液面降至1/2左右时，拉坯速度应降到正常拉坯速度的50%左右。此时开始捞结晶器内的保护渣。捞渣操作时，用捞渣耙沿着钢液表面将粉渣和熔融态渣全部捞净，同时做好关闭水口的准备。

（3）注意观察中间包钢液面。当中间包钢液面降到可能使中间包渣流入结晶器的时候，立即关闭中间包水口，同时将拉坯速度降到“蠕动”。

（4）升起中间包车，将车开离结晶器并停在中间包放渣的位置。

（5）再一次捞尽结晶器内所残余的保护渣和水口碎片，并准备封顶操作。

4.4.7 Operation Steps of Capping

(1) Just before the steel slag of tundish appears, close the tundish, turn the caster to creep speed, or turn to the minimum casting speed and drive the tundish car away. After that, immediately remove the slag on the molten steel in the mold.

(2) After all the slag is removed, the liquid steel can be stirred gently with a fine steel bar or argon blowing tube. This operation should be even and sufficient. Then spray water on the mold copper plate around the end of the billet to accelerate its solidification and form the shell.

(3) After slow stirring, if there is no solidification on the surface of the continuous casting tail billet in the mold, slow cooling and solidification can be achieved by reducing the casting speed.

(4) If the casting speed is reduced and the continuous casting tail billet in the mold is not solidified, it can be solidified by stopping the casting completely, continuing to spray water or sprinkle iron scraps.

(5) From the beginning to the end of capping, the secondary cooling shall be weakened and water distribution shall be carried out according to the regulations of capping operation.

4.4.7 封顶操作步骤

（1）当中间包钢渣刚刚出现之前，关闭中间包，连铸机转为蠕动速度，或转到最低拉速并将中间包车开走，此后，马上捞净结晶器钢液面上的渣子。

（2）当所有的渣除去之后，才能用细钢棒或吹氩管轻轻搅动钢液，这一操作要均匀而充分。然后用喷淋水喷在铸坯尾端周围的结晶器铜板上，加快其凝固形成坯壳。

（3）缓慢搅动后，若结晶器内尾部连铸坯表面没有凝固，可采用减低拉速，使其缓慢冷却凝固。

（4）若减低拉速后，结晶器内尾部连铸坯还未凝固时，可以采用完全停止拉坯，继续喷水或撒铁屑，使其凝固。

（5）从封顶开始到结束过程中，应将二次冷却减弱，按封顶操作的规定进行配水。

4.4.8 Operation Steps of Tail Billet Output

(1) Gradually increase the casting speed to the specified output speed of tail billet. Generally, the pulling out speed of tail billet is about 30% higher than that of normal casting speed.

(2) When the tail billet is pulling out, the water distribution in the secondary cooling section will return to normal.

(3) The output of the billets can only be finished when the end of the continuous casting billets leaves the last pair of pinch rolls of the tension leveler.

4.4.8 尾坯输出操作步骤

（1）将拉坯速度逐渐升到规定的尾坯输出速度。一般尾坯输出速度比正常拉坯速度高30%左右。

（2）尾坯输出时，二次冷却段的配水将恢复到正常。

（3）尾坯输出只有等到连铸坯尾端离开拉矫机最后一对夹辊才算结束。

4.4.9 Cleaning and Inspection after Casting

(1) Lift off the residual protective slag and slag rake on the mold, and wash the cover plate of the mold.

(2) Clean the residual steel and residue of the protective pipe, and move the operating machinery to a safe position.

(3) Clean the residual steel, residue and dirt on the inner wall of the mold, and inspect the mold according to the requirements of 'inspection before casting'.

(4) After the residual steel and residue in the old tundish are completely discharged, it shall be lifted away in time.

(5) Replace the full slag pan and overflow tank.

(6) Clean all casting tools and instruments.

(7) Other inspections shall be carried out according to 'inspection before casting'.

4.4.9 浇铸结束后的清理和检查

（1）抬下结晶器上的剩余保护渣和渣耙，并对结晶器盖板进行冲洗。

（2）清理保护管的残钢、残渣，并将操纵机械移至安全位置。

（3）清理结晶器内壁残钢、残渣和污垢，并按“浇铸前的检查”要求对结晶器进行检查。

（4）将旧中间包内的残钢、残渣放尽后，及时吊走。

（5）更换已装满的渣盘、溢流槽。

（6）清理所有的浇铸工器具。

（7）其他的检查均按“浇铸前的检查”进行。

Exercises:

(1) Describe the operation content of stopping the casting process.

(2) Use the simulation software to complete the stop casting operation according to the steps.

(3) Use simulation software to check the parameters after production and analyze the problems in production.

思考与习题:

(1) 阐述停止浇铸过程的操作内容。

(2) 利用仿真软件按照步骤完成停浇操作。

(3) 利用仿真软件查看生产结束后的各项参数，分析生产中产生的问题。

Task 4.5 Continuous Casting Process System Control
任务 4.5 连铸工艺制度控制

Mission objectives:

任务目标:

(1) Master the key points of stopping casting operation.

(1) 能够阐述连铸工艺制度的种类和内容。

(2) Be able to operate the equipment to complete the stop casting task, and carry out inspection and cleaning.

(2) 能够确定浇铸温度、拉坯速度、冷却制度等工艺参数。

(3) Be able to complete ladle finishing casting, speed reduction, capping, tail billet output and other operations according to the process requirements.

(3) 能够根据工艺要求调整连铸操作，保证产品质量。

Task Preparation 任务准备

Continuous casting process system includes temperature system, speed control and cooling system. In the 'three stability' of continuous casting production, the premise of stable casting speed is stable, which creates conditions for stable liquid level.

连铸工艺制度包括温度制度、拉速控制、冷却制度。连铸生产的“三稳定”中拉速稳定的前提是稳定，而拉速稳定又为液面稳定创造了条件。

4.5.1 Casting Temperature Control

Casting temperature refers to the temperature of molten steel in tundish. The temperature of the first ladle is higher than that of the normal casting, and the starting casting temperature of the tundish should be 20~50℃ above the liquidus of the steel.

According to the requirements of steel quality, the lower superheat should be controlled and the uniform and stable casting temperature should be maintained. Therefore, the tundish heating technology can be used in the initial stage, the final stage and the change of ladle to compensate the temperature loss of molten steel. In the normal casting process, it can also be properly heated to compensate for the natural temperature drop of the molten steel, so as to maintain the constant temperature of the molten steel.

4.5.1 浇铸温度控制

浇铸温度是指中间包钢水温度。第一包钢水温度比正常浇铸温度要高，中间包开浇温度应在钢中液相线以上20~50℃为宜。

根据钢种质量的要求控制较低的过热度，并保持均匀稳定的浇铸温度。为此在浇铸初期、浇铸末期、换包时，可采用中间包加热技术，补偿钢水温度降低损失。在正常浇铸过程也可适当加热，补偿钢水的自然温降，以保持钢水温度的恒定。

The target temperature is determined according to the steel grade, casting billet section, and the total temperature drop T_{Total}.

$$T_{Tapping} = T_{Casting} + T_{Total}$$

For example, the tapping temperature of 120t converter continuous casting Q235 slab in a steel plant is as follows:

(1) Casting temperature, 1530~1535℃.

(2) Process temperature drop, 30~35℃ from the ladle to the tundish.

Temperature drop from argon blowing to starting casting waiting time is 10℃.

Argon blowing cooling, 15℃.

Tapping cooling, 60℃.

Therefore, the tapping temperature is:

$$T_{Tapping} = 1535 + (35 + 10 + 15 + 60) = 1655℃$$

钢种、铸坯断面等确定了中间包钢水温度的目标值，再加上钢水传递过程中总温降$T_{总}$，就可确定出钢温度。

$$T_{出} = T_C + T_{总}$$

例如某钢厂120t转炉连铸Q235板坯的出钢温度：

(1) 浇铸温度，1530~1535℃。

(2) 过程温降，钢包到中间包的温降30~35℃。

吹氩后到开浇等待时间温降10℃。

吹氩降温，15℃。

出钢降温，60℃。

所以，出钢温度为：

$$T_{出} = 1535 + (35 + 10 + 15 + 60) = 1655℃$$

4.5.2 Control of Casting Speed

4.5.2.1 Determination of Casting Speed

The casting speed usually refers to the length of billet pulled out per first-class unit time by the continuous caster (m/min). The following factors shall be considered:

(1) Ensure that there is no steel leakage at the lower outlet of the mold;

(2) The length of liquid core is less than that of metallurgy;

(3) The casting period matches the production capacity of steelmaking and refining;

(4) To ensure the quality of billet.

The casting speed should be determined according to the section size, casting temperature and steel grade. The calculated casting speed is the theoretical maximum casting speed, and then the maximum working casting speed is determined by the theoretical maximum casting speed.

4.5.2 拉坯速度控制

4.5.2.1 拉速的确定原则

拉速通常指连铸机每一流单位时间拉出的铸坯长度（m/min），应考虑以下因素：

（1）保证出结晶器下口不漏钢；

（2）液芯长度小于冶金长度；

（3）浇铸周期与炼钢、精炼生产能力相匹配；

（4）保证铸坯质量。

拉速要考虑铸坯断面尺寸、浇铸温度和浇铸钢种来确定。计算得出的拉速为理论最大拉速，再由理论最大拉速确定最大工作拉速。

4.5.2.2 Calculation of Casting Speed

There are two commonly used calculation methods: one is that the complete solidification is just selected at the straightening point, and the speed at this time is the maximum casting speed, which is calculated according to the metallurgical length of the casting machine. The formula of this method is as follows:

$$\frac{D}{2} = K\sqrt{t} = \sqrt{\frac{L_M}{v_{max}}} \qquad v_{max} = \frac{4K^2}{D^2} \times L_M$$

Where D——The thickness of billet, mm;

K——The solidification constant of billet, $mm/min^{1/2}$, K of billet is generally 28~32, K of slab is generally 26~29;

v_{max}——The theoretical maximum casting speed, m/min;

L_M——The metallurgical length or liquid core length, m;

t——The solidification time of billet, min.

In the actual production, the theoretical maximum casting speed is basically not applicable, so when setting process parameters, the maximum casting speed that may be used is called the maximum working casting speed.

4.5.2.2 拉速的计算

计算方法常用的有两种：一种是完全凝固正好选在矫直点上，此时的速度即是最大拉速，按铸机冶金长度计算，这种方法公式如下：

$$\frac{D}{2} = K\sqrt{t} = \sqrt{\frac{L_M}{v_{max}}} \qquad v_{max} = \frac{4K^2}{D^2} \times L_M$$

式中 D——铸坯厚度，mm；

K——铸坯凝固系数，$mm/min^{1/2}$，方坯 K 一般取 28~32，板坯 K 一般取 26~29；

v_{max}——理论最大拉速，m/min；

L_M——冶金长度或液芯长度，m；

t——铸坯凝固时间，min。

实际生产中基本不适用理论最大拉速，因此设定工艺参数时，将可能使用的最大拉速称为最大工作拉速。

Another method is that when the shell at the bottom of the mold is minimum thickness, it is called safe thickness (δ_{min}), and the corresponding casting speed is maximum casting speed.

$$\delta_{min} = K_m\sqrt{t_m} = K_m\sqrt{\frac{L_m}{v_{max}}} \qquad v_{max} = \left(\frac{K_m}{\delta_{min}}\right)^2 \times L_m$$

Where δ_{min}——The shell safety thickness at the outlet of the mold, mm, generally, the shell safety thickness of the billet is 10mm, and the safety thickness of the slab is 15mm;

K_m——The solidification constant of the steel in the mold, $mm/min^{1/2}$, K_m can be taken as $20mm/min^{1/2}$;

L_m——The effective length of the mold, m, can be used according to the total length of the mold to reduce 80mm;

t_m——The billet residence time in the mold, min;

v_{max}——The maximum casting speed, m/min.

另一种方法是当出结晶器下口的坯壳为最小厚度时，称为安全厚度（δ_{min}），此时对应的拉速为最大拉速。

$$\delta_{min} = K_m\sqrt{t_m} = K_m\sqrt{\frac{L_m}{v_{max}}} \qquad v_{max} = \left(\frac{K_m}{\delta_{min}}\right)^2 \times L_m$$

式中 δ_{min}——结晶器出口处坯壳安全厚度，mm，一般小方坯的坯壳安全厚度为 10mm，板坯的坯壳安全厚度为 15mm；

K_m——结晶器内钢的凝固系数，$mm/min^{1/2}$，可取 $K_m = 20mm/min^{1/2}$；

L_m——结晶器的有效长度，m，可按照结晶器总长度减少 80mm 使用；

t_m——铸坯在结晶器内停留时间，min；

v_{max}——最大拉速，m/min。

4.5.2.3 Factors Affecting Casting Speed

(1) The solidification constant of different steels is different, as shown in Table 4-2. The solidification constant of carbon steel is the largest and that of alloy steel is the smallest. Therefore, the casting speed of carbon steel with the same section is higher than that of alloy steel. This is because the steel with small solidification constant has a large thermal stress in the cooling process, so it can only use a small casting speed. The casting speed of general alloy steel is 20% ~30% lower than that of carbon steel.

4.5.2.3 影响拉速的因素

(1) 不同钢种凝固系数不同，见表4-2，碳素钢凝固系数最大，合金钢凝固系数最小。因此，断面相同的碳素钢拉速要比合金钢的拉速大。这是因为凝固系数小的钢种在冷却过程中产生的热应力大，只能采用较小的拉速。一般合金钢的浇铸速度比碳钢低20%~30%。

Table 4-2 Influence of steel grades and billet sections on casting speed

表4-2 钢种和铸坯断面对拉坯速度的影响

Section 铸坯断面 /mm×mm	Steel grade 钢种					
	Al-killed steel 铝镇静钢 $K=28\text{mm/min}^{1/2}$		Low alloy steel 低合金钢 $K=26\text{mm/min}^{1/2}$		Alloy steel 合金钢 $K=24\text{mm/min}^{1/2}$	
	Metallurgical length 冶金长度 /m	Casting speed 拉速 /m·min^{-1}	Metallurgical length 冶金长度 /m	Casting speed 拉速 /m·min^{-1}	Metallurgical length 冶金长度 /m	Casting speed 拉速 /m·min^{-1}
100×100	10.8	3.4	9.2	2.5	8.7	2.0
150×150	16.5	2.3	15.8	1.9	16.6	1.7
200×200	22.9	1.8	23.7	1.6	22.6	1.3
200×1000	22.9	1.8	23.7	1.6	22.6	1.3
200×2000	20.4	1.6	20.7	1.4	19.1	1.1
300×500	37.3	1.3	36.6	1.1	35.2	0.9
300×1000	37.3	1.3	36.6	1.1	35.2	0.9

(2) Influence of section shape and dimension of billet. For billets with different section shapes, the specific surface of cooling is different with different peripheral dimensions of unit mass. The specific surface area of the round section is smaller than that of the square and rectangle, and the cooling is slower, so the casting speed is smaller. Compared with the square, when the rectangular slab solidifies in the mold, the narrow side solidifies faster than the wide side, and the time for the solidified shell to form the air gap from the mold wall is earlier, so the solidification speed is reduced. Therefore, the casting speed of rectangular slab is smaller than that of billets. On the other hand, billets with different section shapes have different crystallization and

solidification characteristics, so the casting speed is also different.

（2）铸坯断面形状及尺寸的影响。不同断面形状的铸坯，单位质量的周边尺寸不同，冷却的比表面积不同。圆形断面比方形和矩形的比表面积小，冷却慢，故拉坯速度要小一些；矩形与方形相比，矩形坯在结晶器中凝固时，窄边比宽边凝固快，凝固壳脱离器壁形成气隙的时间早，使凝固速度降低，故矩形坯的拉速比方坯小一些。另一方面，不同断面形状的铸坯有不同的结晶凝固特点，因此拉速也不同。

（3）The influence of casting temperature and the content of sulfur and phosphorus in steel. The casting temperature is high, the solidification time is prolonged, the casting speed should be reduced. In the practice of continuous casting, the casting speed should be adjusted according to the temperature of molten steel in tundish. In addition, when the casting temperature is high or the sulfur and phosphorus in the molten steel are high, the casting speed should be reduced appropriately.

（3）浇铸温度及钢中硫磷含量的影响。浇铸温度高，凝固时间延长，拉速应减小，反之亦然。在连铸生产实践中要根据中间包中钢水温度来调整拉坯速度。另外当浇铸温度偏高或钢水中硫磷较高时，都要适当降低拉速。

4.5.3 Cooling System Control

4.5.3.1 Division of Continuous Casting Cooling Area

During the continuous casting process, the molten steel solidification can be divided into three heat transfer and cooling areas, as shown in Figure 4-4.

（1）Primary cooling zone. The mold is cooled to form a billet shell with enough thickness to ensure no steel leakage from the mold.

（2）Secondary cooling zone. After the casting billet is out of the mold, water spray or mist will make the casting billet completely solidified.

（3）Third cooling zone. Air cooling zone, the billet radiates heat to the air, and the final temperature drops to room temperature.

4.5.3 冷却制度控制

4.5.3.1 连铸冷却区域划分

连铸过程中钢液凝固可分3个传热冷却区，如图4-4所示。

（1）一次冷却区。即结晶器冷却，形成足够厚的坯壳来保证铸坯出结晶器不漏钢。

（2）二次冷却区。铸坯出结晶器后，喷水或水雾使铸坯完全凝固。

（3）三次冷却区。空冷区，铸坯向空气中辐射散热，最终温度降到室温。

According to the measurement and calculation of the solidification heat balance of the billet, the heat dissipation ratio of each cooling zone of the continuous casting machine is respectively: the mold accounts for 16%~20%, the secondary cooling zone accounts for 23%~28%, and the radiation zone accounts for 50%~60%.

根据铸坯凝固热平衡测定计算，连铸机各冷却区散热比例分别为：结晶器占 16%~20%，二冷区占 23%~28%，辐射区占 50%~60%。

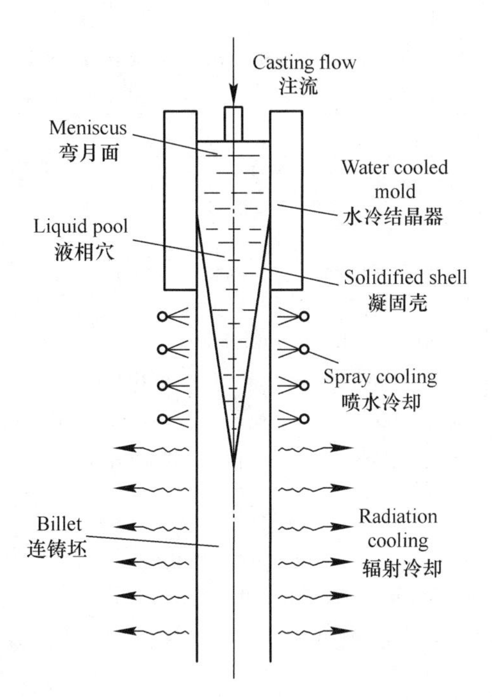

Figure 4-4 Diagram of continuous casting billet cooling and solidification

图 4-4 连铸坯冷却凝固示意图

4.5.3.2 Mold Cooling System

During the casting process, the cooling water flow rate of the mold is usually kept constant. Water supply shall be started 10~20min before the start of casting and stopped 10~20min after the stop casting. The water volume of the mold can be calculated according to the formula (4-1):

$$Q = 0.0036Sv \tag{4-1}$$

Where Q——The cooling water volume of the mold, m^3/h;

S——The total area of water gap in the mold, mm^2;

v——The flow rate of cooling water in the water joint, m/s, 6~12m/s for billet and 3.5~5m/s for slab.

See Table 4-3 for cooling water volume of common billet mold.

4.5.3.2 结晶器冷却制度

在浇铸过程中，结晶器的冷却水流量通常保持不变。在开浇前 10~20min 开始供水，停浇后 10~20min 停水。结晶器水量可根据式（4-1）计算：

$$Q = 0.0036Sv \tag{4-1}$$

式中 Q——结晶器冷却水量，m^3/h；

S——结晶器水缝总面积，mm^2；

v——冷却水在水缝内的流速，m/s，方坯取6~12m/s，板坯取3.5~5m/s。

常见方坯结晶器冷却水量见表4-3。

Table 4-3　Cooling water volume of mold of billet

表4-3　小方坯结晶器冷却水量

Billet section 铸坯断面/mm×mm	150×150	120×120	90×90
Mold water volume 结晶器水量/$m^3 \cdot h^{-1}$	72~108	57.6~86.4	43.2~64.8

4.5.3.3　Secondary Cooling System

In the secondary cooling zone, with the increase of the distance from the mold along the direction of the billet, the heat loss through the surface of the billet is gradually reduced, and the amount of water spray in the secondary cooling zone is also gradually reduced. The secondary cooling water consumption can be calculated according to the formula (4-2):

$$Q = WG \tag{4-2}$$

Where　Q——The water volume in the secondary cooling zone, m^3/h;

W——The secondary cooling strength, m^3/h;

G——The theoretical hourly output of the caster, t/h.

4.5.3.3　二次冷却制度

在二冷区内，沿铸坯方向随着距结晶器距离的增加，通过铸坯表面散失的热量逐渐减少，相应二冷段的喷水量也逐渐减少。二次冷却耗水量可按式（4-2）进行计算：

$$Q = WG \tag{4-2}$$

式中　Q——二冷区水量，m^3/h；

W——二次冷却强度，m^3/h；

G——连铸机理论小时产量，t/h。

Task Implementation 任务实施

4.5.4　Control of Casting Speed

4.5.4.1　Control of Starting Casting Speed

After starting casting, the opening of tundish nozzle is adjusted to maintain stable mold liquid level, so that the casting speed is kept at the starting casting speed. Generally, the casting speed is 80% of the working speed, at the same time, a certain seedling emergence time is needed.

4.5.4　连铸生产中拉速的控制

4.5.4.1　开浇拉速控制

开浇后调整中间包水口开度，保持稳定的结晶器液面，使拉速保持在开浇拉速。开浇拉速一般为工作拉速的 80%，同时还需一定的出苗时间。

4.5.4.2 Control of Normal Casting Speed

According to the specified turning fast speed time of the caster (holding time of casting speed, generally after the dummy bar head enters the secondary cooling), gradually increase the casting speed, adjust the opening of the nozzle, and maintain a stable steel level. The casting speed is adjusted to 0.1m/min for one gear every time, and the casting speed is adjusted to 1~2min for stable transition every time (the stable transition time of small billet can be shorter, and that of slab is longer).

When the casting speed reaches the specified working speed, the opening of the tundish nozzle shall be controlled to ensure the stability of the steel level. The casting speed is required to be stable during normal casting.

(1) When the temperature of molten steel changes, the working speed should be adjusted properly. Within the specified temperature range, the working speed of lower limit should be selected for the higher temperature molten steel, while the working speed of higher limit should be selected for the opposite (generally, the speed formulated by the caster has a certain range).

The target temperature is generally within the range of 15~25℃ above the liquidus (molten steel temperature of tundish). When the liquid steel temperature exceeds the target temperature, the following measures shall be taken:

1) When the temperature of tundish is lower than the lower limit temperature, the casting speed should be increased by 0.1~0.2m/min.

2) When the temperature of tundish is 5℃ higher than the upper limit temperature, the casting speed is reduced by 0.1m/min.

3) When the temperature of tundish is 6~10℃ higher than the upper limit temperature, the casting speed is reduced by 0.2m/min.

4) When the temperature of tundish is 11~15℃ higher than the upper limit temperature, the casting speed is reduced by 0.3m/min. For higher temperature molten steel, tundish should be stopped casting.

(2) When the plasma heating device is installed in the tundish, when the temperature of the molten steel in the tundish is low, the temperature can be raised by heating to maintain the normal casting speed.

(3) When the fluidity of molten steel is poor, the nozzle is clogged, the steel flow cannot be opened, and the casting speed drops to 0.2~0.3m/min below the specified lower limit, the tundish nozzle must be cleaned.

(4) When the oxygen content of molten steel is too high or the nozzle cannot be controlled due to other reasons, and the casting speed is more than 0.3m/min above the upper limit, the nozzle of tundish shall be out of control.

4.5.4.2　正常拉速控制

根据铸机操作规定的转快时间（开浇拉速保持时间，一般待引锭头进入二次冷却后），逐步提高拉速，并调整水口开口度，保持稳定的钢液面，拉速每次调整 0.1m/min 为一档，每调整一次要保持 1~2min 时间作稳定过渡（小方坯的稳定过渡时间可短一些，板坯则长一点）。

拉速达到规定的工作拉速后，控制中间包水口开度，保证钢液面的稳定。铸机正常浇铸时要求拉速稳定不变。

（1）当钢液温度变化时，工作拉速要适当调整，在规定温度范围内，较高温度的钢液选择低限的工作拉速，反之则选择为高限的工作拉速（铸机制订的拉速表一般有一定范围）。

目标温度一般规定在液相线之上 15~25℃范围内（中间包钢液温度）。当钢液温度超过目标温度时，要采取以下措施：

1）当中间包温度低于下限温度时，要提高拉速 0.1~0.2m/min。

2）当中间包温度高于上限温度 5℃之内时，降低拉速 0.1m/min。

3）当中间包温度高于上限温度 6~10℃时，降低拉速 0.2m/min。

4）当中间包温度高于上限温度 11~15℃时，降低拉速 0.3m/min。对于更高温度的钢液，中间包应作停浇处理。

（2）装有中间包等离子加热装置时，当中间包钢液温度偏低时，可进行加热提温，保持正常拉速。

（3）当钢液流动性差，水口发生粘堵，钢流无法开大，拉速下降到规定下限以下 0.2~0.3m/min 时，中间包水口必须进行清洗工作。

（4）当钢液含氧量过高或其他原因造成水口无法控制，拉速高于规定上限 0.3m/min 以上时，中间包水口要做水口失控处理。

4.5.4.3　Control of Capping Speed

When the casting is to be stopped, the capping casting speed must be controlled. With the liquid level of tundish decreasing, the casting speed gradually decreases. After the tundish molten steel is finished pouring, the casting speed can be kept low or the billet withdrawing can be stopped to facilitate the billet capping operation. After the billet is pulled out of the mold, we can recover the working speed or 1.3 times of the working speed to pull out the tail billet, so as to prevent too much temperature loss of the billet, which makes the straightening and cutting operation difficult.

4.5.4.3　封顶拉速控制

当要停止浇铸时，必须控制封顶拉速。随中间包液面降低，拉速逐渐降低，中间包钢水浇完后，可维持低拉速或停止拉出，以利于进行封顶操作。铸坯拉出结晶器后，可恢复工作拉速或 1.3 倍的工作拉速将尾坯拉出，以防止铸坯温度损失太大，造成矫直和切割操作困难。

4.5.5 Control of Cooling Water in Mold

(1) Before the start of casting, the water treatment station shall be informed to start pumping water to make the water volume and pressure within the range specified by the process.

(2) Check the leakage of the mold and the water inlet and outlet pipes. If any abnormality is found, the pump must be stopped and the water supply must be stopped after maintenance to ensure the normal water supply.

(3) In the preparation and casting process of continuous casting, the cooling water of mold is generally not regulated, as long as it is controlled within the specified range of water quantity and water pressure. Otherwise, the valve can be adjusted on site or at the water treatment station.

(4) In addition to monitoring the cooling water quantity and water pressure of the mold, the temperature of the water outlet and the difference temperature of the inlet and outlet water should also be monitored during the casting process. Generally, it is not necessary to control below the specified value. When the temperature exceeds the standard, the water supply and pressure must be increased or the speed must be reduced.

(5) When the water temperature cannot be controlled (lowered) or rises suddenly after adjustment, the caster must stop casting.

(6) Before and during water supply, water quality analysis must be carried out according to regulations to ensure water supply conditions.

(7) At the end of casting, after all the billets are lifted away from the conveying roller table and the cooling bed, the cooling water of the mold shall be shut down, and the water treatment station shall be informed to stop the pump and stop the water.

4.5.5 结晶器冷却水的控制

(1) 开浇前，通知水处理站开泵送水，使水量和水压在工艺规定的范围内。

(2) 对结晶器和进出水管道进行渗漏水检查，发现异常必须停泵停水，待检修后再送水，确保供水正常。

(3) 在连铸准备和浇铸过程中，结晶器冷却水一般不作调节，只要控制在规定的水量和水压范围内。否则可在现场或水处理站调节阀门。

(4) 浇铸过程中除监视结晶器冷却水水量和水压外，还要监视出水温度和进出水温差。凡在规定值以下一般可不作控制，当温度超标时，必须加大供水量和水压或作降速处理。

(5) 经过调节无法控制（降低）水温或水温突然升高时铸机必须作停浇处理。

(6) 供水前和供水过程中，必须按规定作水质分析，保证供水条件。

(7) 浇铸结束，铸坯吊离输送辊道、冷床后，关结晶器冷却水，通知水处理站停泵停水。

4.5.6 Control of Secondary Cooling Water in Different Stages

4.5.6.1 Relationship between Casting Speed and Cooling Water Volume

(1) The starting casting speed is low, and the secondary cooling water supply is small.

(2) The normal casting speed is the working casting speed, and the secondary cooling water supply is large.

(3) At the highest speed, the casting speed is the highest, and the secondary cooling water supply is the largest.

(4) When the tail billet is capped, the casting speed slows down until the withdrawing is stopped, and the secondary cooling water supply decreases accordingly.

4.5.6 不同阶段的二次冷却水控制

4.5.6.1 拉速与冷却水量的关系

(1) 起步拉坯，拉速为起步拉速，速度较低，二冷供水量小。

(2) 正常拉坯，拉速为工作拉速，二冷供水量较大。

(3) 最高速拉坯，拉速为最高拉速，二冷供水量最大。

(4) 尾坯封顶，拉速减慢直至停止拉坯，二冷供水量相应减小。

4.5.6.2 Relationship between Section and Cooling Water Volume

(1) The section of billet is small, the secondary cooling water is small, and the water supply increases with the increase of section.

(2) The section of slab is large, the secondary cooling water is large, and the water supply increases with the increase of setion.

4.5.6.2 断面与冷却水量的关系

(1) 方坯断面较小，其二冷水量小，随断面增大其供水量逐渐增大。

(2) 板坯断面较大，其二冷水量也大，随断面增大其供水量逐渐增大。

Exercises:

(1) What is the relationship between cooling system and billet quality?

(2) How to control the casting speed in production?

(3) Use simulation software to complete continuous casting production, pay attention to the adjustment of casting speed and cooling parameters under different conditions.

思考与习题:

(1) 冷却制度与铸坯质量有什么关系?

(2) 生产中拉速该如何控制?

(3) 利用仿真软件完成连铸生产，注意不同情况下拉速、冷却参数的调整。

Task 4.6 Continuous Casting Protective Casting
任务 4.6 连铸保护浇铸

Mission objectives:

任务目标:

(1) Master the key points of stopping casting operation.

(1) 能够阐述连铸工艺制度的种类和内容。

(2) Be able to operate the equipment to complete the stop casting task, and carry out inspection and cleaning.

(2) 能够确定浇铸温度、拉坯速度、冷却制度等工艺参数。

(3) Be able to complete ladle finishing casting, speed reduction, capping, tail billet output and other operations according to the process requirements.

(3) 能够根据工艺要求调整连铸操作保证产品质量。

Task Preparation 任务准备

Protective casting (shielded casting) is the protection of the whole process of molten steel transmission, which can prevent the reoxidation of molten steel and improve the quality of billet.

保护浇铸就是对钢水传递全过程的保护，可以防止钢液二次氧化，改善连铸坯质量。

4.6.1 Protection Operation of Ladle to Tundish Injection Flow

During the normal casting process, special personnel shall be assigned to monitor on the operation platform. If the nozzle is clogged, the protective pipe shall be removed immediately; when the length of the protective pipe after erosion leaves the normal casting liquid level of the tundish, it shall be replaced immediately; if the vent ring falls off, it cannot be ensured that it is tightly occluded with the ladle nozzle or the protective pipe has penetrating cracks and holes, it shall be replaced immediately; If the loss of the argon pipe joint affects the argon supply, replace it immediately. When the ladle has finished casting for 2~3min, the protective pipe shall be removed to observe the ladle flow and move the operating machinery to a safe position.

4.6.1 钢包到中间包铸流的保护操作

在正常浇铸过程中，应有专人在操作平台上监护，如果水口发生堵塞，立即快速移开保护管；当保护管侵蚀后的长度离开中间包正常浇铸液面时，立即更换；透气环掉块，不能保证与钢包水口咬合严密或保护管出现贯穿性的裂纹及孔洞，立即更换；保护管氩管接头损失，影响通氩时，立即更换。当钢包钢液浇完前2~3min时，应将保护管卸下，以便观察钢包铸流，并将操作机械移至安全位置。

4.6.2 Protection Operation of Tundish

The protection of the liquid level of the tundish is mainly through adding the covering flux to the tundish. The tundish with the protective slag can be insulated, reduce the heat loss of the molten steel, isolate the air, reduce the reoxidation of the molten steel, and absorb the inclusions floating from the molten steel. According to the lining material used in the tundish, the tundish protection slag can also be divided into acid protection slag or alkaline protection slag. According to the requirements of ultra-low carbon steel, the tundish slag can also be divided into non carbon and carbon slag.

4.6.2 中间包的保护操作

中间包液面的保护主要是通过向中间包加入覆盖渣，中间包用保护渣的可以隔热保温，减少钢液的散热损失，隔离空气减少钢液的二次氧化；吸收由钢液中上浮的夹杂物。根据中间包使用的内衬材质，中间包保护渣也可分为酸性保护渣或碱性保护渣。根据超低碳钢的要求，中间包保护渣还可以分为无碳和有碳保护渣。

4.6.3 Protection Operation of casting Flow from Tundish to Mold

The protection casting from tundish to mold is used by the combination of submerged entry nozzle and protective slag casting. In the normal casting process, the erosion of submerged nozzle should be observed frequently. In serious cases, the preparation for tundish replacement should be done well, and the tundish should be replaced timely and accurately.

4.6.3 中间包到结晶器铸流的保护操作

中间包到结晶器的保护浇铸是由浸入式水口完成的，在正常浇铸过程中，应经常观察浸入式水口的侵蚀情况，严重时做好换中间包操作的准备工作，并及时准确地更换中间包。

4.6.4 Protection Operation in Mold

The protection of mold steel level is mainly achieved by adding mold fluxes. In the normal casting process, the protective slag shall be added evenly at any time to ensure that the steel liquid level is not exposed, the slag thickness is generally controlled at about 30mm, and the powder slag thickness is controlled at 10~15mm. When the powder slag agglomerates and is obviously damp, it shall not be used; the 'slag strip' around the mold shall be picked out at any time, and the primary billet shell shall not be touched during the picking, and the action shall be agile to prevent it from getting involved in the molten steel and causing accidents. Therefore, the slag layer in the mold shall be inspected by point and pull method, and the slag layer shall not be stirred as much as possible; if the casting and spreading ability of the protective slag are found to be decreased, agglomerate or seriously bond the submerged nozzle, the slag shall be replaced. During slag replacement, the slag layer shall be removed from both sides of the mold by the slag rakes, and new slag shall be added while fishing to avoid the exposure of molten steel. If the submerged nozzle is found to be broken, the residue shall be removed; if there is slag rolling phenomenon on the liquid surface, the casting speed shall be reduced or the casting shall be stopped; if there is steel hanging on the inner wall of the mold, the casting speed shall be reduced to eliminate the fault or stop the casting in time.

4.6.4 结晶器的保护操作

结晶器钢液面的保护主要是通过加保护渣完成的。在正常浇铸过程中，要随时均匀地添加保护渣，保证钢液面不暴露，渣厚一般控制在30mm左右，粉渣厚度控制在10~15mm，当粉渣结块成团，明显发潮时，不得使用；应随时将结晶器周边的"渣皮条"挑出，挑时不要触及初生坯壳，动作要敏捷，以防卷入钢液，造成事故；采用点、拨方法检查结晶器内渣层情况，尽可能不搅动渣层；若发现保护渣浇铸性、铺展性下降，成团结块或严重黏结浸入式水口时，应进行换渣。换渣时，用捞渣耙将渣层从结晶器两侧捞除，边捞边添加新渣，以避免钢液暴露，如果发现浸入式水口断裂或破裂，应将残体捞出；如果观察液面有卷渣现象时，应降低拉坯速度或停浇；若结晶器内壁产生挂钢现象，应降低拉坯速度及时排除故障或停浇。

4.6.4.1 Function of Mold Powder

(1) Provide thermal insulation and heat preservation.
(2) Isolate air and protect the steel meniscus from oxidation.
(3) Clean the interface of slag and steel, absorb the inclusions into the molten slag.
(4) Lubricate the strand and improve the solidification heat transfer.

4.6.4.1 保护渣的作用

(1) 绝热保温。
(2) 隔绝空气，防止二次氧化。
(3) 净化钢渣界面，吸收钢水中的非金属夹杂物。
(4) 润滑坯壳并改善凝固传热。

4.6.4.2 Working Process of Mold Powder

As shown in Figure 4-5, the melting process of the mold powder is composed of molten slag flux layer, sintering layer and original slag layer (powder slag layer). The low melting point (1050~1100℃) slag powder or particles added to the high temperature steel liquid surface of the mold form a certain thickness of liquid slag cover layer (10~15mm) on the steel liquid surface. The mold powder on the molten flux layer receives the heat transmitted from the steel liquid, and the temperature can reach 800~900℃, but it is already soften and formed sintering layer. A solid powder or granular original slag layer (also known as powder slag layer) is on the sintered layer. With the development of the billet withdrawing process, the liquid powder on the steel surface is consumed continuously, and the sintered layer drops to the steel surface to melt into the molten flux layer, and the powder slag layer becomes the sintered layer, so the production needs to add new mold powder in time.

4.6.4.2 保护渣的工作过程

保护渣熔化过程如图 4-5 所示，由液渣层、烧结层、原渣层（粉渣层）组成，添加到结晶器高温钢液面上的低熔点（1050~1100℃）渣粉或颗粒，在钢液面上形成一定厚度的液渣覆盖层（10~15mm），在液渣层上面的保护渣受到钢液传过来的热量，温度可达800~900℃，但已软化烧结在一起，形成一层烧结层，在烧结层上面是固态粉状或粒状的原渣层（也称粉渣层）。随着拉坯的进行，钢液面上液渣不断被消耗掉，而烧结层下降到钢液面熔化成液渣层，粉渣层变成烧结层，所以生产需要及时添加新的保护渣。

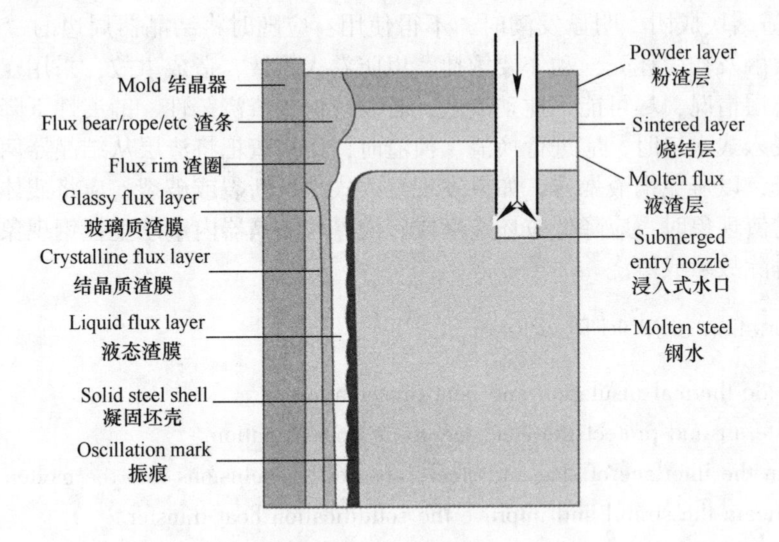

Figure 4-5 Schematic diagram of the melting process of the mold powder

图 4-5 保护渣熔化过程结构示意图

4.6.4.3 Composition of Mold Powder

The main components of the mold powder are CaO and SiO_2, containing a small amount of Al_2O_3. Therefore, the physical properties of the protective slag are based on the composition of the ternary phase diagram of $CaO-SiO_2-Al_2O_3$. In order to adjust the melting point and viscosity of the flux, various additives can be added, such as Na_2O, CaF_2, K_2O, etc. In order to control the melting rate, a certain amount of carbon was added. The composition of general mold powder is controlled in the following range: $w(CaO)=30\%\sim40\%$; $w(SiO_2)=30\%\sim40\%$; $w(Al_2O_3)=3\%\sim7\%$; $w(CaO)/w(SiO_2)=0.85\sim1.25$; $w(C)=2\%\sim5\%$.

4.6.4.3 保护渣的成分

保护渣的主要成分是 CaO 和 SiO_2，含有少量的 Al_2O_3。因此，保护渣的物理特性是依据 $CaO-SiO_2-Al_2O_3$ 三元相图的组成。结合经验和工艺条件的要求可以加入各种添加剂，如为了调整保护渣的熔点和黏度加助熔剂 Na_2O、CaF_2、K_2O 等；为了控制熔化速度适当添加一定的碳元素。一般保护渣成分控制在如下范围：$w(CaO)=30\%\sim40\%$；$w(SiO_2)=30\%\sim40\%$；$w(Al_2O_3)=3\%\sim7\%$；$w(CaO)/w(SiO_2)=0.85\sim1.25$；$w(C)=2\%\sim5\%$。

Task Implementation 任务实施

4.6.5 Operation of Mold Flux Feeding

(1) Prepare relevant tools, such as slag pusher, slag raking, etc., and debug the automatic slag feeding machine.

(2) Prepare the mold powder required by the requirement of the steel grade and billet section.

(3) Before tundish casting, the mold powder is added into the automatic slag feeder or removed and stacked around the upper inlet of the mold.

(4) After the tundish is casting, the mold powder can be added after the steel level rises to the outlet of the submerged nozzle. It should be noted that in some plants, especially slab caster, different mold powder with different physical and chemical properties should be used in starting casing and normal casting in tundish.

(5) After the starting casting powder is basically consumed, the slag surface turns from black to red, and the mold powder for normal casting is feeding, or the automatic slag feeding machine is started.

(6) Generally, the slag pusher is used to push the slag into the mold for manual feeding of the protective slag. The slag pusher should have certain strength so that the slag can be evenly scattered on the liquid surface.

(7) When feeding the mold powder, it is required to be less, frequent and even.

(8) During the normal casting process, the liquid level of the mold shall be kept from being exposed, the slag surface shall be red and bright on the wall of the mold, and the other places shall be the natural color of the protective slag. It is better to control the thickness of powder layer (liquid powder plus powder slag) to 35~40mm.

(9) When using the automatic slag feeding machine, pay attention to the fine adjustment of the feeding speed and position after starting to ensure that the slag feeding requirements are met.

(10) If it is found that the mold powder on the mold liquid level is caked and there is slagging ring near the wall of the mold, it must be removed in time with slagging scraper, otherwise it will affect the quality or cause casting accident.

(11) Before stop casting, slag adding can be stopped in advance according to experience.

(12) If the continuous casting technology of tundish changing is adopted, if the mold is equipped with connecting parts, the residual protective slag in the mold shall be removed in advance, or the adding of protective slag shall be stopped in advance, so as to ensure that the liquid steel can be seen when the tundish is stopped for tundish changing operation.

4.6.5 加结晶器保护渣操作

（1）准备相关工具，如推渣棒、捞渣耙等，调试好自动加渣机。

（2）根据所浇的钢种、铸坯断面准备好工艺要求的保护渣。

（3）中间包开浇前，保护渣加入自动加渣器内，或拆除包装堆放在结晶器上口四周。

（4）中间包开浇后，待钢液面上升到淹没水口出孔后，才能开始加入保护渣。应注意某些工厂，特别是板坯连铸机，中间包开浇时采用与正常浇铸理化性能不同的开浇渣。

（5）待开浇渣基本消耗之后即渣面由黑转红，开始加正常浇铸保护渣，或自动加渣。

（6）保护渣的手工加入一般用推渣棒推入结晶器，推渣要有一定力度，使渣能在液面上均匀洒落。

（7）保护渣加入时要求少、勤、匀。

（8）结晶器液面在正常浇铸过程中应保证钢液不裸露，渣面在结晶器壁呈红亮色，其他地方呈保护渣本色。控制渣层厚度（液渣加粉渣）以35~40mm为宜。

（9）当采用自动加渣机时，开动后要注意加入速度和加入位置的微调，确保符合加渣要求。

（10）若发现结晶器液面上的保护渣有结块、近结晶器壁有结渣圈时，必须用捞渣耙及时捞出清除，否则会影响质量或造成浇铸事故。

（11）在浇铸结束前，根据经验可提前停止加渣。

（12）采用换中间包连浇技术时，如结晶器加连接件，则结晶器内残留保护渣事前应捞出清除干净，或提前停止加保护渣，保证中间包停浇进行换包操作时结晶器液面能看见钢液。

4.6.6 Operation of Slag Dragging

In the no protective casting process, the large inclusions in the molten steel float up and gather, or the scum in the tundish floats up with the steel flow into the mold, which forms the scum on the steel surface of the mold. If these dross are not removed in time, it may enter into the billet and cause slag inclusion defect, or it may be involved in the shell and cause steel leakage accident after leaving the mold.

4.6.6 捞渣操作

在敞开浇铸过程中，钢液中的大型夹杂物上浮聚集，或中间包内的浮渣随钢流进入结晶器而上浮等形成了结晶器钢液面上的浮渣。这些浮渣不及时捞除，可能进入铸坯造成夹渣缺陷，也可能卷入坯壳内从而造成出结晶器后的漏钢事故。

4.6.6.1 Tool Preparation

(1) The slag rod is made of ϕ6.5mm wire, with a length of about 1 ~ 1.5m. A certain amount shall be prepared before casting.

(2) Number of small slag packages.

4.6.6.1 工具准备

(1) 捞渣棒。以 ϕ6.5mm 线材制成，长度约 1~1.5m，浇铸前要准备一定数量。

(2) 小渣包数只。

4.6.6.2 Operation Process

(1) Monitor the liquid level of the steel, and prepare to scoop up the dross if there is dross.

(2) Extend the slag rod to the liquid level of the mold, and stick the scum on the liquid level with the head of the slag rod.

(3) Quickly lift the slag rod, because the cold steel rod head contacts with the slag, the slag will instantly and quickly stick to the steel rod, lift the slag rod and lift the scum at the same time.

(4) Tap the slag rod on the edge of the small slag ladle to knock off the solidified scum and drop it into the slag ladle. If the volume of slag is small, continue to remove the slag.

(5) The slag rod with larger slag sticking can be temporarily put aside, and the slag rod without slag sticking can be replaced to continue the slag fishing. The slag rod with slag sticking can be cleaned with oxygen cutting torch after casting.

4.6.6.2 操作过程

(1) 监视钢液面，发现有浮渣存在就准备捞渣。

(2) 将捞渣棒伸至结晶器液面上，用捞渣棒头部粘住液面上的浮渣。

(3) 迅速提起捞渣棒，因冷的钢棒头与渣子接触，渣子会瞬时迅速粘住钢棒，提起捞渣棒同时提起浮渣。

(4) 在小渣包边上敲击捞渣棒，用力把已凝固的浮渣敲掉并落入渣包。一时敲不掉粘渣的捞渣棒，如渣子体积较小可继续去捞渣。

(5) 粘住较大渣子的捞渣棒可暂时放在一边，另换无粘渣的捞渣棒继续捞渣。带有粘渣的捞渣棒待浇铸结束可用氧气割炬清理后再用。

Exercises:

(1) What are the main links of continuous casting protection casting?

(2) What is the function of mold powder feeding?

(3) Select the appropriate mold powder and use the simulation software to complete the continuous casting production.

思考与习题：

(1) 连铸保护浇铸主要包括哪些环节的保护？

(2) 结晶器加入保护渣有何作用？

(3) 选择合适的保护渣，利用仿真软件完成连铸生产。

Project 5 Accident Treatment of Continuous Casting

项目 5 连铸生产事故处理

Task 5.1 Ladle Accident and Treatment

任务 5.1 钢包事故及处理

Mission objectives:

任务目标:

(1) Be able to deal with accidents caused by ladle and related equipment in continuous casting production to ensure normal production and safety.

(1) 能够处理连铸生产中钢包及相关设备产生的事故，保证生产的正常进行和安全。

(2) Master the causes of ladle accidents and prevent them.

(2) 掌握钢包事故发生的原因并进行预防。

5.1.1 Ladle Sliding Nozzle Blocked

5.1.1 钢包滑动水口堵塞

Task Preparation 任务准备

Tool preparation:

(1) Several oxygen tubes (ϕ8mm).

(2) A pair of oxygen acetylene cutting torch.

工具准备:

(1) 氧气皮管氧气管数根 (ϕ8mm)。

(2) 氧-乙炔割炬 1 副。

Task Implementation 任务实施

Processing method:

(1) Usually, the flow guiding sand does not flow down automatically. The slide plate can be

quickly closed and reopened once, and the flow guiding sand can flow down automatically depending on the vibration of the slide plate movement to reach the normal starting casting, or the oxygen pipe can be used to dip into the ladle nozzle (the slide gate is opened) without oxygen to poke sand, so that the flow guiding sand can be loose and flow down automatically for starting casting.

(2) If the above operation is invalid, first confirm that the sliding nozzle is open.

(3) Clean the ladle nozzle with oxygen tube under normal oxygen pressure (0.3~0.5MPa) to help the flow guiding sand flow down completely and achieve the purpose of staring casting.

(4) If the above operation is invalid, ignite the oxygen pipe under the condition of small oxygen pressure, then carefully move the ignited oxygen pipe into the ladle nozzle, and then adjust the large oxygen pressure to clean the ladle nozzle for flow guiding sand or blocked cold steel slag.

(5) The previous operation can be carried out 3~4 times until the molten steel flow is normal. If the temperature of molten steel is too low or the repeated use of oxygen is ineffective, the accident molten steel can be returned to the furnace.

(6) Pay close attention to the liquid level of tundish in the process of dealing with the clogging accident of nozzle.

处理方法：

(1) 通常情况下，引流砂不会自动流下。可以快速关闭和重开滑板一次，依靠滑板运动的振动使引流砂自动流下，达到正常开浇，或可用氧气管在不带氧的情况下浸入钢包水口（滑板打开）捅砂，使引流砂松动自动流下开浇。

(2) 在进行上项操作无效情况下，先确认滑动水口在打开状态。

(3) 用氧气管在一般氧压条件下（0.3~0.5MPa）清洗水口，帮助引流砂全部流下，达到开浇目的。

(4) 在进行上项操作无效情况下，先在小氧压条件下点燃氧气管，然后小心把点燃的氧气管移入水口内，再开大氧压清洗水口内引流砂或堵塞的冷钢渣等。

(5) 以上操作可进行3~4次，直到钢流正常。若钢液温度过低或多次用氧不见效的情况下，事故钢包钢液可作回炉处理。

(6) 在处理水口堵塞事故过程中密切注意中间包液面高度。

5.1.2 Steel Channeling of Ladle Sliding Nozzle

5.1.2 钢包滑动水口窜钢

Task Preparation 任务准备

5.1.2.1 Causes of Steel Channeling of Ladle Sliding Nozzle

(1) Installation quality problems. Poor installation quality is the main reason for the leakage of sliding nozzle.

(2) Refractory quality problems.

(3) Operation error.

(4) Failure of sliding nozzle mechanism.

5.1.2.1 滑动水口窜钢的原因

(1) 安装质量问题。安装质量不佳是滑动水口穿漏的主要原因。

(2) 耐火材料质量问题。

(3) 操作失误。

(4) 滑动水口机构故障问题。

5.1.2.2 Tool Preparation

(1) Oxygen belt and oxygen pipe (ϕ8mm);

(2) One set of oxyacetylene cutting torch.

5.1.2.2 工具准备

(1) 氧气皮管氧气管数根(ϕ8mm);

(2) 氧-乙炔割炬1副。

Task Implementation 任务实施

Steel channeling occurs in sliding nozzle mechanism during tapping:

(1) No matter where steel channeling occurs, steel tapping shall be stopped immediately.

(2) Move the ladle car immediately, whether the steel channeling stops or not, start the ladle car immediately.

(3) Command the crane to lift the accident ladle. If steel channeling continues, command the nearby personnel to avoid it.

(4) Lift the ladle that is channeling steel to the top of the standby ladle to let the liquid steel flow into the standby ladle.

(5) If the factory adopts the process of crane hoisting and steel tapping, the steel tapping shall be stopped immediately if steel channeling occurs in the sliding nozzle mechanism during steel tapping. If the accident ladle doesn't stop channeling, it must be lifted above the standby ladle to let the liquid steel flow into the standby ladle.

(6) If the steel channeling is stopped and there is still liquid steel in the accident ladle, the furnace returning operation must be taken immediately. During operation, the possibility of steel liquid channeling again shall be considered, and the operator shall avoid it reasonably.

Steel channeling is found in the sliding nozzle mechanism during refining or transportation to the casting position from the completion of ladle tapping to the opening of sliding nozzle:

(1) If the amount of steel channeling is very small, and immediately stop the channeling and leakage, continue refining or transport to the steel casting position, the chance may not affect the normal casting. However, the cold steel affecting sliding action of sliding plate shall be cleaned with cutting torch in advance.

(2) If the casting is not normal, the accident ladle shall be returned to the furnace.

(3) If there is a large amount of steel channeling, the accident ladle must be lifted to the top of the standby ladle immediately, so that the accident ladle liquid steel flows out, and the standby ladle liquid steel is returned to the furnace for treatment.

(4) If the amount of steel channeling is very small, but it is not stopped immediately, the accident ladle must be returned to the furnace through ladle.

In the casting position, the sliding nozzle has just opened or steel channeling occurs between sliding plates or between nozzles during casting:

(1) If the amount of steel leakage is small and the steel channeling is stopped immediately, the casting can be continued.

(2) If the amount of steel leakage is very small, but it cannot be stopped immediately, and there is a trend of expansion, the sliding nozzle must be closed to stop casting. Turn from the rotary table or hoist to the top of the standby ladle.

(3) The liquid steel in the accident ladle is poured into the standby ladle for treatment.

Precautions:

(1) In the process of ladle transportation, if the steel channeling at the sliding nozzle is continuing, it is likely to burn off the entire sliding plate mechanism, resulting in large leakage. It is necessary to pay attention to the avoidance of surrounding personnel and equipment.

(2) Sometimes, the amount of steel channeling in the sliding nozzle is very small, and the channeling will be stopped immediately, but the sliding plate mechanism is often stuck, so the operation cannot continue. For the stuck sliding plate mechanism, the oxygen acetylene cutting torch can be properly used on site to deal with the cold steel channeling out, but the pressing screw or the door closing device cannot be loosened, so as to prevent more accidents.

(3) Due to various reasons, a small amount of steel channeling occurs in the sliding nozzle. When oxygen is used to clean the nozzle during the opening process, it must be ensured that the nozzle is in the fully open position.

(4) In order to deal with the steel channeling accident of the sliding nozzle, the site must be dry to prevent the explosion and splashing of the liquid steel and slag.

出钢过程中滑动水口机构有窜钢现象发生：

(1) 不论窜钢发生在什么部位，应立即停止出钢。

(2) 立即开出钢包车，无论窜钢停止与否，都要立即开出钢包车。

(3) 指挥吊车吊起事故钢包，如窜钢在继续应指挥附近人员避让。

(4) 把正在窜钢的钢包，吊至备用钢包上方，让钢液流入备用钢包。

(5) 如工厂采用吊车吊包配合出钢的工艺，出钢过程中发现滑动水口机构有窜钢发生，也应立即停止出钢。事故钢包如窜钢不止，则必须吊至备用钢包上方，让钢液流入备用钢包。

(6) 如窜钢中止，事故钢包中仍有钢液，必须立即采取回炉操作。操作时应考虑钢液再次窜钢的可能，操作人员要合理避让。

钢包出钢完毕至滑动水口打开前，在精炼或运送到浇铸位置过程中发现滑动水口机构有窜钢现象：

（1）如果窜钢量很小，并马上停止窜漏，继续精炼或运至浇钢位置，偶然的机会可能不影响正常浇铸。但事先应将影响滑板滑动动作的冷钢用割炬清理干净。

（2）如果不能正常浇铸，事故钢包作回炉处理。

（3）如果窜钢量很大，则必须立即把事故钢包吊至备用钢包上方，让事故钢包钢液流完，备用钢包钢液作回炉处理。

（4）如果窜钢量很小，但没有马上停止，事故钢包必须进行过包回炉。

在浇铸位置，滑动水口刚打开或在浇铸过程中发生滑板之间或水口之间窜钢：

（1）如漏钢量很少，并立即停止窜钢，则可继续浇铸。

（2）如漏钢量很小，但不能立即停止，并有扩大趋势，则必须关闭滑动水口，停止浇铸。由回转台转至或吊车吊运至备用钢包上方。

（3）事故钢包内钢液倒入备用钢包作回炉处理。

注意事项：

（1）在钢包运送过程中如滑动水口窜钢正在继续，则很可能会烧脱整个滑板机构，造成大窜漏，必须注意周围人员和设备的避让。

（2）滑动水口窜钢有时窜钢量很小，并马上停止窜漏，但往往会卡住滑板机构，无法继续操作，对于滑板机构卡死，现场可适当用氧-乙炔割炬处理窜出的冷钢，但不能松动压紧螺丝，或松动门关闭装置，以防止更大事故的发生。

（3）因各种原因造成滑动水口少量窜钢，在开浇过程中用氧气清洗水口时，必须保证水口在全打开位置。

（4）处理滑动水口的窜钢事故，必须保证场地干燥，防止钢液、渣液爆炸飞溅。

5.1.3 Steel Flow Out of Control of Ladle Sliding Nozzle

5.1.3 钢包滑动水口钢流失控

Task Implementation 任务实施

Treatment of steel flow out of control in ladle sliding nozzle：

（1）Notify the tundish steelworker to pay attention to prevent the tundish from overflowing or the steel flow splashing when the tundish is lifted（transferred）from the tundish, resulting in the injury accident.

（2）Command the crane or turret operator to be ready to move the accident ladle away from the top of the tundish at any time.

（3）Monitor the liquid level of tundish steel.

（4）Check whether the switch parts are flexible and complete, and take repair measures in time. If necessary, ask the fitter to repair on site.

（5）The steel flow of the sliding nozzle mechanism is out of control, which is mainly caused by the power failure of the sliding nozzle control. Therefore, first check the operation direction and eliminate the misoperation factors.

（6）Check whether the sliding nozzle control power system, namely the hydraulic system, is normal: first check the operation of the hydraulic pump, then check the hydraulic pressure, then check the oil level of the oil tank, then check the action of the reversing valve, and take corresponding maintenance measures.

(7) If the hydraulic pump stops running or the hydraulic pressure is insufficient, but the oil level in the oil tank is normal, it can be quickly turned into manual pump operation to control the steel flow.

(8) In the process of handling the accident, if it is found that the liquid level of the steel in the tundish has reached the emergency maximum position, and there is no overflow tank in the tundish, it is necessary to immediately dismantle the oil pump or the hydraulic pipe on the oil pump that controls the sliding nozzle, and command the crane or the rotary turret operator to lift (rotate) the ladle to the position of the standby ladle.

(9) The molten steel flowing into the standby ladle shall be recycled.

钢包滑动水口钢流失控的处理方法：

(1) 通知中间包浇钢工注意，防止中间包满溢或钢包吊离（转离）中间包时发生的钢流飞溅而产生伤害事故。

(2) 指挥吊车或回转台操作人员，随时准备把事故钢包运离中间包上方。

(3) 监视中间包钢液面高度。

(4) 检查开关机件是否灵活、齐全，并及时采取修复措施，必要时请钳工到现场修复。

(5) 滑动水口机构发生钢流失控，主要由滑动水口控制动力故障所造成，所以先得检查操作方向，排除误操作因素。

(6) 检查滑动水口控制动力系统即液压系统是否正常：先检查液压泵的运转，后检查液压压力，再检查油箱油位，然后检查换向阀动作，并针对性地采取相应检修措施。

(7) 凡液压泵停转或液压压力不足，但油箱油位正常，可迅速转为手动泵操作来控制钢流。

(8) 在处理事故过程中，如发现中间包钢液面已到紧急最高位置，该中间包又没有溢流槽，则必须立即拆卸控制滑动水口的油泵或油泵上的液压管，指挥吊车或回转台操作工将钢包吊离（转离）到备用钢包位置。

(9) 流入备用钢包内的钢液进行回炉处理。

5.1.4 Ladle Stopper Nozzle Blocked

5.1.4 钢包塞棒水口堵塞

Task Preparation 任务准备

Tool preparation:

(1) Several oxygen tubes (ϕ8mm);

(2) A pair of oxygen acetylene cutting torch.

工具准备：

（1）氧气皮管氧气管数根（ϕ8mm）；

（2）氧-乙炔割炬1副。

Task Implementation 任务实施

Processing method：

（1）If it is considered that the blockage in the nozzle has not been cleaned up，the nozzle must be cleaned again when the stopper is closed. If necessary，the oxygen pipe shall be used for cleaning once，and then the test run shall be conducted.

（2）If it is considered that the stopper head block of stopper rod causes the plug block，the stopper rod can be closed by increasing the force to make the stopper head stick to the stopper block，so as to open the nozzle gently and slowly. In this case，the stopper rod can be closed by increasing the force for 2~3 times，or it may fail. In the case of successful sticking of the stopper block，the casting operation must be more slow and careful.

（3）In the case of failure in the operation of adhesive chip dropping，the ladle nozzle can be cleaned with oxygen pipe under the condition of high pressure oxygen.

（4）If it is a cold ladle，there is a cold ladle bottom in the ladle or the tapping temperature is low，which may be affected by the cold steel near the nozzle，resulting in the blockage of the nozzle. Therefore，under the condition that the above operation is invalid，the molten steel can be recycled.

（5）If the steel flow cannot be controlled，the liquid level of the tundish must be paid close attention to，and the mold pulling speed should be controlled within the allowable range. Once the liquid level of the tundish is close to the upper overflow tank（some tundish do not have overflow tank），the liquid steel can flow into the standby ladle or slag ladle through the overflow tank. The slag ladle must be kept dry and protected by some bottom garbage.

Once the liquid level of the tundish is close to the upper port and there is a danger of overflow，the ladle must be lifted away by a crane or the revolving table must be operated to place the ladle in the accident position so that the molten steel can flow into the standby ladle or slag ladle. After the liquid level of the tundish drops，the ladle can be reentered into the casting position to continue to supply molten steel to the tundish，which can be repeated many times until the molten steel is casted out.

The molten steel flowing into the standby ladle can be returned to the furnace for treatment. The molten steel in the ladle is transported to the slag yard for overturning after cooling. The large cold steel is divided（oxygen cutting）and then returned to the furnace.

（6）If the molten steel in tundish has been casting out in the process of dealing with the clogging accident of ladle nozzle，tundish casting can be used for tundish changing operation：the billet in the mold is wriggled or the section is reduced to wait. If the waiting time exceeds the specified time of tundish changing operation，the caster can be used for stopping casting operation. At this time，the accident treatment of ladle nozzle blockage can be stopped，and the molten steel in the accident ladle can be returned to the furnace. If the tundish is not prepared，when the ladle steel flow is reopened, the tundish used in the accident can be continuously casting (also as the operation of changing ladle to start casting) on trial under the condition that no slag is found in the molten steel. The success rate of recasting with used tundish is low, so it is necessary to prepare for the accident that the nozzle of tundish cannot be closed (stopper mechanism).

处理方法:

(1) 若考虑是水口内堵塞物没有清理干净，必须在塞棒关闭状态下重新清理水口，必要时用氧气管清洗一次，再行试开。

(2) 若考虑是塞棒的塞头砖掉片造成水口堵塞，可加大力量关闭塞棒，使塞头砖粘起塞头掉片，从而轻缓打开水口，在此情况下，塞棒可加大力度关闭 2~3 次，也可能失败，在成功粘起塞头掉片的情况下浇铸操作必须更缓慢和小心。

(3) 在试行粘合掉片的操作失败的情况下，可用氧气管在大压力氧气的条件下清洗水口，清洗时塞棒要在打开状态。

(4) 如果是冷钢包，钢包内有冷钢包底或出钢温度偏低，都有可能因水口附近冷钢影响，造成水口堵塞。为此，在上述操作无效的条件下，钢液可作回炉处理。

(5) 经水口堵塞事故处理钢流打开后，若发生钢流无法控制时，必须密切关注中间包液面，结晶拉速在允许的范围内以偏上限操作控制。一旦中间包液面接近上口溢流槽时(有些中间包没有)，可让钢液通过溢流槽流入备用钢包或渣包。渣包必须保持干燥并带有一些垫底的垃圾保护底部。

一旦中间包液面接近上口有满溢危险时，必须用吊车吊走该钢包或操作回转台把该钢包置于事故位置，让钢液流入备用钢包或渣包。在中间包液面下降后，钢包可重新进入浇铸位置继续向中间包供钢液，这样可反复多次直至把钢液浇完。

流入备用钢包内的钢液可作回炉处理，渣包内钢液冷却后运往渣场倒翻，大块冷钢分割（氧气切割）后回炉。

(6) 在处理水口堵塞事故过程中若中间包钢液已浇完，中间包浇铸可作换中间包操作：结晶器内铸坯蠕动或缩小断面进行等候，如果等候时间超过换包操作规定时间，铸机可作停浇操作处理。此时水口堵塞事故处理可以中止，事故钢包内钢液可回炉。如没有准备好中间包，当钢包钢流重新打开后，事故时使用的中间包在钢液未见渣的条件下，可以试行继续浇铸（也作换包开浇操作)。用使用过的中间包再行浇铸成功率较低，必须做好发生中间包水口关不死事故的准备（塞棒机构)。

5.1.5 Ladle Leakage Accident

5.1.5 钢包穿漏事故

Task Preparation 任务准备

5.1.5.1 Cognition of Ladle Leakage Accident

From the beginning of high temperature molten steel entering the ladle to the whole process operation stage when the molten steel flows out from the sliding nozzle, the molten steel leaks out from the bottom or wall outside the sliding nozzle, which is called the ladle penetration accident.

There are four types of ladle penetration accidents according to the location of leakage:

(1) Slag line leakage.

(2) Ladle wall leakage.

(3) Bottom of the ladle leakage.

(4) Purging plug leakage.

5.1.5.1 钢包穿漏事故认知

从高温钢液进入钢包开始，直至钢液全部从滑动水口流出的整个工艺操作阶段发生钢水从滑动水口以外的底部或壁部漏出，则称为发生穿包事故。

钢包穿包事故根据穿漏的部位不同可分四种：

(1) 穿渣线。

(2) 穿包壁。

(3) 穿包底。

(4) 穿透气塞。

5.1.5.2 Hazards of Ladle Leakage Accident

(1) In addition to the ladle slag line accident sometimes can continue to cast, the vast majority of leakage accidents will cause the end of casting.

(2) There is a great loss of liquid steel.

(3) It is also possible to damage the equipment and increase the production cost.

(4) Leakage accidents often cause personal injury.

5.1.5.2 钢包穿漏事故的危害

(1) 除钢包穿渣线事故有时可继续浇铸外，绝大多数穿漏事故都会造成浇铸终止。

(2) 钢液有较大的损失。

(3) 处理冷钢或钢液回炉，也有可能损坏设备，要增加生产成本。

(4) 穿漏事故往往会造成人员伤害。

5.1.5.3 Signs and Prevention of Wearing Bags

As long as we carefully check the ladle in the production process, we can find the signs of ladle leakage (or the omen of ladle penetration accident), and immediately take emergency measures to avoid the accidents as much as possible.

The main symptom of ladle leakage is the temperature change of outer steel shell of ladle, which results in the change of steel shell color.

Generally speaking, the ladle leakage accident is that the molten steel in the ladle penetrates into the ladle steel shell gradually and transfers heat, which makes the temperature of the steel shell rise continuously until the molten steel penetrates into the steel shell, thus heating the steel shell to the melting temperature. When the steel shell melts, the molten steel (slag liquid) in the ladle will leak from the leakage to the large leakage, resulting in the ladle leakage accident. Therefore, when the steel shell of a certain part of the ladle starts to turn red, it is a sign of ladle penetration.

5.1.5.3 穿包的征兆和预防

只要在生产过程中我们认真检查钢包，就可以发现穿包征兆（或称为先兆穿包事故），并立即采取紧急措施以尽可能避免事故的发生。

穿包的征兆主要是钢包外层钢壳的温度变化从而造成钢壳颜色的变化。

一般穿包事故是包内钢液逐渐向包的钢壳渗透，并传递热量，使该处钢壳的温度不断升高，直到钢液渗到钢壳从而使钢壳加热到熔化温度。钢壳熔化，包内的钢液（渣液）就会从渗漏到大漏造成穿包事故。为此当钢包的某一部位钢壳开始发红时，则是穿包的征兆。

5.1.5.4 Causes of Ladle Leakage Wearing Accidents

(1) Liquid steel temperature too high.

(2) Over oxidation of liquid steel.

(3) Poor quality of refractory.

(4) Poor quality of ladle masonry.

(5) Over use causes ladle leakage.

5.1.5.4 造成穿包事故的原因

(1) 钢液温度过高。

(2) 钢液氧化性过强。

(3) 耐火材料质量不好。

(4) 钢包砌筑质量不好。

(5) 过度使用造成穿包。

Task Implementation 任务实施

5.1.5.5 Inspection before Ladle Use

In order to prevent ladle leakage accident and forewarning ladle leakage accident in the process of tapping, refining and casting, the inspection of ladle must be strengthened before use.

For refractory materials, including refractory bricks, refractory mud, casting materials, etc., quality acceptance shall be carried out according to the standard requirements. Refractory materials that do not meet the requirements shall not be used. Measures shall also be taken in time in case of quality problems found in the process of ladle construction. Materials in doubt shall be reaccepted and those with problems confirmed shall be stopped in whole batch.

For the ladle with new lining, check the masonry quality of the working layer. If cracks are found in the ladle with integral casting lining, it shall be repaired or recasted and tied. If the bricks of the masonry ladle do not meet the requirements, they shall also be rebuilt.

Before the used ladle is used again, the residual slag must be removed in the red hot state, and then the corrosion condition shall be checked in detail. If the brick joint of the brick masonry package is enlarged and the liquid steel has penetrated, the working layer erosion is more than 1/3 in the liquid steel area and more than 1/2 in the slag line area, the ladle cannot be reused. The working lining of the ladle impact zone must be replaced and cannot be reused. The ladle to be reused can be repaired in the red hot state where local erosion or peeling is serious. The repair method can be patched with mud (the same as the lining material) or by spray (coating) machine.

When the ladle is ready, including the completed baking operation to be baked, the lining must be checked again before entering the tapping position. If lining collapse, large area peeling and other phenomena are found, the ladle must be abandoned and prepared again.

5.1.5.5 钢包使用前的检查

为了防止出钢、精炼、浇铸过程中出现穿包事故和先兆穿包事故，使用前必须加强对钢包的检查。

对于耐火材料，包括耐火砖、耐火泥、浇铸材料等，应该按标准要求进行质量验收。不符合要求的耐火材料不可投入使用。在钢包修砌过程中发现质量问题也要及时采取措施，有疑问的材料应重新验收，确认有问题的要整批停用。

对于更换新衬的钢包要检查工作层的砌筑质量。整体浇铸内衬的钢包，如发现有裂纹，则需修补或重新浇铸、打结，砌筑包的砖不合要求的也要重砌。

使用过的钢包再次使用前必须在红热状态下清除残余钢渣，再详细检查侵蚀情况。如砖砌包砖缝扩大并已有钢液渗入，工作层侵蚀在钢液区超过1/3，在渣线区超过1/2，则该钢包不可再次使用。钢包冲击区的工作衬必须更换，不可重复使用。要重复使用的钢包在红热状态下可对局部侵蚀或剥落严重的部位进行修补。修补方法可用泥料（与内衬材质相同）贴补，也可用喷补（涂）机喷补。

当钢包准备完毕，包括需烘烤的已完成烘烤操作将进入出钢位前必须再次检查内衬。如发现内衬坍塌、大面积剥落等现象则必须弃用，重新准备钢包。

5.1.5.6 Treatment of Ladle Leakage Accident

Generally speaking, the ladle slag line leakage occurs before ladle casting. If the ladle is not hoisted above the casting platform, it can be hoisted to the place to be casted after stopping the leakage through in the slag plate area of the casting site. If the leakage occurs above the casting position, the ladle shall be hoisted away from the slag plate area to stop the leakage before casting. If casting has started, it depends on whether it affects personal safety and the quality of the slab, it decides to stop casting or continue casting. In general, continue casting, when the liquid level drops, it will stop leakage.

The normal refining or casting operation can only be interrupted due to the leakage accident at the middle and lower wall and bottom of the ladle. The breakout liquid can be put into the standby ladle or into the slag pan for heat recovery or cooling treatment.

5.1.5.6 钢包穿漏事故处理

钢包穿渣线一般发生在钢包开浇之前，如钢包未吊到浇铸平台上方，则可在浇铸场地的渣盘区让其停止穿漏后再吊到待浇位，继续准备浇铸；如在浇铸位上方发生穿漏，则要将钢包吊离到渣盘区让其停止穿漏后再浇铸；如已经开始浇铸，则视是否影响人身安全和铸坯质量情况，决定停浇或继续浇铸。一般情况下继续浇铸，当其液面下降后会停止穿漏。

发生在中、下部包壁、包底的穿漏事故，只能中断正常精炼或浇铸操作。穿漏钢液可放入备用钢包或流入渣盘后热回炉或冷却处理。

5.1.5.7 Handling of Foreshadowing Ladle Leakage Accidents

In case of the pre-warning ladle leakage accident, in addition to the possible slag line leakage accident, the refining or casting must be stopped immediately, and the ladle molten steel shall be returned to the furnace after ladle leakage.

5.1.5.7 先兆穿包事故处理

发生先兆穿包事故，除可能穿渣线事故以外，必须立即终止精炼或浇铸，钢包钢液作过包回炉处理。

5.1.5.8 Precautions

In case of leakage accident, the relevant personnel around the operation area must be informed to pay attention to the operation direction of the accident package and avoid it in time. The operator shall pay attention to prevent the splashing of molten steel from hurting people.

5.1.5.8 注意事项

发生穿漏事故必须通知操作区周围的有关人员注意事故包的运行方向，并及时避让，操作人员要注意预防钢液飞溅伤人。

Exercises:

(1) During the continuous casting production, it is found that a small strand of molten steel flows out from the bottom of the ladle. How to deal with it?

(2) How to deal with steel flow out of control or clogging of ladle nozzle?

思考与习题：

(1) 某连铸机生产时，发现钢包底部有小股钢水流出，应如何处理？

(2) 钢包水口钢流失控或水口堵塞该如何处理？

Task 5. 2 Tundish Accident and Treatment
任务 5. 2 中间包事故及处理

Mission objectives:

任务目标:

(1) Be able to deal with accidents caused by tundish and related equipment in continuous casting production to ensure normal production and safety.

(1) 能够处理连铸生产时中间包及相关设备产生的事故，保证生产的正常进行和安全。

(2) Master the causes of tundish accidents and prevent them.

(2) 掌握中间包事故发生的原因并进行预防。

5. 2. 1 Tundish Stopper Accident

5. 2. 1 中间包塞棒事故

Task Preparation 任务准备

5. 2. 1. 1 Cognition of Tundish Stopper Rod Accident

In the process of casting, the accident of abnormal operation or casting suspension caused by stopper rod fault of ladle or tundish is stopper rod accident.

In the process of continuous casting, there are several kinds of stopper accidents:

(1) Steel leakage of nozzle caused by stopper head brick not in place.

(2) The stopper walking. In order to control the steel flow, the center line of the stopper rod has a certain dip angle difference with the nozzle brick, and some production is called 'rubbing' . If the inclination angle changes during the casting process, the steel flow at the nozzle will be out of control, unable to close or unable to control the steel flow.

(3) The stopper rod is broken.

(4) When the stopper head brick turns falls off, sometimes the stopper brick will block the nozzle and cause the nozzle to be blocked; sometimes the falling off due to the buoyancy of the liquid steel will cause the liquid steel out of control.

There are also four cases mentioned above in the tundish stopper rod accident, but the tundish stopper rod is usually composed of ventilating and cooling steel pipes. Before the stopper is broken, there is gas leakage. If the casting is not stopped in time, it will cause the stopper rod breaking accident.

5.2.1.1 认知塞棒事故

在浇铸过程中，因钢包或中间包塞棒故障造成操作不正常或浇铸中止的事故则为塞棒事故。

连铸过程中的塞棒事故有以下几种：

（1）塞头砖没有就位造成的水口漏钢。

（2）塞棒走位。为控制钢流，塞棒中心线与水口砖有一定倾角差位，有的生产称为“擦身”。如安装时的倾角在浇铸过程中有变化往往会造成水口钢流失控、关不死或无法控制钢流大小。

（3）塞棒折断。

（4）塞头砖掉头或掉片，塞头砖掉头或掉片有时会堵死水口造成水口堵塞；有时因钢液浮力作用脱落物浮出钢液面则造成钢液失控。

中间包塞棒事故也有上述4种情况，但中间包塞棒往往由通气冷却的钢管组成，在断塞棒前先有气体穿漏，如不及时停止浇铸则会造成断棒事故。

5.2.1.2 Causes of Stopper Accidents

The main causes of stopper accidents are as follows:

(1) Refractory quality problems.

(2) The influence of liquid steel quality. The temperature of the molten steel is too high (exceeding the requirements of the regulations), the liquid steel is over oxidized and the steel slag contains too much FeO.

(3) Improper operation when laying stopper rod. If the stopper brick is installed with excessive force during the construction of stopper rod, it will cause hidden danger of internal crack.

(4) Installation quality problems of stopper rod. For example, when installing the stopper rod, it will cause damage to the refractory material by striking hard, and the sleeve brick compression nut is not loosened (expansion joint is not left) after installation, which will cause stopper rod accident.

(5) The cooling of the stopper rod center is not in place.

(6) Impact by external force.

(7) Improper operation.

5.2.1.2 造成塞棒事故的原因

造成塞棒事故的原因主要有以下几点：

（1）耐火材料质量问题。

（2）钢液质量的影响。钢液温度过高（超过规程要求），钢液过氧化及钢渣含FeO过高等。

（3）砌筑塞棒时操作不当。砌筑塞棒时如安装塞头砖过度用力会造成内裂留下隐患。

（4）塞棒安装质量问题。如安装塞棒时用力敲击使耐火材料受损，安装完毕后袖砖压紧螺母未松开（未留胀缝）等都会造成塞棒事故。

（5）塞棒中心冷却不到位。

（6）外力撞击。

（7）操作不当。

Task Implementation 任务实施

Treatment of stopper rod accident: When the steel leakage and clogging of the nozzle are caused by the ladle stopper rod accident and cannot be recovered, the only way to deal with it is to stop casting immediately. The molten steel in the ladle is returned to the furnace and the molten steel leaked out becomes the cold steel scrap.

If emergency measures are taken to restore the control of steel flow at the nozzle, casting can be continued as the case may be; the molten steel that has not been cast shall be turned to other casters or furnaces.

In case of high temperature of molten steel, the refractoriness of refractories (stopper brick, nozzle brick) is not enough to withstand such severe conditions or the refractoriness and load softening point of refractories themselves are not up to the standard, both of which may cause deformation and chipping of stopper brick or nozzle brick, resulting in nozzle blockage. At this time, the residual head of the stopper rod can be used to bond the missing piece so as to make the nozzle unobstructed. This kind of bonding is not firm again, it is likely to fall off again and cause the next nozzle to be blocked, so the stopper rod must move slowly after opening the steel flow.

塞棒事故的处理：因钢包塞棒事故造成水口漏钢、水口堵塞，且无法恢复时，只能立即采用停浇的处理办法，钢包内钢液回炉，漏出钢液成冷钢废品。

如采取应急措施以恢复水口控制钢流的话，视情况可继续浇铸；未开浇的钢液应转向其他铸机或回炉。

钢液高温，耐火材料（塞头砖、水口砖）的耐火度不足以承受这种恶劣条件或耐火材料本身耐火度和荷重软化点达不到标准，这两种情况都有可能造成塞头砖或水口砖变形掉片，造成水口堵塞。此时可用塞棒残头用力去粘合掉片从而使水口通畅。这种粘合又是不牢固的粘合，很有可能再次脱落造成下一次水口堵塞，所以打开钢流后塞棒动作必须缓慢。

5.2.2 Submerged Nozzle Clogged

5.2.2 浸入式水口堵塞

Task Preparation 任务准备

The clogging accidents of submerged nozzle are mostly caused by the low temperature of molten steel, the deviation of fluidity or the low speed of casting. Therefore, after handling the accident, if it is caused by low temperature or low casting speed, the casting speed must be raised and controlled at the upper limit of casting to prevent the nozzle from freezing again due to low temperature or casting speed not keeping up.

The integral submerged nozzle is blocked, because the nozzle can't be cleaned, it can only be stopped for casting. Other casting streams of the multi flow casting machine can be produced normally. If the production rhythm is affected, the casting can be stopped after the molten steel of the furnace finished casting.

The technology of replacing the nozzle on line is to replace the nozzle quickly without stopping casting after the nozzle is blocked.

Tool preparation:

(1) Several oxygen belts and oxygen pipes (ϕ8mm).

(2) A pair of oxygen acetylene cutting torch.

浸入式水口的堵塞事故，大都是因为钢液温度偏低、流动性偏差或拉速过低造成的。所以在处理事故后，如是低温或低拉速所造成的堵塞，则拉速必须提上去并控制在偏上限浇铸，防止水口再一次因低温或拉速未跟上而冻结。

整体式浸入式水口堵塞，因水口无法清洗，只能作该钢流停浇处理。多流铸机其他铸流可正常生产，如因此影响生产节奏，可待该炉钢液浇完后停浇。

在线更换水口，是在水口堵塞后快速换上新水口而不停浇的技术，国内外均已有工厂采用。

工具准备：

(1) 氧气皮带及氧气管数根（ϕ8mm）。

(2) 氧-乙炔割炬 1 副。

Task Implementation 任务实施

5.2.2.1 Integral Submerged Nozzle Clogging

(1) The submerged nozzle clogging during the starting casting:

1) Inform the main control room, and through the main control room, inform all posts of the caster and the production scheduling that there is a clogged nozzle accident.

2) In case of an accident, the casting flow is no longer drawn, and other casting flows of the multi flow casting machine can be produced normally.

3) If other casting streams of the multi flow caster are in normal production, the dummy bar of the accident caster stream can be pulled out and recycled at the same time.

4) When the casting machine of single unit and single flow is clogged, the ladle can be stopped, and the ladle can also be transferred to other casters to continue casting. After the ladle stops casting, the tundish shall be operated as follows:

① After the steel liquid in the accident tundish is slightly frozen, the tundish car can be started to the tundish hoisting position.

② Carefully lift the tundish to the turnover site, and pay attention to prevent the liquid steel from tipping during lifting.

③ After the molten steel in the tundish is completely frozen, the tundish can be turned over and the frozen steel can be poured out.

④ Cut frozen steel and return to furnace.

5) After casting the molten steel in the furnace, the multi flow casting machine can decide whether to continue continuous casting, stop casting or change the tundish according to the situation.

(2) The nozzle is clogged in the middle period of casting:

1) Inform the main control room, and through the main control room, inform all posts of the caster and the production scheduling that there is a clogged nozzle accident.

2) The caster is in the state of stop casting, and the billet in the mold is treated as the tail billet.

3) The ladle shall be stopped casting and the molten steel in the ladle shall be returned to the furnace. The ladle can also be transferred to other casters for continuous casting.

5.2.2.1 整体式浸入式水口堵塞

(1) 开浇时发生浸入式水口堵塞：

1) 通知主控室，并通过主控室通知铸机所有岗位和生产调度发生了水口堵塞事故。

2) 发生事故的铸机流，不再拉坯，多流铸机的其他铸流可正常生产。

3) 如多流铸机的其他铸流正常生产，可把事故铸机流的引锭同时拉出回收。

4) 单机单流的铸机发生开浇堵塞，钢包作停浇处理，钢包也可转到其他铸机上去继续浇铸。钢包停浇后，中间包做如下操作：

① 待事故中间包内存钢液稍微冻结后，才能开动中间包车到中间包吊运位置。

② 小心吊运中间包至翻包场地，注意吊运过程中防止钢液倾翻。

③ 待中间包内钢液全部冻结后，才能翻转中间包倒出冻钢。

④ 切割冻钢回炉。

5) 多流铸机待浇铸完该炉钢液后，可视情况决定是否继续连浇、停浇或更换中间包。

(2) 浇铸中期发生水口堵塞：

1) 通知主控室，并通过主控室通知铸机所有岗位和生产厂调度发生了水口堵塞事故。

2) 铸机进入停浇状态操作，结晶器内铸坯作尾坯处理。

3) 钢包作停浇处理，钢包内钢液作回炉处理。有多台铸机的生产厂，钢包也可转到其他铸机上去继续浇铸。

5.2.2.2 The Split Submerged Nozzle Clogging

(1) Inform the main control room, and through the main control room, inform all posts of the caster and the production scheduling that there is a clogged nozzle accident.

(2) The submerged nozzle is clogged during the staring casting:

1) Remove the submerged nozzle and check the blockage of the submerged nozzle. If the submerged nozzle is not clogged, but the upper nozzle is blocked, then use oxygen to clean the nozzle, etc.

2）If the submerged nozzle is clogged, the submerged nozzle shall be cleaned with oxygen pipe and reinstalled immediately, or the standby submerged nozzle can be immediately installed to continue the casting operation.

3）After treatment, the nozzle will continue to be clogged or the molten steel temperature in the tundish is confirmed to be too low, so the start-up casting can be stopped.

（3） clogging in the middle period of casting：

1）Stop the casting flow in the clogging of submerged nozzle steam and replace the tundish.

2）Remove the submerged nozzle, check the clogging situation, clean the submerged nozzle, upper nozzle or both clean.

3）Start casting operation for tundish replacement again, install submerged nozzle and turn to normal casting.

4）The casting can be stopped if the nozzle is clogged or the temperature of molten steel is confirmed to be too low.

5.2.2.2 分体式浸入式水口堵塞

（1）通知主控室，并通过主控室通知铸机所有岗和生产厂调度发生了水口堵塞事故。

（2）开浇时发生浸入式水口堵塞：

1）卸下浸入式水口，检查浸入式水口堵塞情况。如浸入式水口没有堵塞，而是上水口堵塞，则用氧气清洗水口等。

2）如是浸入式水口堵塞，应立即用氧气管清洗浸入式水口重新装上，也可立即装上备用浸入式水口，继续开浇操作。

3）经处理后水口继续堵塞，或确认中间包钢液温度过低，可停止开浇。

（3）浇铸中期发生堵塞：

1）堵塞水口的铸流停车，作换中间包操作。

2）卸下浸入式水口，检查堵塞情况，清洗浸入式水口，上水口或都进行清洗。

3）重新作中间包更换的开浇操作，装上浸入式水口，转入正常浇铸。

4）处理后继续发生水口堵塞，或确认钢液温度过低，可停止浇铸。

5.2.2.3 Precautions

（1）In the process of casting, if the submerged nozzle is clogged, attention must be paid to the treatment time. If the time exceeds the time allowed for tundish replacement, casting must be stopped and forced casting is not allowed.

（2）When cleaning the submerged nozzle, pay special attention to the outlet angle（slab caster）. Generally, a standby submerged nozzle that can be put into use immediately should be prepared.

（3）Special attention shall be paid to the treatment of the full tundish after the accident：the tundish shall be kept still for a period of time, the tundish car shall be started and the tundish shall be hoisted stably, so as to absolutely prevent the greater accident caused by the overflow of the tundish tipping over liquid steel.

5.2.2.3 注意事项

(1) 浇铸过程中发生浸入式水口堵塞事故，必须注意处理的时间，凡超过换中间包允许的时间，必须停浇，不可强行浇铸。

(2) 清洗浸入式水口时特别要注意出口倾角（板坯连铸机），一般应备好立即能投入使用的备用浸入式水口。

(3) 特别要注意事故发生后的满包中间包的处理：中间包要静等一段时间，中间包车开动和中间包吊运都要平稳，绝对防止因中间包倾翻钢液溢出而造成更大的事故。

Exercises:

(1) During the continuous caster production, it is found that the stopper bar breaks and falls into the molten steel. How to deal with it?

(2) How to deal with the clogging of submerged nozzle?

思考与习题：

(1) 某连铸机生产时，发现塞棒破裂并掉到钢水中，应如何处理？

(2) 浸入式水口发生堵塞该如何处理？

Task 5.3 Mold Accident and Treatment
任务 5.3 结晶器事故及处理

Mission objectives:

任务目标：

(1) Be able to deal with accidents caused by mold and related equipment in continuous casting production to ensure normal production and safety.

(1) 能够处理连铸生产时结晶器及相关设备产生的事故，保证生产的正常进行和安全。

(2) Master the causes of mold accidents and prevent them.

(2) 掌握结晶器事故发生的原因并进行预防。

5.3.1 Tundish Stopper Accident

5.3.1 结晶器断水事故

Task Preparation 任务准备

The accident water tower of the caster is designed to deal with the sudden water failure of the caster, especially the water failure caused by the power failure. The water tower of the large-scale caster should supply water for mold, secondary cooling spray, equipment cooling, etc. at the same time, so the caster can't continue casting generally, only stop casting in case of emergency, and pull the casting billet out of the tension leveler.

The secondary cooling zone of small casters, especially some small billet casters, is relatively simple, and the cooling water pressure requirement of the mold is relatively low. Therefore, under the permission of water quantity and water pressure, the speed can be reduced for a period of time (several minutes), and then the casting and shutdown can be stopped (generally only used when the pump room switches the water pump or power supply).

The height of the emergency water tower is generally 20~50m, and the water capacity is 20~40m^3. The water quality requirement is the same as that of mold cooling water. The accident water tower shall be spot checked and commissioned regularly, and the water storage shall be replaced regularly.

铸机的事故水塔是为应对铸机突然停水事故而设计的，特别是停电后造成的停水事故。大型铸机的水塔要同时供应结晶器、二冷喷淋、设备冷却等用水，所以铸机一般不能继续浇铸，只有应急停浇，把铸坯拉出拉矫机。

小型铸机，特别是有的小方坯铸机二冷区比较简单，结晶器冷却水压要求也相对较低。所以在水量、水压许可的条件下可继续降速拉坯一段时间（几分钟），然后停浇、停机（一般仅限泵房切换水泵或电源后即可供水情况下采用）。

事故水塔高度一般为20~50m，水容量20~40m^3。水质要求同结晶器冷却水质。事故水塔应定期抽查调试，储水应定期更换。

Task Implementation 任务实施

5.3.1.1 Water Break Accident of Mold with Accident Water Tower

(1) Inform the main control room and other posts that the water supply of the mold enters the water supply status of the emergency water tower.

(2) Accurately judge the remaining molten steel in the tundish and ladle, control the time when the emergency water tower starts to supply water, and take the measures of ladle stop casting and tundish stop casting. After tundish stop casting, take tail billet pulled out operation and caster stop.

(3) The remaining molten steel in the tundish shall be driven into the tundish car after the surface solidifies. Lift the tundish to the tundish turning site, cool and solidify, turn over the tundish, cut the cold steel and return to the furnace; the remaining molten steel in the ladle shall be returned to the furnace for treatment or casting (other casters or ingot molds).

(4) When there is an accident water tower but the water supply is out of order, the water cut-off accident of the mold is treated as if there is no accident water tower.

5.3.1.1 有事故水塔的结晶器断水事故

（1）通知主控室和其他岗位，结晶器供水进入事故水塔供水状态。

（2）准确判断中间包和钢包内剩余钢液，在事故水塔开始供水时控制好时间，采取钢包停浇，中间包停浇措施。中间包停浇后采取尾坯操作、铸机停机。

（3）中间包剩余钢液待表面凝固后开中间包车。吊运中间包至翻包场地，冷却凝固，翻包，切割冷钢后回炉；钢包剩余钢液作回炉处理或转浇（其他铸机或锭模）。

（4）有事故水塔但供水失灵状态下，结晶器断水事故视同无事故水塔结晶器断水事故一样处理。

5.3.1.2 Water Break Accident of the Mold without the Accident Water Tower

(1) When the cooling pressure or flow alarm light of the mold is on or the bell rings, it can be immediately judged that the mold is out of water, i. e. the tundish steel flow is closed, the ladle steel flow is closed, the tundish car is driven away, and the casting speed is controlled at the upper limit of the allowable casting speed.

(2) If the copper plate at the top of the mold changes color suddenly and the steel level in the mold turns abnormally, it can also be judged that the mold is out of water, and the above operations shall be taken immediately.

(3) The mold capping operation is adopted for the steel level in the mold, which can spray water on the copper plates around to accelerate the cooling and solidification of the liquid level.

(4) After the caster stops casting, the molten steel in the tundish and ladle shall be treated according to the furnace return.

Other accidents (out of control of nozzle, steel leakage, steel overflow, etc.) caused by stopping of casting speed and closing of steel flow in tundish shall be handled according to other accident operation requirements. After the water cut-off, the mold should be replaced, and then the casting preparation should be made again.

5.3.1.2 无事故水塔的结晶器断水事故

（1）当结晶器冷却压力或流量报警灯亮或铃响后，则可立刻判断为结晶器断水，即关闭中间包钢流，关闭钢包钢流，开走中间包车，拉速控制在允许拉速上限操作。

（2）结晶器上口铜板突然变色，结晶器内钢液面翻动异常，也可判断为结晶器断水，并立即采取上述操作。

（3）结晶器内钢液面采用封顶操作，可在四周铜板上淋水加速液面冷却凝固。

（4）铸机停浇后，中间包和钢包内钢液按回炉处理。

在处理结晶器断水中，因拉速停止、中间包钢流关闭等而引发其他事故（水口失控、漏钢、溢钢等），应再按其他事故操作要求处理。断水后的结晶器应更换，再重新做浇铸准备。

5.3.1.3 Precautions

(1) It is easy to cause explosion accident if the water break accident is not handled well. Therefore, operators in the casting area should pay attention to avoid. In case of emergency water supply to the water tower, the casing shall be stopped immediately. If the casting is maintained, the casting time shall not exceed the specified time.

(2) After the mold is cut off (flow and water pressure are close to zero), the mold can't supply water immediately, only after the casting slab is out of the mold or all the casting slabs are cooled can the water supply be restored.

(3) If the cooling water flow and water pressure of the mold alarm and continue to drop, it shall be deemed as if the mold is cut off and the casting shall be stopped immediately (the steel flow of the tundish is closed and the straightening continues).

(4) It is not allowed to pull the mold copper plate by force when it is burned out and bonded with the casting billet.

(5) The mold after water cut-off can not be used any more, so it should be disassembled, checked and assembled again.

(6) No matter which first-class discovered the water break in the mold, other casting flows must take the same treatment steps.

5.3.1.3 注意事项

(1) 结晶器断水事故处理不好很容易造成爆炸事故。所以浇铸区域的操作人员应注意避让。在事故水塔供水情况下，通常立即作停浇处理，若维持浇铸，浇铸时间不得超过规定时间。

(2) 结晶器断水（流量、水压接近于零）后，结晶器不能立即供水，只能待铸坯出结晶器或铸坯全部冷却后才能恢复供水。

(3) 结晶器冷却水流量、水压报警，并在继续下降，应视同结晶器断水立即采取停浇措施（中间包钢流关闭，拉矫继续）。

(4) 结晶器铜板烧坏与铸坯黏结不可强行拉坯。

(5) 断水后的结晶器不可继续使用，应拆卸重新检查组装。

(6) 一机多流或多机多流铸机，结晶器断水多数同时发生，无论哪一流先发现结晶器断水，其他铸流必须采取相同的处理步骤。

5.3.2 Mold Water Leakage Accident

5.3.2 结晶器漏水事故

Task Implementation 任务实施

5.3.2.1 Treatment of Mold Water Leakage Accident

(1) Check the cooling water pressure of the mold immediately after water leakage.

(2) Before casting, if there is a large amount of water leakage at the support steel plate or water pipe joint of the mold, and there is leakage on the surface of the copper plate of the mold or the water at the upper entrance leaks into the mold, immediately replace the mold or take maintenance steps.

(3) Water leakage such as mold support plate or water pipe joint occurs during casting:

1) As long as the water leakage is small, it will not affect the cooling of billet, and the water leakage is not toward the mold steel level, the casting can continue, but pay close attention to whether the water leakage will expand.

2) If the uncontrollable water leakage affects the billet cooling or the steel level (slag level) in the mold, and the copper plate leakage in the mold flows into the mold, the casting flow or caster shall be stopped.

3) During the casting process, it is found that there is abnormal turnover of the steel level near the copper plate, or there is steam bubble emerging from the slag surface, then it can be judged as the water leakage fault of the mold. In case of such a situation, the operator around must be informed immediately to pay attention, and the main control room must be informed to stop casting immediately.

5.3.2.1 处理方法

(1) 发生漏水后立即检查结晶器冷却水压。

(2) 在浇铸前发生结晶器支撑钢板或水管接头大量漏水，结晶器铜板表面有渗漏或上口渗水漏入结晶器内部，则立即采取更换结晶器操作或采取检修步骤。

(3) 在浇铸过程中发生结晶器支撑板或水管接头等漏水：

1) 只要漏水量少，不影响铸坯冷却，漏水不向着结晶器钢液面则可继续浇铸，但要密切关注漏水情况是否会扩大。

2) 无法控制的漏水，影响铸坯冷却或影响结晶器内钢液面（渣液面），发现结晶器内铜板渗漏流入结晶器，则该铸流或铸机采取停浇措施。

3) 在浇铸过程中发现结晶器内钢液面靠近铜板有不正常的翻腾，或渣面翻腾并有气泡冒出，则可判断为结晶器漏水故障。发现该种情况必须立即通知周围操作工注意，并通知主控室，立即停浇。

5.3.2.2 Precautions

(1) In case of equipment leakage (mold, secondary cooling equipment, secondary cooling equipment, etc.) during casting, water supply shall not be reduced or stopped, otherwise greater accidents will be caused. Water cut-off of mold means water cut-off of mold, which results in burning out or explosion accident of mold; water cut-off of secondary cooling zone may result in steel leakage of slab or burning out of equipment, etc.

(2) Water leakage (seepage) in the mold is likely to cause explosion in the mold, so operators nearby must be informed to avoid it. The casting flow or caster must be stopped immediately.

(3) When fastening screws during casting, it is necessary to pay attention to the personal in-

jury caused by other accidents or splashing of molten steel. Therefore, it can only be operated when there is no sign of accident, other operations are normal, and the operation point is possible to avoid retreat.

5.3.2.2 注意事项

（1）在浇铸过程中发生设备漏水（结晶器、二冷、二冷设备等）千万不能降低供水或停止供水，否则会造成更大事故。结晶器停水即结晶器断水，其结果是结晶器烧坏或爆炸事故；二冷区停水则铸坯可能漏钢，或烧坏设备等。

（2）结晶器内漏水（渗水）很有可能造成结晶器内爆炸，所以必须通知附近操作工注意避让，铸流或铸机必须立即停浇。

（3）在浇铸过程中紧固螺丝，必须注意其他事故的发生造成人身伤害，或钢液飞溅的人身伤害。所以必须在做到万无一失的情况下（没有事故迹象、其他操作正常、操作点又有避让后退可能等）才可操作。

5.3.3 Mold Overflow Accident

5.3.3 连铸结晶器溢钢

Task Preparation 任务准备

Tool preparation:

(1) Oxygen belt and oxygen pipe (ϕ8mm).

(2) One oxygen acetylene belt and cutting torch.

(3) Several small crowbars (ϕ12mm).

工具准备：

（1）氧气皮带和氧气管若干（ϕ8mm）。

（2）氧-乙炔皮带及割炬1副。

（3）小撬杠数根（ϕ12mm）。

Task Implementation 任务实施

5.3.3.1 Treatment Method

(1) Close the tundish nozzle and stop the tension leveler.

(2) Close the ladle molten steel flow, lift the tundish.

(3) Inform the main control room, and inform all posts of the caster through the main control room. The caster has steel overflow accident.

(4) Immediately find out the cause of steel overflow and quickly eliminate the operation failure of slab:

1) If it is caused by the billet against secondary cooling roller, it shall be handled immediately as the billet against secondary cooling roller accident.

2) If the power of the tension leveler is tripped, the power shall be restored immediately or the standby power shall be supplied.

3) As a result of the hydraulic system of the tension leveler, handle the hydraulic system fault immediately and recover the pressure of the lower oil pump of the tension leveler.

(5) At the same time of eliminating the operation failure of the casting billet, use oxygen or oxygen acetylene cutting torch to clean up the cold steel overflowing from the upper entrance of the mold. Because the liquid level of the steel in the solidification shrinkage mold will be lower than the upper entrance of the mold, clean up the cold steel overflowing from the upper entrance of the mold, make all the copper plates around the upper entrance of the mold bare to ensure that the shell in the mold does not hang from the upper entrance of the mold.

5.3.3.1 处理方法

(1) 关闭中间包水口，拉矫机停车。

(2) 关闭钢包钢流，升起中间包车。

(3) 通知主控室，并通过主控室通知铸机所有岗位，铸机发生了溢钢事故。

(4) 立即寻找溢钢的原因，迅速排除铸坯运行故障：

1) 因顶坯造成，立即以顶坯事故操作处理。

2) 因拉矫机跳电造成，立即恢复送电，或送上备用电。

3) 因拉矫机液压系统造成，立即处理液压系统故障，恢复拉矫机压下油泵压力。

(5) 在排除铸坯运行故障的同时，用氧气或氧-乙炔割炬清理结晶器上口溢出的冷钢，因凝固收缩结晶器内的钢液面会低于结晶器上口，清理上口溢出的冷钢，必须使结晶器上口四面铜板全部裸出，保证结晶器内坯壳不与结晶器上口悬挂。

5.3.3.2 Precautions

(1) If the tundish nozzle is out of control after the overflow accident, it must be dealt with decisively, and the tundish car shall be driven away in time according to the steps, and the caster shall be stopped for treatment, otherwise it may cause more accidents; if the overflow liquid steel burns the water pipe, it will cause explosion, etc.

(2) In order to prevent other accidents, the main control room must inform all post operators in time, make emergency preparation, and prepare to evacuate the dangerous area in time.

(3) When dealing with the molten steel overflowing from the upper entrance of the mold (generally frozen into cold steel immediately), it is necessary to pay attention to burn out the equipment when using oxygen pipe or oxygen acetylene cutting torch, especially to prevent water leakage of the mold caused by burning out the copper plate or steel plate of the mold.

5.3.3.2 注意事项

(1) 溢钢事故发生后的中间包水口失控事故，必须果断处理，及时按步骤开走中间包车，铸机作停车处理，否则可能造成更大事故；溢出钢液烧坏水管引起爆炸等。

(2) 溢钢事故发生后为防止引发其他事故，主控室一定要及时通知所有岗位操作人员，做好应急准备，并准备及时，迅速撤离危险区域。

（3）处理结晶器上口溢出的钢液时（一般立即冻结成冷钢），必须注意在使用氧气管或氧-乙炔割炬时烧坏设备，特别是防止烧坏结晶器铜板或钢板造成结晶器漏水。

5.3.4 Other Continuous Casting Mold Accidents

5.3.4 其他连铸结晶器事故

Task Implementation 任务实施

5.3.4.1 Mold Oscillation Failure

Causes: the electrical system of oscillation device trips; the resistance of blank drawing suddenly increases; power failure, etc.

Treatment: put it into the swing groove, set the speed to zero, and inform the electrician to deal with it. If the oscillation is restored, the casting can continue, otherwise, the flow casting will be stopped.

Only by strengthening the equipment management, combining the regular inspection with the spot inspection, and ensuring the equipment in good condition, can such accidents be reduced or even eliminated.

5.3.4.1 结晶器振动故障

原因：振动装置电气系统跳闸；拉坯阻力突然增大；停电等。

处理：摆入摆槽，拉速调零，通知电工，若恢复振动可继续浇铸，否则停止该流浇铸。

只有加强设备管理，定检和点检相结合，确保设备状态良好，才能减少甚至杜绝此类事故。

5.3.4.2 Liquid Level Automatic Control Failure

When the liquid level control of the mold breaks down in the process of putting into operation, the operator immediately turns the liquid level control mode to manual mode and quickly reduces the casting speed to below 0.6m/min, controls the liquid level of the mold to prevent the liquid level of the mold from falling down, carefully observes whether there is residual steel around the mold, if there is residual steel, remove it with a steel bar or a steel drill rod, and pick out a large slag skin. After confirming that the mold is in normal condition, press standard speed up, and finally contact with the treatment of mold level control.

5.3.4.2 液位自动控制故障

当结晶器液位控制在投入过程中出现故障时，操作工应立即将液位控制方式转为手动并快降拉速至0.6m/min以下，控制好结晶器液面防止结晶器液面下降，仔细观察结晶器四周有无残钢，如有残钢则用点钢棒或钢钎去除，同时挑出大块渣皮，确认结晶器状况正常后，按标准升速，最后联系处理结晶器液位控制。

5. 3. 4. 3 The Slag Involved in the Mold Accident

The slag involved in the mold accident is the slag from the tundish entering the mold. Generally, it is difficult to drain and wait for a long time at the end of casting or continuous casting for ladle replacement.

In order to prevent slag from falling from tundish, the liquid level of tundish should be controlled at the end of casting. When the liquid level of tundish drops to about 150mm, close the tundish water inlet immediately.

In case of slag falling, the casting speed shall be reduced and the slag shall be fished out quickly. If the slag cannot be fished out within three minutes, the casting shall be stopped.

5. 3. 4. 3 结晶器下渣故障

下渣是中间包渣进入结晶器。一般出现在浇铸末期或连浇换钢包时引流困难、等待时间较长。

为防止中间包下渣，浇铸末期应控制中间包液面高度。当中间包液面降至约 150mm 时，立即关闭中间包水口。

出现下渣时，应降低拉速，将渣迅速捞出，3min 内不能捞净，应停浇。

5. 3. 4. 4 Mold Deformation

(1) Causes of mold deformation:

1) The temperature gradient of the inner and outer walls of the mold is large during the casting process, and the mold will deform rapidly due to cooling during the casting stop.

2) The surface of meniscus fluctuates, the temperature fluctuates greatly and the deformation is serious.

3) In the later period of use, the mold has serious wear and deformation tendency.

4) Insufficient water pressure and water volume, poor cooling effect, recrystallization on the inner wall of the mold.

Using mold with serious deformation will cause serious defects of continuous casting slab, such as pit, longitudinal crack, etc.

(2) Precautions: establish the use and maintenance files of the mold, determine the reasonable use cycle, and strictly check and confirm the work before pouring.

5. 3. 4. 4 结晶器变形

(1) 结晶器变形的原因:

1) 结晶器内外壁在浇铸过程中温度梯度大，停浇期间结晶器因冷却会迅速变形。

2) 弯月面处液面波动，温度波动大，变形严重。

3) 结晶器使用后期，磨损严重，变形倾向大。

4) 水压、水量不足，冷却不良，结晶器内壁产生再结晶。

使用变形严重的结晶器，会造成严重的连铸坯缺陷，如凹坑、纵裂等。

(2) 预防措施：建立结晶器使用维护档案，确定合理的使用周期，严格浇铸前的检查确认工作。

Exercises:

(1) How to deal with the overflow of molten steel in the mold during the continuous casting?
(2) How to deal with the accident of water break or water leakage in the mold?

思考与习题:

(1) 某连铸机生产时，发现结晶器钢水溢出，应如何处理?
(2) 生产时结晶器出现断水或漏水事故该如何处理?

Task 5.4 Secondary Cooling System Accident and Treatment
任务5.4 二冷事故及处理

Mission objectives:

任务目标:

(1) Be able to deal with accidents caused by secondary cooling system and related equipment in continuous casting production to ensure normal production and safety.
(1) 能够处理连铸生产时二次冷却及相关设备产生的事故，保证生产的正常进行和安全。
(2) Master the causes of secondary cooling system accidents and prevent them.
(2) 掌握二次冷却事故发生的原因并进行预防。

Task Implementation 任务实施

5.4.1 Treatment of Abnormal Situation of Secondary Cooling Water System

(1) If the cooling water flow or water pressure of the equipment is abnormal, the slab is partially overcooled during the normal casting process, and there is abnormal water flow in the secondary cooling area during the normal casting process or the preparation for commissioning, it can be judged that there may be cooling water leakage of the secondary cooling equipment or abnormal water leakage of the secondary cooling spray system.

(2) If the above phenomenon is found before casting, measures must be taken to find out the cause of water leakage.

(3) If the above phenomenon is found in the casting process, the source of water leakage must be found out. If the spray causes local over cooling on the casting slab, it can try to block the water leakage stream with a proper size of steel plate. If the partition is effective, the casting can

be continued.

(4) The above phenomena were found in the casting process, but no local over cooling was found in the slab, and the surface quality of the slab was normal, so the caster could continue to cast.

(5) In the process of casting, if the above phenomena are found, and no water leakage can be found, or the water leakage stream cannot be blocked, thus causing abnormal cooling of the slab and abnormal surface quality, the casting stream or caster shall be stopped for casting.

5.4.1 二冷水系统异常的处理

(1) 凡设备冷却水流量或水压异常，在正常浇铸过程中铸坯局部过度冷却，在正常浇铸过程中或准备调试过程中二冷区有异常水流股出现，则可判断有可能存在二冷设备冷却漏水或二冷喷淋系统异常漏水。

(2) 在浇铸前发现上述现象，必须找出漏水原因采取措施。

(3) 在浇铸过程中发现上述现象，必须找出漏水根源，如喷淋在铸坯上造成局部过度冷却，则可设法用适当大小的钢板隔断漏水流股。隔断有效可继续浇铸。

(4) 在浇铸过程中发现上述现象，但铸坯未发现有局部过度冷却，铸坯表面质量又没有异常，铸机可继续浇铸。

(5) 在浇铸过程中发现上述现象，又找不到漏水处，或漏水流股无法阻隔，从而造成铸坯冷却异常，表面质量异常，则该铸流或铸机作停浇处理。

5.4.2 Pull off Operation Accident

The precondition for recovering the drawing steel after pulling off is to be able to quickly handle the frozen billet of the mold. Therefore, after pulling off, the mold should continue to oscillate and drip proper amount of mold lubricating oil on the four walls of the mold to quickly remove the frozen billet in the mold. After the frozen billet treatment is completed, the inner wall of the mold and the secondary cooling equipment are checked to be free of problems, the dummy bar can be sent for secondary casting.

5.4.2 拉脱操作事故

拉脱以后恢复拉钢的前提条件是能快速处理结晶器的冻坯，因此拉脱以后，结晶器应继续保持振动，并在结晶器四壁滴适量的结晶器润滑油，以快速脱除结晶器内的冻坯，冻坯处理完毕，检查结晶器内壁以及二次冷却设备没有问题之后，可以二次送引锭开浇。

Exercises:

(1) How to deal with the leakage accident of the secondary cooling device in production?

(2) Use simulation software to simulate water leakage accident and deal with it.

思考与习题：

(1) 生产时二次冷却装置器出漏水事故该如何处理？

(2) 利用仿真软件模拟漏水事故并进行处理。

Task 5.5 Breakout Accident and Treatment
任务 5.5 连铸漏钢事故及处理

Mission objectives:

任务目标:

(1) Be able to deal with steel leakage accident during continuous casting production to ensure normal production and safety.

(1) 能够处理连铸生产时漏钢事故，保证生产的正常进行和安全。

(2) Master the causes of steel leakage during continuous casting production and prevent it.

(2) 掌握连铸生产时漏钢发生的原因并进行预防。

Task Preparation 任务准备

5.5.1 Cognition of Steel Leakage Accident

Breakout refers to the phenomenon that the solidification condition of billet shell is not good or the billet shell breaks or breaks due to other external forces in the early stage of continuous casting or in the casting process, which makes the internal molten steel flow out. Breakout is one of the malignant accidents in continuous casting production. Serious breakout not only affects the normal production of continuous casting machine and reduces the operation rate, but also damages the equipment of the caster and causes equipment damage. It can be divided into:

(1) Steel leakage during start-up. Steel leakage caused by poor start-up of starting casting.

(2) Hanging steel leakage. Large corner seam, concave corner base plate or scratch of copper plate in the mold, which increases the resistance of blank drawing in the mold, and it is easy to start hanging steel leakage.

(3) Crack steel leakage. Serious longitudinal crack, angular crack or square breaking occurs in the shell of the mold, which causes steel leakage after leaving the mold.

(4) Slag inclusion steel leakage. Steel leakage is caused by the thickness of the shell is too thin due to the inclusion of mold slag or foreign matters in the local area of the solidified shell.

(5) Cut off steel leakage. When the casting speed is too fast and the secondary cooling water is too weak, the liquid phase cavity is too long. After cutting the slab, the central liquid flows out.

(6) Bonding steel leakage. The steel leakage caused by the casting billet sticking to the mold wall and breaking.

5.5.1 漏钢事故认知

漏钢是指连铸初期或浇铸过程中，铸坯坯壳凝固情况不好或因其他外力作用引起坯壳断裂或破漏使内部钢水流出的现象。漏钢是连铸生产中恶性事故之一，严重的漏钢事故不仅影响连铸机的正常生产，降低作业率，而且还会破坏铸机设备，造成设备损坏。其可以分为：

（1）开浇漏钢。开浇起步不好而造成漏钢。

（2）悬挂漏钢。结晶器角缝大，角垫板凹陷或铜板划伤，致使在结晶器中拉坯阻力增大，极易发生起步悬挂漏钢。

（3）裂纹漏钢。在结晶器坯壳产生严重纵裂、角裂或脱方，出结晶器后造成漏钢。

（4）夹渣漏钢。由于结晶器渣块或异物裹入凝固壳局部区域，使坯壳厚度太薄而造成漏钢。

（5）切断漏钢。当拉速过快，二次冷却水太弱，使液相穴过长，铸坯切割后，中心液体流出。

（6）黏结漏钢。铸坯黏结在结晶器壁而拉断造成的漏钢。

5.5.2 Causes of Steel Leakage Accident

5.5.2.1 Steel Leakage during Starting Casting

(1) Billet emergence time is not enough.

(2) Too fast starting speed and acceleration.

(3) Too little or uneven distribution of iron nail scraps.

(4) The steel plate of dummy bar head is not placed properly, falling into the grid and stuck between the shell.

(5) Too early and large amount of protective slag is pushed into the mold, resulting in slag entrapment.

(6) The first section of the mold and the second cooling zone is not aligned with the arc, which may lead to steel leakage in the starting casting.

5.5.2.2 Steel Leakage during Casting

The basic reason is that the local solidified shell is too thin after casting out of the mold, which can't bear the static pressure of the molten steel and causes the breakout of the steel. It includes equipment factors, process operation factors, foreign matters or cold steel biting into solidified shell and other reasons.

5.5.2.3 Breakout by Sticking

Sticking breakout refers to the fact that there is no lubricant between the solidified shell of the slab on the meniscus of the mold and the copper plate of the mold, which increases the drawing resistance, breaks the bond and causes breakout when it reaches the mold outlet.

(1) The formed slag ring blocks the passage of liquid slag into the inner wall of copper tube and the shell.

(2) The high Al_2O_3 content, high viscosity and liquid surface crusting of mold flux make the fluidity of mold powder poor and it is not easy to flow into the shell and copper plate to form lubricating slag film.

(3) High casting speed under abnormal conditions. Such as the high casting speed when the liquid level fluctuates and the high casting speed when the molten steel temperature is low.

(4) The liquid level of the mold fluctuates too much, such as the submerged nozzle is clogged, the nozzle deflects seriously, and the nozzle condenses when replacing the ladle, which will cause the liquid level fluctuation.

5.5.2 漏钢事故产生的原因

5.5.2.1 开浇时发生漏钢

(1)“出苗”时间不够。

(2)起步升速过快。

(3)铁钉屑撒得太少或撒放不均匀。

(4)堵引锭头钢板未放好，落入格栅内与坯壳间卡住。

(5)保护渣加得过早且大量推入造成卷渣。

(6)结晶器与二冷区首段不对弧等都可能导致开浇漏钢。

5.5.2.2 浇铸过程中发生漏钢

根本原因在于铸坯出结晶器后局部凝固壳过薄，承受不住钢水静压力而破裂导致漏钢。包括设备因素、工艺操作因素和异物或冷钢咬入凝固壳等原因。

5.5.2.3 黏结漏钢

黏结漏钢是指结晶器弯月面的铸坯凝固壳与结晶器铜板之间无润滑液，使拉坯阻力增大，黏结处被拉断，到达结晶器下口就发生漏钢。

(1)形成的渣圈堵塞了液渣进入铜管内壁与坯壳间的通道。

(2)结晶器保护渣 Al_2O_3 含量高、黏度大、液面结壳等，使渣子流动性差，不易流入坯壳与铜板之间形成润滑渣膜。

(3)异常情况下的高拉速，如液面波动时的高拉速、钢水温度较低时的高拉速。

(4)结晶器液面波动过大，如浸入式水口堵塞，水口偏流严重、更换钢包时水口凝结等会引起液面波动。

5.5.3 Tool Preparation for Steel Leakage Accident Treatment

(1) Oxygen belt and oxygen pipe (ϕ8mm).

(2) Oxygen acetylene belt and cutting torch (special long cutting torch).

(3) Wire rope and other lifting appliances.

5.5.3 漏钢事故处理的工具准备

(1) 氧气皮带和氧气管若干（ϕ8mm）。

(2) 氧-乙炔皮带及割炬（特制长割炬）。

(3) 钢丝绳等吊具。

Task Implementation 任务实施

5.5.4 Treatment of Steel Leakage in Small Section Billet

(1) For small section continuous casting billets, sometimes the amount of steel leakage is small, and the casting can be continued after salvage treatment, but generally it is only limited to the case that the steel level has not dropped or there is no obvious steel shell, and it is suggested to adopt the treatment method of stopping and filling to avoid the accident of frozen billets.

(2) In case of steel leakage (such as the drop of steel level or spark splashing at the outlet of the mold), the tundish steel flow shall be closed immediately and the tension leveler shall be stopped.

(3) Judge the drop of the liquid level in the mold, the amount of steel leakage at the outlet of the mold and the damage extent of the secondary cooling equipment.

(4) Try to pull the billet at a lower casting speed (generally the starting speed of the starting casting), and pay attention that the motor current of the tension leveler shall not exceed the allowable value.

(5) If it is judged that the liquid level drop of the mold is small, that is, the steel leakage is small, the influence on the equipment is small, and the trial casting slab can be pulled normally, the continuous casting can be tried out.

(6) Pull the slab to the casting position of the dummy bar head.

(7) Open the tundish nozzle according to the requirements of dummy bar head casting, and pay close attention to whether there is steel leakage again under the mold, and whether the steel level rises normally.

(8) Under normal conditions, the continuous casting machine can be used to start casting and transfer to normal casting.

(9) Pay attention to the surface quality of the normal casting billet after steel leakage rescue treatment. If there is no problem in the surface quality and secondary cooling water spray, the next continuous casting can be carried out. Otherwise, when the casting accident occurs, the casting of furnace of molten steel in the ladle will be stopped.

(10) In case of trial drawing of billets, under the rated current condition of the tension level-

er, those who fail to pull the billets shall be subject to heat treatment after steel leakage.

(11) If there is any sign of continuous steel leakage when the above operation (7) is taken, the steel flow shall be closed immediately, and the pulling straightener shall be started to pull the billet to end the casting of the billet. If the billet can't be pulled, stop casting and stop, and heat the billet.

(12) After stopping casting due to steel leakage, if it is a multi-caster multi flow casting machine, other casting flows can continue to cast (at the same time, the casting flow of tension straightening may be affected and can't continue to cast); if it is a single machine single flow casting machine, the tundish car can be moved to the accident position after the liquid level of tundish steel is frozen, and then it can be lifted to the tundish turning position stably and turned over after all freezing, and the frozen steel can be cut back to the furnace. After the caster stops casting, the remaining molten steel in the ladle shall be returned to the furnace for treatment.

(13) If the steel leakage is found to be large and the equipment has a large influence range, the accident steel flow can be closed or blocked immediately after the steel leakage occurs, and the slab can be pulled down at the original speed (the oscillation can be closed after the slab is out of the mold). Pay attention to that the tension straightening current cannot exceed the value, and strive to pull out the hot slab.

5.5.4 小方坯发生漏钢时的抢救处理

(1) 对小断面的连铸坯有时漏钢量较少，经过挽救处理后可以继续浇铸，但一般仅限于钢液面基本未下降或未见明显钢壳的情况下，并建议采用停车补注的处理方法，避免冻坯。

(2) 发现漏钢现象（钢液面下降或结晶器下口发现火花飞溅等），立即关闭中间包钢流，拉矫机停车。

(3) 判断结晶器内液面下降情况、结晶器下口的漏钢量和对二冷设备的损坏程度。

(4) 以较低的拉速（一般为开浇起步拉速）试拉铸坯，注意拉矫机马达电流不得超过允许值。

(5) 凡判断为结晶器液面下降量较小，即漏钢量较少，对设备影响又较小，试拉铸坯又能正常拉动的情况，可以试行继续浇铸。

(6) 试拉铸坯到引锭头开浇位置。

(7) 以引锭头开浇的要求打开中间包水口，并密切注意结晶器下是否有再次漏钢发生，同时注意钢液面是否正常上升。

(8) 在正常情况下可以用连铸机开浇操作步骤，进行开浇和转入正常浇铸。

(9) 注意漏钢挽救处理后正常浇铸的铸坯的表面质量，凡表面质量没有问题，二次冷却喷水也没有问题，即可连续下一炉浇铸。否则浇完事故发生时钢包内一炉钢液后即停浇。

(10) 试拉铸坯，在拉矫机额定电流条件下，如未能拉动铸坯则作漏钢后的热坯处理。

（11）如采取上述第（7）项操作时，发现有继续漏钢迹象则立即关闭钢流，并启动拉矫机拉动铸坯，结束该铸坯浇铸。如铸坯无法拉动，则停浇停车，铸坯作热坯处理。

（12）因漏钢铸流停浇后，如为多机多流铸机，其他铸流可继续浇铸（同时拉矫的铸流有可能受影响也不能继续浇铸）；如为单机单流铸机，待中间包钢液面冻结后才能开中间包车到事故位置，然后平稳吊至翻包位置待全部冻结后才能翻包，将冻钢切割回炉。铸机停浇后钢包内的剩余钢液作回炉处理。

（13）如发现漏钢量大，设备影响范围大，发生漏钢后可立即关闭或堵塞事故钢流，铸坯以原速拉下（铸坯出结晶器后可关闭振动），注意拉矫电流不能超值，力争把热坯拉出。

5. 5. 5 Precautions

(1) After steel leakage accident, it is not allowed to force over load drawing, otherwise it is easy to damage the equipment.

(2) The mold after steel leakage must be checked in detail from cooling water volume, surface quality of copper plate, mold size, especially taper, otherwise it cannot be reused.

(3) In case of steel leakage, if the steel flow of the tundish is not closed tightly, the tundish car shall be driven away immediately.

(4) In order to protect the equipment, the cooling water of mold after steel leakage must be kept normal.

(5) When the caster is casting, for the sake of safety, it is not allowed to handle the accident billet in the secondary cooling zone.

5. 5. 5 注意事项

（1）漏钢事故发生后不能强行超负荷拉坯，否则容易损坏设备。

（2）漏钢后的结晶器必须对冷却水量、铜板表面质量、结晶器尺寸（特别是锥度）做详细检查，否则不得重新使用。

（3）发生漏钢后，中间包钢流又关不紧密时应立即开走中间包车。

（4）为保护设备，漏钢后结晶器冷却水，设备冷却水一定要保持正常。

（5）铸机在浇铸时，为安全起见，不能到二冷区去处理事故坯。

Exercises:

(1) How to deal with steel leakage accident in production?

(2) Use simulation software to simulate steel leakage accident and deal with it.

思考与习题：

（1）生产时发生漏钢事故该如何处理？

（2）利用仿真软件模拟漏钢事故并进行处理。

Task 5.6 Treatment of Frozen Slab and Roller Blocking Billet Accident
任务 5.6 连铸冻坯和顶坯处理

Mission objectives:

任务目标:

(1) Be able to deal with the accidents of frozen slab and roller blocking billet in continuous casting production to ensure the normal operation and safety of production.

(1) 能够处理连铸生产的冻坯和顶坯事故，保证生产的正常进行和安全。

(2) Master the reasons for the difficulty of billet pulling out during continuous casting production and prevent it.

(2) 掌握连铸生产时钢坯拉出困难的原因并进行预防。

Task Preparation 任务准备

5.6.1 Cognition of Frozen Slab Accident and Roller Blocking Billet Accident

The frozen billet and roller blocking billet can't be pulled under the action of a certain pulling force of dummy bar, mainly because of the shrinkage deformation during the cooling process of the billet, or the hanging of steel after the spilled steel, which makes the billet contact or connect with the roller in the secondary cooling zone to form the pulling resistance. However, the lifting distance of the pressing cylinder in the secondary cooling zone is limited, which can not eliminate the drag of the blank drawing. The reason for the top roller blocking billet is that it can not be pulled out normally because of the formation of obstacles after deformation.

Both of them have certain damage to the equipment due to the above cooling deformation, and the secondary cooling section is often damaged when dealing with the accidents of frozen billet and roller blocking billet. Therefore, after other accidents of continuous casting, it is generally necessary to seize the time to pull out in the hot state of billet, rather than cause the accidents of frozen billet and roller blocking billet.

5.6.1 冻坯和顶坯事故认知

冻坯和顶坯都是引锭杆一定拉矫力作用下不能拉的情况，主要是因为铸坯冷却的过程中收缩变形，或溢漏钢后挂钢，使铸坯与二冷区辊子强力接触或连接形成拉坯阻力。而二冷区压紧缸的升起的距离有限，不能消除这股拉坯阻力。顶坯是因为变形后形成障碍导致铸坯无法正常拉出。

两者因上述的冷却变形对设备有一定的损坏，处理冻坯和顶坯事故又往往会造成二冷段损坏，所以在连铸发生其他事故后，一般应抓紧时间，在钢坯较热状态下拉出，避免造成冻坯和顶坯事故。

5. 6. 2 Tool Preparation

(1) Oxygen acetylene belt and cutting torch;

(2) 2~3 crowbars;

(3) Wire rope and other lifting appliances.

5. 6. 2 工具准备

(1) 氧-乙炔皮带及割炬；

(2) 撬杠 2~3 根；

(3) 钢丝绳等吊具。

Task Implementation 任务实施

5. 6. 3 Treatment of Frozen Billet when Dummy Bar Still in the Tension Leveller

(1) Loosen the hydraulic cylinder of the secondary cooling sector to make the billet have a certain degree of freedom.

(2) The cutting torch is used to remove the steel hanging on the connecting casting billet and the secondary cooling roller or the equipment (such as the frozen billet caused by steel leakage and overflow). Try to pull the dummy bar and frozen billet.

(3) When the dummy bar head is out of the cutting point of the tension leveler, the oxygen acetylene cutting torch is used to cut the billet in front of the dummy head to separate the frozen billet from the dummy bar, or the automatic device is used for disconnecting the dummy bar.

(4) Recover dummy bar head and chain. The dummy bar head can be reused after treatment.

(5) After the frozen billet is pulled out for a certain length, the steel wire rope is used to hang the frozen billet outside the cutting point roller of the tension leveler, and the frozen billet is cut outside the cutting point roller, that is, the frozen billet is processed in sections. After cutting, the broken billet is lifted away and used as scrap steel for melting return. Continue to pull the billet, hang the frozen billet, cut and lift the other billet, repeat the operation until all the frozen billet are pulled out of the tension leveller.

(6) When the secondary cooling sector is loosened, the frozen billet cannot be pulled out, which may be caused by the arc deformation of the billet and the resistance of the secondary cooling guide roller. At this time, part of the frozen billet can be cut in the thickness direction, but do not cut off, about 2/3~3/4 of the thickness of the frozen billet, so that the frozen billet has a certain amount of deformation in the arc direction, and then pull the frozen billet again. If it is possi-

ble to pull out, operate according to (3)~(5) above until the whole frozen billet is pulled out of the tension leveler.

(7) If the above (1)~(6) operations still fail to pull out the frozen billet, the mold, foot roller section and section 0 can be removed. The frozen billet is lifted from the upper part by steel wire rope and lifted off in sections above the secondary cooling section. Then send the dummy bar out and cut and lift the frozen billet section by section.

(8) For the above operations (1)~(7), if the frozen billet cannot be pulled out, the frozen billet can only be cut in front of the dummy bar head to separate the frozen billet from the dummy bar, the dummy bar is pulled out for recycling, and the dummy bar head can be reused if it is in good condition after treatment. The frozen billet is cut between the secondary cooling sector sections, lifted out the sector section, processed offline, and then reinstalled the sector section, arc regulating, centering, commissioning, inspection, and ready to start casting again.

(9) If the horizontal section of the tension leveler is long, it can be handled according to item (7) or item (8) above.

5.6.3 引锭未出拉矫机的冻坯处理

(1) 松开二冷扇形段液压缸，使铸坯有一定自由度。

(2) 用割炬去除连接铸坯与二冷辊或设备上挂钢（如漏、溢钢造成的冻坯），可试拉引锭和冻坯。

(3) 当引锭头出拉矫机切割点后，用氧-乙炔割炬，切割引锭头前铸坯，使冻坯和引锭脱开，或采用自动脱锭装置脱锭。

(4) 回收引锭头和引锭链。引锭头处理后可重复使用。

(5) 在冻坯拉出一定长度后，用钢丝绳吊住拉矫机切点辊外的冻坯，并在切割点辊外切割冻坯，即冻坯分段处理。切割后的断坯吊走作废钢回炉。继续拉坯，吊住冻坯，切割，吊走分段坯，重复操作直至冻坯全部出拉矫机。

(6) 松开二冷扇形段后，冻坯无法拉动，可能因为铸坯弧向变形，顶住二冷导辊造成阻力。这时可在冻坯厚度方向作部分切割，但不要切断，约割开坯厚的2/3~3/4，使冻坯在弧度方向有一定变形量，再试拉冻坯。如果可以拉动按上述（3）~（5）项操作，直至冻坯出拉矫机。

(7) 作上述第（1）~（6）项操作仍未能拉出冻坯，可拆除结晶器、足辊段及0号段，冻坯从上部用钢丝绳吊住，在二冷段上面分段吊走。然后送引锭顶坯再一段段切割吊走冻坯。

(8) 作上述第（1）~（7）项操作，冻坯还不能顶动者，只能在引锭头前切割冻坯，使冻坯与引锭分离，引锭拉出回收，引锭头处理后完好则可重复使用。冻坯在二冷扇形段之间切割分段，吊出扇形段、离线处理冻坯，再重新安装扇形段、调弧、对中、调试、检查、准备重新开浇。

(9) 拉矫机水平段较长，可按上述第（7）项或第（8）项操作处理。

5. 6. 4 Treatment of Roller Blocking Billet Accident

(1) In case of roller blocking billet accident, the tundish steel flow shall be closed immediately, the ladle steel flow shall be closed, and the tension leveller shall be commanded to stop.

(2) The steel level in the mold shall be treated according to the operation requirements of changing tundish, the protective slag shall be removed and the preparation for capping the billet shall be made.

(3) The tundish nozzle shall be cleaned.

(4) The cooling of the mold and the secondary cooling zones remain unchanged, and the secondary spray cooling is kept at the lowest level.

(5) Using oxyacetylene torch to quickly deal with the obstacles at the position of roller blocking the billet.

(6) Try to pull the casting billet, pay attention to the guide during the casting process, and correct it with crowbar if necessary.

(7) If it is found that the cold steel at the tundish nozzle is clogged after closing the tundish nozzle, it can be used to clean the nozzle.

(8) If the accident of roller blocking billet is found later or the nozzle is not closed tightly, it may cause steel overflow. In case of steel overflow, it shall be treated as steel overflow.

5. 6. 4 连铸顶坯处理

(1) 发现顶坯事故后，立即关闭中间包钢流，关闭钢包钢流，指挥拉矫停车。

(2) 结晶器内钢液面按换中间包操作要求进行处理，撇除保护渣，并做好封顶的准备。

(3) 中间包水口作一定清理。

(4) 二次冷却区的设备冷却和结晶器冷却保持不变，二次喷雾冷却保持在最低水平。

(5) 用氧-乙炔割炬迅速处理顶坯处的障碍。

(6) 试拉铸坯，注意铸坯行进中的导向，必要时用撬杠纠正。

(7) 如发现因中间包水口关闭后造成水口冷钢堵塞，可以作清洗水口操作。

(8) 顶坯事故如发现稍晚或水口关闭不严可能造成溢钢。出现溢钢则按溢钢处理。

5. 6. 5 Precautions

(1) If the processing time of the roller blocking billet exceeds the specified time of changing the ladle, the strand shall be stopped after the part of roller blocking billet is processed to ensure the smooth progress of the casting billet.

(2) If it takes too long to process the accident, resulting in frozen billet, it shall be treated as frozen billet.

(3) When using the crowbar, pay attention to the rebound of the crowbar or the injury caused by the irregular deformation of the crowbar when the billet is running.

(4) If the tundish nozzle is not closed tightly, which may cause steel overflow, it shall be immediately driven away from the tundish car.

5.6.5 注意事项

(1) 处理顶坯时间超过规定的换包时间，则处理好顶坯部位保证铸坯顺利行进后，该铸流作停浇处理。

(2) 处理顶坯时间过长，造成冻坯，则按冻坯操作处理。

(3) 使用撬杠时注意撬杠反弹或铸坯运行时撬杠受力不规则变形后伤人。

(4) 如中间包水口关闭不严可能造成溢钢应立即开离中间包车。

Exercises:

(1) How to deal with the accidents of frozen slab and roller blocking billet during production?

(2) Use simulation software to simulate the accident of frozen slab and roller blocking billet, and deal with it.

思考与习题:

(1) 生产时发生冻坯和顶坯事故该如何处理?

(2) 利用仿真软件模拟顶坯和冻坯事故并进行处理。

Project 6　Defect Control of Continuous Casting Billet

项目6　连铸坯缺陷控制

Task 6.1　Quality Control of Continuous Casting Billet

任务6.1　连铸坯质量控制

Mission objectives:

任务目标:

(1) Master the factors that affect the quality of continuous casting billet, and ensure the product quality by targeted control in production.

(1) 掌握影响连铸坯质量的因素，在生产中能对其针对性控制保证产品质量。

(2) Master the classification of billet defects and understand the common control methods

(2) 掌握铸坯缺陷的分类，了解常用的控制手段。

Task Preparation 任务准备

6.1.1　Cognition of Billet Quality

The quality of continuous casting billet refers to the severity of billet defects allowed by qualified products. In the solidification process of liquid steel, due to many factors, there will be a variety of defects. Some of these defects can be treated by certain means, which can not affect the quality of the billet, while some of the defects can't be treated and eventually make the billet scrapped. The quality of billet generally refers to the condition of billet purity, surface defect, internal defect and shape defect.

6.1.1　铸坯质量认知

铸坯质量是指得到合格产品所允许的铸坯缺陷的严重程度。液态钢水在连铸凝固过程中，由于受诸多因素影响，会产生各种各样的缺陷，这些缺陷有的通过一定手段的处理，可以不造成对铸坯质量的影响，而部分的缺陷是无法处理并最终使铸坯报废。铸坯质量一般是指铸坯纯净度、铸坯表面缺陷、铸坯内部缺陷和形状缺陷四方面的状况。

6. 1. 2 Classification of Billet Defects

There are many kinds of defects in continuous casting billet, and the causes are complicated. However, there are quite a number of defects which are often repeated in production, which are called common defects.

Common defects of continuous casting billet can be divided into three categories: surface defects, internal defects and shape defects, which can be further distinguished. As shown in Table 6-1, these defects may occur at the same time.

Table 6-1 Classification of continuous casting billet defect

Defect categories	Defect name
Surface defects	longitudinal surface crack; longitudinal crack of the billet head; longitudinal crack near the corner; transverse surface crack; transverse corner crack; narrow transverse crack; star crack; pinhole on the billet head; pinhole and point slag inclusion on the surface; pinhole and flocculent slag inclusion distributed on the longitudinal straight line; pinhole on the whole surface; carburization; deep oscillation mark
Internal defect	internal crack (segregation crack, intermediate crack); triangle crack (side crack); diagonal crack (corner crack); centerline crack; center segregation and central porosity; slag inclusion; spherical inclusion; clusters inclusion; hypodermic inclusion; hypodermic bubble
Shape defect	longitudinal pit; transverse pit; billet bulge; width deviation; thickness deviation; deflection; straightness; rhombus deformation; ellipse

6. 1. 2 铸坯缺陷分类

连铸坯的缺陷种类繁多，产生的原因也较复杂，但是有相当一部分缺陷是生产中经常重复出现的，称为常见缺陷。

连铸坯常见缺陷分为三大类，即表面缺陷、内部缺陷和形状缺陷，各类缺陷还可以进一步予以区分，见表 6-1，这些缺陷有可能同时出现。

表 6-1 连铸坯缺陷分类

缺陷类别	缺 陷 名 称
表面缺陷	纵向表面裂纹；头坯的纵向裂纹；靠近角部的纵向裂纹；横向表面裂纹；横向角部裂纹；窄面横裂纹；星状裂纹；头坯上的针孔；面上的针孔和点状夹渣；分布在纵向直线上的针孔和团絮状夹渣；整个表面上的针孔；渗碳；深的振痕
内部缺陷	内部裂纹（偏析裂纹、中间裂纹）；三角区裂纹（侧边裂纹）；对角线裂纹（角部裂纹）；中心线裂纹；中心偏析和中心疏松；保护渣夹杂；球状夹杂；团絮状夹杂；皮下夹杂；皮下气泡
形状缺陷	纵向凹坑；横向凹坑；铸坯鼓肚；宽度偏差；厚度偏差；挠曲；不直；菱形变形；椭圆

Task Implementation 任务实施

There is a certain relationship between billet quality requirements and continuous casting process, as shown in Figure 6-1. The purity of continuous casting liquid steel is determined by the liquid steel above the mold. The control of the purity of liquid steel before entering the mold should be focused on. The surface defect quality of billet is mainly affected by the solidification process of molten steel in mold. The internal defect quality of billet is determined by the solidification process below the mold, which mainly depends on the billet cooling process and the billet support system in the secondary cooling zone. The section shape and dimension of the billet are related to the billet cooling and the equipment status.

铸坯质量要求和连铸的工艺过程有一定的对应关系，如图 6-1 所示。连铸钢液的纯净度是由结晶器之上的液态钢所决定的，应着眼于未进入结晶器前钢水纯净度的控制。铸坯的表面缺陷质量主要是受结晶器内钢水凝固过程所影响。铸坯的内部缺陷质量则是由结晶器以下的凝固过程所决定的，主要取决于在二次冷却区的铸坯冷却过程和铸坯支撑系统。关于铸坯的断面形状和尺寸则和铸坯冷却以及设备状态有关。

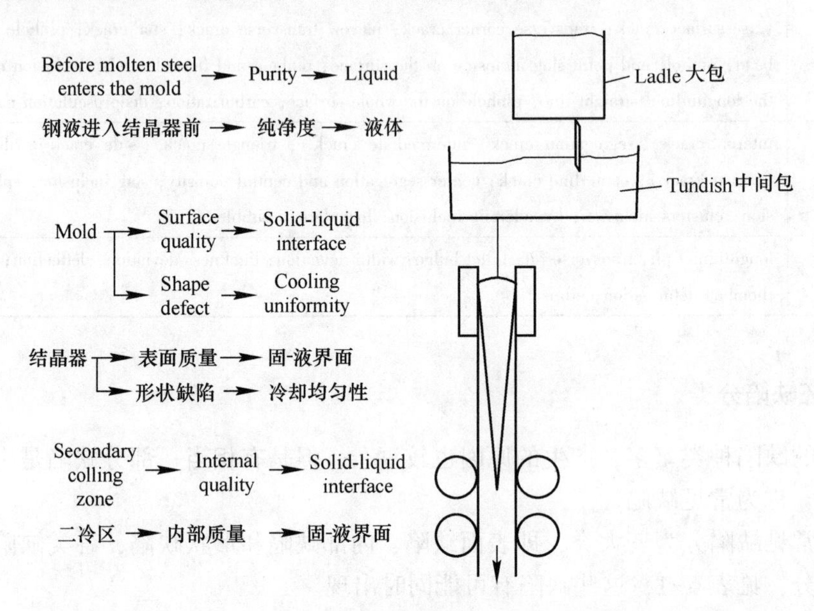

Figure 6-1 Schematic diagram of continuous casting billet quality control

图 6-1 连铸坯质量控制示意图

Exercises:

(1) Which aspects should be considered to evaluate the quality of continuous casting billet?

(2) Analyze the relationship between billet quality and continuous casting process.

思考与习题：

(1) 评价连铸坯质量应从哪几个方面考虑?

(2) 分析铸坯质量和连铸工艺过程的关系。

Task 6.2 Surface Defects of Continuous Casting Billet and Control
任务 6.2 连铸坯表面缺陷及控制

Mission objectives:

任务目标:

(1) Be able to identify various defects on the surface of continuous casting billet.

(1) 能够识别连铸坯表面各种缺陷。

(2) Master the causes of surface defects of billet.

(2) 掌握铸坯表面缺陷产生的原因。

(3) Master the control means to reduce the surface defects of the billet.

(3) 掌握减少铸坯表面缺陷的控制手段。

Task Preparation 任务准备

6.2.1 Cognition of Billet Surface Defects

The surface quality of continuous casting billet or slab determines whether the billet needs to be finished before hot processing, and also the precondition of billet hot delivery and direct rolling. The causes of surface defects of continuous casting billet are complex, but generally speaking, they are mainly controlled by the solidification of molten steel in the mold. The common surface defects of continuous casting billet are shown in Figure 6-2.

6.2.1 铸坯表面缺陷认知

连铸坯表面质量的好坏决定了铸坯在热加工之前是否需要精整，也是影响金属收得率和成本的重要因素，还是铸坯热送和直接轧制的前提条件。连铸坯表面缺陷形成的原因较为复杂，但总体来讲，主要是受结晶器内钢液凝固所控制，连铸坯常见的表面缺陷如图 6-2 所示。

6.2.2 Surface Cracks

Continuous casting billet crack is one of the most common and numerous defects. The reason for its formation depends on the stress of shell and solidification interface on the one hand, and on the other hand, the plasticity and strength of steel at high temperature.

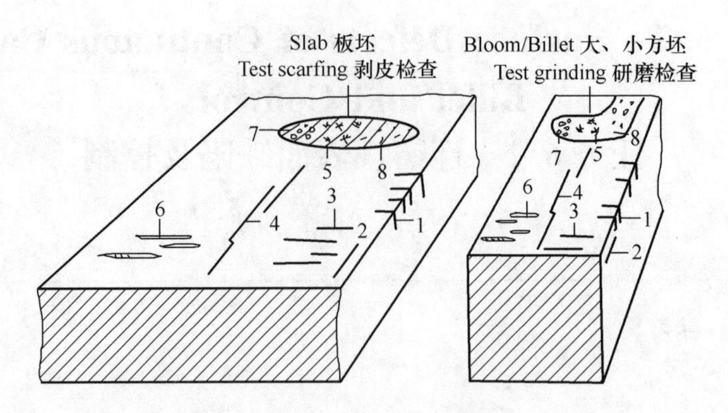

Figure 6-2 Surface defects of continuous casting billet

1—Transverse corner cracks; 2—Longitudinal corner cracks; 3—Transverse cracks; 4—Longitudinal facial cracks; 5—Star cracks; 6—Deep oscillation marks; 7—Pinholes; 8—Macro inclusions

图 6-2 连铸坯表面缺陷

1—角部横向裂纹；2—纵向角部裂纹；3—横向裂纹；4—纵向裂纹；5—星状裂纹；6—深振痕；7—针孔；8—宏观夹杂

6.2.2.1 Longitudinal Facial Cracks

The longitudinal facial cracks is the crack formed on the billet surface along the axial direction (drawing direction), which mostly occurs in the center of the slab wide surface, and the square slab appears on the face. When serious, the crack will be more than 10mm. The longitudinal facial cracks are located on the face of the slab. The longitudinal cracks near the corner are called the longitudinal corner cracks. The crack length direction is consistent with the axis direction of the slab, and the distribution is irregular. The longitudinal cracks are caused by the uneven thickness of the primary shell and the stress concentration at the place where the shell is thin. When the stress exceeds its tensile strength, the main reasons are as follows:

(1) The wear and deformation of the mold lead to the uneven solidification shell, and the crack occurs in the weak part.

(2) The viscosity of the mold powder does not match the casting speed. The excessive inflow of slag along the meniscus makes the slag circle locally thickened, reduces the heat conduction and hinders the development of solidified shell.

(3) The fluctuation of the liquid level in the mold is too fast and too large, which directly affects the uniformity of the solidified shell and easily forms the longitudinal cracks.

(4) The inclined installation of the submerged nozzle sleeve causes local scour, which makes the condensing shell thin.

(5) Too high casting temperature also has a great influence on the uniform growth of solidified shell.

6.2.2 表面裂纹

连铸坯裂纹是最常见和数量最多的一种缺陷，其形成原因一方面取决于坯壳和凝固界面的受力情况，另一方面取决于钢在高温下的塑性和强度。

6. 2. 2. 1　表面纵裂纹

表面纵裂纹是铸坯表面沿轴向（拉坯方向）形成的裂纹，多发生在板坯的宽面中央部位，方坯出现于面部，严重时裂纹将高达10mm以上。面部纵裂位于铸坯面部处，靠近角部出现的纵裂纹称为角部纵裂纹，裂纹长度方向与铸坯轴心方向一致，分布不规则。发生纵裂纹是由于初生坯壳厚度不均匀，在坯壳薄的地方应力集中，当应力超过其抗拉强度时产生裂纹，主要原因有：

（1）结晶器的磨损、变形，导致凝固壳不均匀，裂纹产生于薄弱部位。

（2）保护渣的黏度与拉速不匹配，渣子沿弯月面过多流入使渣圈局部增厚，降低了热传导，阻碍了凝固壳的发展。

（3）结晶器内液面波动过快、过大，也直接影响凝固壳形成的均匀性并易形成纵裂纹。

（4）浸入式水口套安装偏斜，造成局部冲刷，使该部位凝壳变薄。

（5）过高的浇铸温度也对凝固壳的均匀生长有较大的影响。

6. 2. 2. 2　Transverse Facial Cracks

The cracks on the surface of the billet along the wave trough are called transverse facial cracks. The surface transverse cracks include face transverse cracks and corner transverse cracks. This kind of crack is caused by tensile stress when the slab is straightened, or continues to expand after cracks appear on the surface of the slab. If the crack is in the corner, the corner transverse crack will be formed. It is especially pointed out that the surface transverse cracks occur in the brittle temperature range of 700~900℃. The continuous transverse facial cracks are caused by abnormal oscillation, especially after the steel leakage accident, they are not cleaned and repaired in time; while the intermittent transverse cracks, if the caster and process parameters are not adjusted, should first check whether the content of sulfur and other impurities in the liquid steel is too high.

6. 2. 2. 2　表面横裂纹

在铸坯表面沿波动纹波谷所发生的开裂，称为表面横向裂纹。表面横裂纹包括面部横裂纹和角部横裂纹。这种裂纹是铸坯矫直时产生拉应力造成的，或在连铸坯表面出现龟裂后继续扩展而成。如果裂纹在角部，就形成角部横裂纹。特别指出，表面横裂纹均发生在700~900℃的脆性温度范围内。连续的表面横向裂纹是由于震动异常造成的，特别是在漏钢事故后未及时清理及检修；而断续的横向裂纹，若铸机和工艺参数均未调整，应首先检查钢液中硫元素等杂质的含量是否过高。

6. 2. 2. 3　Star Cracks

The small cracks distributed on the surface of the slab in the shape of stars are called star cracks (chap), whose depth is generally 1~3mm. In serious cases, they may cause transverse cracks after straightening the slab, or tear the slab surface during rolling, which is a defect on the steel surface. The causes are as follows:

(1) The fall off of copper in the mold. The copper adsorbed on the surface of the shell is preferentially diffused along the austenite grain boundary.

(2) The content of aluminum and sulfur in molten steel is too high. Aluminum nitride and sulfide infiltrate into the surface of the slab, causing high temperature embrittlement of the steel.

(3) Selective oxidation of slab surface. Weak cooling makes the surface temperature of slab high, residual elements (copper and tin) are easy to form cracks along the grain boundary, while strong cooling will increase the temperature gradient of slab and promote the precipitation of aluminum and niobium at the grain boundary.

6.2.2.3 星状裂纹

呈星状分布在铸坯表面的细小裂纹，称为星状裂纹（龟裂），其深度一般为1~3mm，严重时可能在连铸坯矫直后导致横裂纹，或在轧制时造成连铸坯表面撕裂，使钢材表面产生缺陷。其产生原因如下：

（1）结晶器上铜的脱落。坯壳表面吸附了结晶器上的铜，而铜优先沿奥氏体晶界扩散。

（2）钢液中铝和硫含量过高。氮化铝和硫化物渗入铸坯表面，引起钢的高温脆化。

（3）铸坯表面的选择性氧化。弱冷使得铸坯表面温度高，残余元素（铜、锡）沿晶界富集易形成裂纹，强冷则会增大铸坯的温度梯度，促使铝铌等在晶界沉淀。

6.2.3 Oscillation Marks

In order to avoid the bond between shell and mold, the concept of mold oscillation has been put forward for a long time. However, the results of the mold moving up and down on the surface of the slab cause periodic marks of transverse pattern along the whole periphery, which is called oscillation marks. It is considered to be caused by the periodic process of shell pulling and rewelding. If the oscillation marks is very shallow and regular, it will not cause any defects in further processing. However, if the oscillation marks is too deep (more than 3mm), or the transverse cracks, slag inclusion, pinhole and other defects lurk in the vibration mark, it will cause harm to the subsequent processing and the finished product. The main causes are as follows:

(1) Poor mold oscillation;

(2) The liquid level of steel fluctuates violently;

(3) Improper selection of mold powder;

(4) The oscillation marks of low carbon steel is deep and that of high carbon steel is shallow

6.2.3 振动痕迹

为了避免坯壳与结晶器之间黏结，很早就提出了结晶器振动的概念。但结晶器上下运动的结果在铸坯表面上造成了周期性的沿整个周边的横纹模样的痕迹，称为振动痕迹。它被认为是周期性的坯壳拉破和重新焊合过程造成的，如果振痕很浅而且又很规则的话，它在进一步加工时不会引起什么缺陷。但过深的振痕（大于 3mm），或在振痕处潜伏横裂纹、夹渣和针孔等缺陷时，会对后续加工及成品产生危害。其产生原因主要为：

（1）结晶器振动状况不佳；

（2）钢液面波动剧烈；

（3）渣粉选择不当；

（4）低碳钢振痕较深、高碳钢较浅。

6.2.4 Gasholes and Bubbles

During the solidification of molten steel, the escape of CO or H_2 from the C-O reaction, and the gasholes near the surface of the billet in the growth direction of the columnar crystal are called blowholes, which are generally 1mm in diameter and about 10mm in depth. The gasholes, relatively small and densely distributed near the slab surface, is called pinholes. According to the position of the gasholes those hidden under the skin without exposure are called subsurface gasholes. subsurface gasholes need to be grinded and pickled to be found.

In actual production, the common causes of gasholes are poor deoxidation. When the dissolved aluminum in steel is more than 0.008%, CO bubbles can be prevented; the superheat of molten steel is large; secondary oxidation, water and gas in the air are inhaled; the moisture content of the mold powder exceeds the standard; water seeps from the upper entrance of the mold; excessive lubricating oil in the mold; and wet lining of the tundish.

6.2.4 气孔和气泡

钢液凝固时 C-O 反应生成的 CO 或 H_2逸出，在柱状晶生长方向接近于铸坯表面形成的孔洞称为气孔，直径一般为 1mm，深度为 10mm 左右。接近于铸坯表面，相对较小且密集分布的称为表面气孔（皮下针孔）。按照气孔的位置，藏于皮下没有裸露的称为皮下气泡。皮下气泡需要对铸坯修磨、酸洗才能发现。

实际生产中，产生气泡的常见原因有：脱氧不良，当钢中溶解铝大于 0.008%就可防止 CO 气泡产生；钢液过热度大；二次氧化，空气中水汽被吸入；保护渣水分超标；结晶器上口渗水；结晶器润滑油过量；中间包衬潮湿。

6.2.5 Surface Slag Inclusion

The scum or inclusions on the steel surface in the mold are drawn into the slab, and the spots formed on the surface of the slab are called surface slag inclusion. Due to the slow solidification of the crust under the slag inclusion, there are often fine cracks and gasholes.

The common causes of slag inclusion in actual production are not timely bailing, unstable liquid level of mold, poor quality of refractory, poor fluidity of molten steel, too slow casting speed or too low casting temperature.

6.2.5 表面夹渣

结晶器中钢液面上的浮渣或上浮的夹杂物被卷入铸坯内，在连铸坯表面形成的斑点称为表面夹渣。由于夹渣下面的凝壳凝固缓慢，故常有细裂纹和气泡伴生。

实际生产中常见夹渣原因主要有：捞渣不及时；结晶器液面不稳定；耐火材料质量差；钢液流动性差；拉速过慢或浇注温度过低。

6. 2. 6 Other Surface Defects

(1) Double skin. When the shell breaks slightly in the mold, a small amount of molten steel will flow out to close the crack. It seems that there is a layer of skin on the shell surface. Usually, the defect is called double skin. Serious heavy double skin will cause steel leakage. The defects such as low casting temperature or low casting speed, slag inclusion, poor lubrication and incorrect arc alignment of the mold may cause double skin defects. The secondary oxidation of molten steel and the double skin defects can be reduced by using the mold powder casting of the submerged nozzle.

(2) Superposition. Superposition defect refers to the condensation of shell at meniscus due to the interruption of molten steel casting (sudden stop of tension leveler), which can't be fused with the molten steel for recasting, resulting in a joint mark around the slab surface, also known as double casting. The longer the casting time is interrupted, the deeper the joint mark is.

(3) Scratch. The mechanical damage caused by the friction between the continuous casting slab and the roller and guide device with poor rotation is called scratch. The preventive measure is to frequently repair the parts that are easy to contact the continuous casting slab and cause scratches, and find out the causes of scratches and deal with them in time.

(4) Surface depression of continuous casting slab or billet. Surface depression of continuous casting slab or billet refers to the irregular depress on the surface of continuous casting slab or billet. Most of the depressions appear on the surface of the billet of austenitic stainless steel and low carbon steel (carbon content is 0. 10%~0. 15%). The transverse depression is easy to occur at the corner of wide surface or any part of wide surface. There are cracks in the serious depressions. If the cracks are deep, there will be steel bleeding and scarring, double skin or even steel leakage on the surface.

6. 2. 6 其他表面缺陷

(1) 重皮。当坯壳在结晶器内发生轻微破裂时，会有少量钢液流出来弥合裂口，在坯壳表面上好像贴一层皮似的，通常把这种缺陷称为重皮。重皮严重时会造成漏钢事故。铸温低或拉速偏低、有夹渣、润滑不良以及结晶器对弧不准等都可能引起重皮缺陷。采用浸入式水口保护渣浇铸，可减少钢液的二次氧化，有助于消除重皮缺陷。

(2) 重接。重接缺陷是指由于钢水浇铸中断（拉矫机突然停止）而在弯月面处产生凝壳，不能与再浇的钢水相融，导致在铸坯表面产生了一个环绕铸坯的接痕，也称为双浇。中断浇铸时间越长，重接部位的接痕越深。

(3) 划伤。连铸坯与转动不良的辊子、导向装置等摩擦而产生的机械损伤称为划伤。防止措施是经常检修容易接触连铸坯并造成划伤的部件，同时一发现划伤立即查找原因，并及时处理。

（4）凹陷。连铸坯表面凹陷是指连铸板坯或方坯表面呈现的不规则的凹坑。多为横向凹坑，也有纵向凹坑。凹陷多出现在奥氏体不锈钢和低碳钢（含碳质量分数在 0.10%~0.15%）的铸坯表面。横向凹陷易产生在铸坯宽面角部或宽面任何部位。严重的凹陷处有裂纹，裂纹深者出现表面渗钢结疤、重皮甚至漏钢。

Task Implementation 任务实施

6.2.7 Prevention of Surface Cracks

6.2.7.1 Longitudinal Facial Cracks

（1）Control the composition of liquid steel.

（2）Choose the appropriate mold flux with proper melting performance and melting speed.

（3）The reverse taper and the thickness of copper wall are designed correctly.

（4）Adopt submerged nozzle and strictly align, adopt reasonable casting temperature, casting speed, mold wide and narrow surface cooling balance.

（5）Secondary cooling strength and the influence of foot roller. The secondary cooling is mainly manifested in the longitudinal depression caused by local subcooling.

6.2.7 表面裂纹的预防

6.2.7.1 表面纵裂纹

（1）控制钢液成分。

（2）选用熔化性能和熔化速度合适的保护渣。

（3）正确设计结晶器的倒锥度和铜壁厚度。

（4）采用浸入式水口并严格对中，采用合理的铸温、铸速，结晶器宽、窄面冷却平衡。

（5）二次冷却强度及足辊的影响。二冷主要表现在局部过冷产生纵向凹陷。

6.2.7.2 Transverse Facial Cracks

（1）Control the composition of molten steel, reduce the content of sulfur, oxygen, nitrogen and other elements in steel, add trace titanium, zirconium, calcium and other elements, and inhibit the precipitation of carbide, oxide and sulfide at the grain boundary.

（2）Control the straightening temperature and avoid the brittle zone.

（3）Slow cooling is used to control the precipitates at the grain boundary.

（4）Strengthen the spot inspection of the mold oscillation device.

（5）The inner arc clamping roller is added in the secondary cooling zone.

（6）Improve the quality of mold lubrication and cooling water.

6.2.7.2 表面横裂纹

（1）控制钢液成分，降低钢中硫、氧和氮等元素的含量，加入微量钛、锆和钙等元素，抑制碳化物、氧化物和硫化物在晶界析出。

（2）控制矫直温度，避开脆性区。

（3）采用缓冷，控制晶界析出物。

（4）加强对结晶器振动装置的点检。

（5）在二冷窗口增加内弧夹持辊。

（6）改善结晶器润滑和冷却水质。

6.2.7.3 Star Cracks

(1) Cr and Ni were plated on the mold.

(2) Proper amount of secondary cooling water.

(3) Select raw materials and control residual elements (Cu, Sn, etc.) in steel.

(4) Control $w(\mathrm{Cu})<0.2\%$ in molten steel.

(5) Control $w(\mathrm{Mn})/w(\mathrm{S})>40$ in molten steel.

6.2.7.3 星状裂纹

（1）结晶器镀 Cr 和 Ni。

（2）合适的二冷水量。

（3）精选原料，控制钢中残余元素（Cu、Sn 等）。

（4）控制钢中 $w(\mathrm{Cu})<0.2\%$。

（5）控制钢中 $w(\mathrm{Mn})/w(\mathrm{S})>40$。

6.2.8 Prevention of Oscillation Marks

(1) Mold oscillation mode with small amplitude and high frequency.

(2) Keep the steel level stable.

(3) Select appropriate mold powder.

(4) A plug-in made of poor thermal conductivity material is added near the liquid level of the mold by using a hot top mold.

6.2.8 振动痕迹的预防

（1）小幅高频的结晶器振动模式。

（2）保持钢液面平稳。

（3）选择合适保护渣。

（4）采用热顶结晶器，在结晶器液面附近加入导热性差材料做的插件。

Exercises:

(1) Describe the causes and preventive measures of longitudinal cracks in continuous casting.

(2) What technical measures should be taken to prevent large slag from 2~10mm below the surface of slab in a factory?

(3) Using simulation software to observe the surface quality of the final product.

思考与习题：

(1) 阐述连铸纵裂纹产生的原因及防止措施。

(2) 某厂铸坯表皮下 2~10mm 处有大块的渣子，应采取哪些技术措施进行预防？

(3) 利用仿真软件观察铸坯最终产品的表面质量。

Task 6.3 Internal Defects of Continuous Casting Billet and Control
任务 6.3 连铸坯内部缺陷及控制

Mission objectives:

任务目标：

(1) Be able to identify various defects inside the continuous casting billet

(1) 能够识别连铸坯内部各种缺陷。

(2) Master the causes of internal defects of billet.

(2) 掌握铸坯内部缺陷产生的原因。

(3) Master the control means to reduce the internal defects of the billet.

(3) 掌握减少铸坯内部缺陷的控制手段。

Task Preparation 任务准备

6.3.1 Cognition of Internal Defects of Billet or Slab

The internal quality of continuous casting billet refers to whether the billet has correct solidification structure, segregation degree, internal crack, inclusion content and distribution. During the operation of high-temperature slab with liquid core, various forces act on the shell of high-temperature slab to produce deformation, exceeding the allowable strength and strain of steel is the external cause of cracks, the internal cause of cracks is the sensitivity of steel to cracks, and the conditions of cracks are the equipment and process of caster. The internal defects of the slab or billet are shown in Figure 6-3.

6.3.1 铸坯内部缺陷认知

连铸坯的内部质量是指铸坯是否具有正确的凝固结构、偏析程度、内部裂纹、夹杂物含量及分布状况等。带液芯的高温铸坯在连铸机运行过程中，各种力作用于高温坯壳上产生变形，超过了钢的允许强度和应变是产生裂纹的外因，钢对裂纹敏感性是产生裂纹的内因，而连铸机设备和工艺因素是产生裂纹的条件。铸坯的内部缺陷如图 6-3 所示。

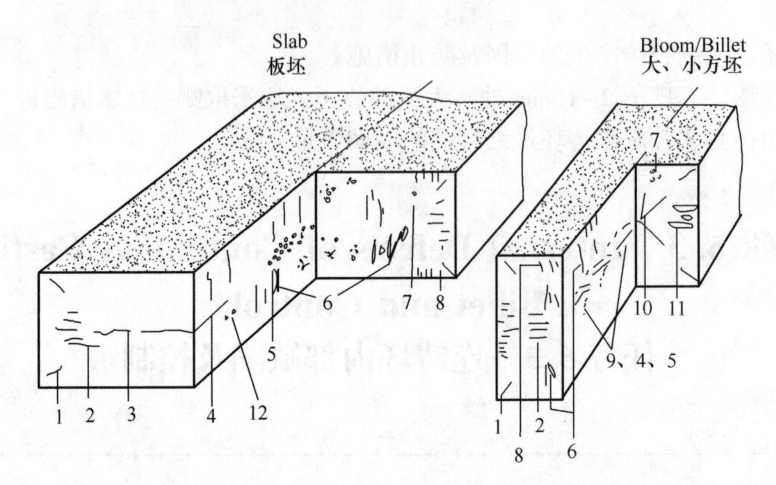

Figure 6-3 Internal defects of continuous casting slab or billet

1—Internal crack; 2—Side intermediate crack; 3—Centerline crack; 4—Centerline segregation; 5—Loose; 6—Intermediate crack; 7—Inclusion; 8—Subcutaneous crack; 9—Shrinkage cavity; 10—Star crack, diagonal crack; 11—Pinhole; 12—Semi macro segregation

图 6-3 连铸坯内部缺陷

1—内部裂纹；2—侧面中间裂纹；3—中心线裂纹；4—中心线偏析；5—疏松；6—中间裂纹；7—非金属夹杂；8—皮下裂纹；9—缩孔；10—中心星状裂纹、对角线裂纹；11—针孔；12—宏观偏析

6.3.2 Internal Cracks

The cracks from the subsurface to the center of the slab or billet are all internal cracks, which are also called solidification cracks because they are generated in the solidification process. Internal cracks are mainly caused by various stresses (including thermal stress, mechanical stress, etc.) acting on the fragile solidification interface. The internal crack of slab originates from solid-liquid interface and accompanied with segregation line. Even if it can be welded during rolling, it will also affect the mechanical properties and service properties of steel.

(1) Intermediate crack. In a certain position between the outside and the center of the slab, the crack produced between the columnar crystals is generally in the middle, so it is called the middle crack or intermediate crack.

(2) Corner crack. The corner crack is produced within 250mm below the meniscus of the mold, and the crack forms at the solid-liquid interface and then propagates.

(3) Extrusion (straightening) crack. When the liquid core of continuous casting slab is bent and straightened, the pressure on it exceeds the limit of slab itself, which leads to the formation of cracks. The cracks are concentrated in the columnar crystal area on the inner arc side, and the cracks are filled with residual mother liquor. In recent years, new technologies such as multi-point straightening, multi-point bending, compression casting, continuous straightening and air-water atomization have been developed to reduce the occurrence of this crack.

(4) Center crack. The cracks in the center of slab cross section are called center cracks, accompanied by positive segregation of S, P and C. It is produced by the bridging of columnar crystals or the bulging of billets at the end of solidification.

The center crack of slab can't be welded in rolling, and there will be serious delamination defects on the section of steel plate. The surface of steel coil or sheet will be wavy defect in the middle, and there will be belt breaking accidents in rolling, which will affect the rolling and use of finished products.

(5) Triangular area crack. In the low power observation state, the triangular columnar crystal area formed from the slab narrow surface production is called triangular area, and the cracks in the area are generally called triangular area cracks.

(6) Subsurface crack. The small cracks (3~10mm) from the surface of the slab are mainly caused by the multiple phase transformation of the slab surface temperature rising repeatedly, and the cracks propagate along the interface between the two structures.

(7) Diagonal crack. It often occurs at the interface of solidification structure of two different cooling surfaces. The rhombic deformation of billet, uneven cooling of mold and asymmetric cooling of secondary cooling all lead to such cracks.

6.3.2 内部裂纹

铸坯从皮下到中心出现的裂纹都是内部裂纹，由于是在凝固过程中产生的裂纹，也称为凝固裂纹。内部裂纹主要是由于各种应力（包括热应力、机械应力等）作用在脆弱的凝固界面上产生的。铸坯内部裂纹起源于固液界面并伴随有偏析线，即使轧制时能焊合，也会影响钢的力学性能和使用性能。

(1) 中间裂纹。在铸坯外侧和中心之间的某一位置，在柱状晶间产生的裂纹，其位置一般在中间，所以称为中间裂纹。

(2) 角部裂纹。角部裂纹是在结晶器弯月面以下 250mm 以内产生的，裂纹在固液交界面形成然后扩展。

(3) 挤压（矫直）裂纹。连铸坯带液芯弯曲和矫直时，所受的压力超过铸坯本身的极限而导致裂纹的产生。裂纹集中在内弧侧柱状晶区，裂纹内充满残余母液。近年来开发了多点矫直、多点弯曲、压缩浇铸、连续矫直及气水雾化新技术可以减少此裂纹的产生。

(4) 中心裂纹。铸坯横断面中心区域可见的缝隙称为中心线裂纹，并伴随有 S、P、C 的正偏析。它是由柱状晶搭桥或凝固末期铸坯鼓肚而产生的。

铸坯中心裂纹在轧制中不能焊合，在钢板的断面上会出现严重的分层缺陷，在钢卷或薄板的表面呈中间波浪形缺陷，在轧制中还会发生断带事故，给成品材的轧制和使用带来影响。

(5) 三角区裂纹。在低倍观察状态下，从板坯窄面生产形成的三角区柱状晶区域称为三角区，在区域内出现的裂纹统称三角区裂纹。

(6) 皮下裂纹。皮下裂纹是与铸坯表面距离不等（3~10mm）的细小裂纹，主要是由于铸坯表层温度反复回升所发生的多次相变，裂纹沿两种组织交界面扩展而形成的。

(7) 对角线裂纹。它常发生在两个不同冷却面凝固组织交界面，小方坯的菱变、结晶器冷却不均匀及二冷不对称冷却都会导致此种裂纹的产生。

6.3.3 Center Segregation

The phenomenon that the content of carbon, phosphorus, sulfur, manganese and other elements in the center of slab is higher than the edge of slab is called center segregation. The reason of central segregation is that the columnar crystal of continuous casting slab is relatively developed. Especially if the columnar crystal is too developed, the center segregation will be more serious. In addition, the mechanical cause (bulging deformation of the shell) will also cause the center segregation of the continuous casting slab.

6.3.3 中心偏析

铸坯中心部位的碳、磷、硫、锰等元素含量高于铸坯边缘的现象称为中心偏析。中心偏析形成的原因是由于连铸坯柱状晶比较发达。特别是柱状晶过分发达，连铸坯会出现“搭桥”现象时，中心偏析更严重。此外机械原因（坯壳的鼓肚变形）也会引起连铸坯的中心偏析。

6.3.4 Center Porosity

In the section of continuous casting slab or billet, there are scattered small voids in different degrees, which are called porosity or loosen. There are three kinds of porosities: general looseness scattered on the whole section, dendrite looseness in the dendrite and central looseness along the slab axis.

The central density of slab determines the degree of center porosity and segregation, while the density mainly depends on the ratio of columnar crystal to equiaxed crystal. It is related to steel grade, cooling system and casting temperature.

6.3.4 中心疏松

在连铸坯剖面上可看到不同程度的分散的小空隙，称为疏松。疏松有 3 种情况，即分散在整个断面上的一般疏松，在树枝晶内的枝晶疏松和沿铸坯轴心产生的中心疏松。

铸坯中心致密度决定了中心疏松和偏析程度，而致密度主要取决于柱状晶与等轴晶比例。它与钢种、冷却制度和浇铸温度有关。

Task Implementation 任务实施

6.3.5 Prevention of Internal Cracks

(1) The reasonable secondary cooling system is determined according to the steel grade, continuous casting billet section and casting speed. When casting medium high carbon steel and large section continuous casting billet, the cooling strength should be reduced accordingly.

(2) Reasonable casting temperature and casting speed. Under the same conditions, the high casting temperature is conducive to the development of columnar crystal, and the increase of cast-

ing speed will increase the center porosity and segregation, thus promoting the generation of internal cracks.

(3) Equipment. The guide roller of the secondary cooling section shall be rigid and strong enough to avoid deformation. The bending radius of continuous casting slab is not easy to be too small, and the tension straightening force should not be too large.

6.3.5 内部裂纹的预防

(1) 根据钢种、连铸坯断面和拉速确定合理的二次冷却制度。对于浇铸中高碳钢、大断面连铸坯时，冷却强度要相应降低。

(2) 合理的铸温和拉速。在其他条件相同的情况下，铸温高有利于柱状晶的发展，增大拉速会增加中心疏松和偏析，从而促进内部裂纹的产生。

(3) 设备方面。二冷段的导辊要有足够的刚性和强度，以免产生变形。连铸坯的弯曲半径不宜过小，拉矫力不要过大，可采用压缩浇铸、多点矫直方法。

6.3.6 Prevention of Center Segregation

(1) Adopt low temperature casting and low casting speed.

(2) The equiaxed crystal is increased by ultrasonic oscillation and electromagnetic stirring.

(3) Reduce the content of sulfur, phosphorus and other impurities in steel, and reduce the segregation degree of sulfide and other easily segregated elements.

(4) In the vicinity of the end of the liquid phase cavity, the distance between rollers should be reduced and the surface of the continuous casting slab should be cooled evenly to improve the shell strength and prevent bulging deformation.

(5) In order to prevent the residual liquid steel from flowing downward due to solidification shrinkage, the so-called ‘clamping’ method, in which the roller spacing is gradually narrowed, is used to expand the proportion of equiaxed crystal.

6.3.6 中心偏析的预防

(1) 采用低温浇铸、低拉速。

(2) 采用超声波振动、电磁搅拌等方法，增加等轴晶。

(3) 减少钢中硫和磷等杂质含量，减少硫化物等易偏析元素的偏析程度。

(4) 在接近液相穴末端位置附近，缩小辊子间距和使连铸坯表面均匀冷却，提高坯壳强度，防止鼓肚变形。

(5) 防止凝固收缩而发生的残留钢液向下方流动，使用辊子间距逐渐变窄的所谓“夹紧”方式，扩大等轴晶比例。

6.3.7 Prevention of Central Porosity

(1) According to the steel grade, liquid steel temperature and secondary cooling strength, determine the appropriate casting speed. For steel grades with a wide solidification temperature range, the casting speed should also be slowed down to reduce the secondary cooling strength.

(2) Select the appropriate casting temperature. The high casting temperature will increase

the center porosity, but the temperature should not be too low. The middle temperature limit should be selected to ensure that the liquid steel has enough fluidity to supplement the volume shrinkage during the solidification of continuous casting slab.

(3) Increasing compression ratio, adding rare earth elements, electromagnetic stirring technology.

6.3.7 中心疏松的预防

(1) 根据钢种、钢液温度和二次冷却强度，确定合适的拉速。对于凝固温度范围宽的钢种，要降低二次冷却强度，拉速也要减慢。

(2) 选择合适的浇铸温度。浇铸温度高将增加中心疏松，但温度不能过低，应选择温度中限，以保证钢液有足够的流动性来补充连铸坯凝固中的体积收缩。

(3) 加大压缩比，添加稀土元素，电磁搅拌技术。

Exercises:

(1) Describe the prevention measures of central porosity and segregation in the billet.

(2) Use the simulation software to improve the internal quality of billet and reduce the internal defects through reasonable control.

思考与习题:

(1) 阐述连铸中心疏松和偏析的防止措施。

(2) 利用仿真软件进行操作，通过合理控制提高铸坯内部质量，减少内部缺陷的产生。

Task 6.4 Shape Defects of Continuous Casting Billet and Control

任务 6.4 连铸坯形状缺陷及控制

Mission objectives:

任务目标:

(1) Be able to identify shape defects of continuous casting billet.

(1) 能够识别连铸坯的形状缺陷。

(2) Master the causes of shape defects of billet.

(2) 掌握铸坯形状缺陷产生的原因。

(3) Master the control means to reduce the shape defects of the billet.

(3) 掌握减少铸坯形状缺陷的控制手段。

Task Preparation 任务准备

6.4.1 Cognition of Billet Shape Defects

The main shape defects of continuous casting billet or slab are bulging and out of square (rhomboidity). Slight shape defects of continuous casting billet can be rolled directly as long as the allowable error is not exceeded, which has little impact on product quality, but serious shape defects are often accompanied by other defects, even after treatment, they can't bite in smoothly.

6.4.1 铸坯形状缺陷认知

连铸坯的形状缺陷主要是鼓肚、脱方（菱变）。轻微的连铸坯形状缺陷，只要不超过允许误差都是可以直接轧制的，对产品质量影响不大，但严重的形状缺陷往往伴有其他缺陷，即使经过处理也不能顺利咬入。

6.4.2 Bulging

Bulging defect refers to the phenomenon that the solidified shell on the surface of the billet bulges into convex surface under the action of the static pressure of the molten steel. When the continuous casting billet has bulging defect, the drawing resistance increases. In serious cases, the billet drawing is difficult to carry out, and the production is forced to be interrupted, even the equipment is damaged. Bulging makes the center segregation of slab more serious and forms the center crack, which seriously endangers the product quality. The causes are as follows:

(1) The rigidity of the caster itself is not enough, the distance between the secondary cooling pinch rolls is too large, or the rigidity is not enough, or the adjustment of the roll diameter center is not correct.

(2) When the casting speed is too high, the secondary cooling control is improper, the reverse cone of the mold is too small, or the lower outlet of the mold is worn seriously, and the billet is separated from the mold wall too early.

(3) The fluidity of powder slag is too good, the cooling strength is too low, the shell is thinner and the possibility of bulge is increased.

6.4.2 鼓肚变形

鼓肚缺陷是指铸坯表面凝壳受到钢液静压力的作用而鼓胀成凸面的现象。当连铸坯有鼓肚缺陷后，拉坯阻力增大，严重时拉坯难以进行，生产被迫中断，甚至损坏设备。鼓肚使板坯中心偏析加重，形成中心裂纹，严重危害了产品质量。产生的原因如下：

（1）铸机本身刚度不够，二冷夹辊间距过大，或刚度不够，或辊径中心调整不准；

（2）当拉速过高，二冷控制不当，结晶器倒锥度过小或结晶器下口磨损严重，铸坯过早脱离结晶器壁。

（3）粉渣流动性过好，冷却强度过低，坯壳减薄，产生鼓肚可能性增大。

6.4.3 Out of Square (Rhomboidity)

The length of the two diagonals on the cross section of the billet is not equal, that is, the angle of two diagonals on the section is greater than or less than 90°, is called rhomboidity, also called out of square. The angle between the adjacent sides of the billet is not a right angle. The reason for this is the uneven cooling of the shell in the mold, which causes the uneven shrinkage of the shell in the mold and the secondary cooling zone.

6.4.3 脱方（菱变）

在方坯横断面上两个对角线长度不相等，即断面上两对角线角度大于或小于 90°，称为菱变，俗称为脱方。铸坯邻边之间的交角不为直角。其发生的原因是坯壳在结晶器内冷却不均匀，引起结晶器和二冷区内的坯壳不均匀收缩。

Task Implementation 任务实施

6.4.4 Prevention of Bulging

(1) The superheat of molten steel is appropriate.

(2) Regularly check and maintain the distance between rollers in the secondary cooling zone to avoid the increase of the distance between rollers, and replace the deformed rollers in time.

(3) Ensure the cooling strength and uniformity of the secondary cooling water.

(4) It is suitable to control the pressure of the tension leveller, especially for the leveler with liquid core, the bulging of the straightening area should be avoided.

6.4.4 鼓肚的预防

(1) 钢水过热度适当。

(2) 定期检查、维护二冷区的夹辊辊间距，避免辊间距增大，同时应及时更换掉发生变形的夹辊。

(3) 保证二冷水的冷却强度和均匀性。

(4) 控制拉矫机的压力合适，尤其对于带液芯拉矫时，应避免铸坯产生矫区鼓肚现象。

6.4.5 Prevention of Rhomboidity

(1) Chromium plating mold is used to enhance the wear resistance of the mold and keep its reverse taper as much as possible.

(2) Reasonably arrange the nozzles in the secondary cooling area to ensure the smooth spray. The shape of the billet can be controlled and the shell can be evenly cooled by adding foot roller or cooling grid at the outlet of mold.

(3) Control the superheat of the liquid steel not to exceed the specified requirements, and try to stabilize the casting speed.

(4) The cooling effect of the shell can be alleviated and even by adopting the technology of submerged nozzle slag.

(5) Strengthen the maintenance and management of equipment. Regularly check the size of the inner cavity of the mold, establish a reasonable waste judgment standard; regularly check the alignment between the mold and the secondary cooling section to prevent shaking during mold oscillation.

6.4.5 脱方的预防

(1) 采用镀铬结晶器，增强结晶器的耐磨性尽量保持其倒锥度。

(2) 合理布置二冷区喷嘴，保证喷水畅通，在结晶下口加设足辊或冷却格栅，可以控制铸坯形状，使坯壳均匀冷却。

(3) 控制钢液的过热度不超过规定的要求，尽量稳定拉坯速度。

(4) 采用浸入式水口保护渣工艺，缓和及均匀坯壳的冷却效果。

(5) 加强设备的维护和管理。定期检查结晶器内腔尺寸，制定合理的判废标准；经常检查结晶器与二次冷却段对中程度，防止结晶振动时摇晃。

Exercises:

(1) Describe the prevention measures of bulging in the slab.

(2) Use the simulation software is to operate to ensure that the billet does not have shape defects through reasonable control.

思考与习题：

(1) 阐述连铸坯鼓肚的防止措施。

(2) 利用仿真软件进行操作，通过合理控制保证铸坯不发生形状缺陷。

Task 6.5 Control of Continuous Casting Billet Purity
任务 6.5 连铸坯纯净度控制

Mission objectives:

任务目标：

(1) Master the expression method of purity of continuous casting billet.

(1) 掌握连铸坯纯净度的表示方法。

(2) Master the source and classification of inclusions in continuous casting billet.

(2) 掌握连铸坯中夹杂物的来源和分类。

(3) Control the inclusions in continuous casting production to reduce the influence of inclusions on the quality of continuous casting billet.

(3) 对连铸生产中的夹杂物进行控制，减少夹杂物对连铸坯质量的影响。

Task Preparation 任务准备

6.5.1 Cognition of Billet Purity

Billet purity refers to the quantity, shape and distribution of non-metallic inclusions in steel. Inclusions in continuous casting have a wide range of sources and complex composition. Inclusions destroy the continuity and compactness of steel matrix. Large inclusions larger than 50μm are often accompanied by cracks, resulting in unqualified macro structure of continuous casting billet, delamination of plate, and damage to the surface of cold-rolled steel plate. The influence of the size, shape and distribution of inclusions on the quality of steel is also different. If the inclusions are small, spherical and dispersive, the influence on the quality of steel is smaller than that existing concentrated; if the inclusions are large, they are accidental distribution, and though the quantity is small, the damage to the quality of steel is greater.

6.5.1 铸坯纯净度认知

铸坯纯净度是指钢中非金属夹杂物的数量、形态和分布。连铸夹杂物的来源范围广，组成也较为复杂。夹杂物的存在破坏了钢基体的连续性和致密性。大于50μm的大型夹杂物往往伴有裂纹出现，造成连铸坯低倍结构不合格，板材分层，并损坏冷轧钢板的表面等，对钢危害很大。夹杂物的大小、形态和分布对钢质量的影响也不同，如果夹杂物细小，呈球形，弥散分布，对钢质量的影响比集中存在要小些；当夹杂物大，呈偶然性分布，数量虽少对钢质量的危害也较大。

6.5.2 Type of Inclusions

The type of inclusions in continuous casting billet is determined by the steel grade and deoxidation method. The most common inclusions in continuous casting slab are Al_2O_3, SiO_2 containing silicate with MnO and CaO, and Al_2O_3 containing aluminate with SiO_2, CaO and CaS. In addition, there are sulfides such as MnS, FeS, etc.

Inclusions smaller than 50μm are called micro inclusions, and larger than 50μm are called macro inclusions. The micro inclusions are mostly deoxidized products, while the macro inclusions are mainly formed by the melting loss of refractories and the reoxidation of molten steel.

6.5.2 夹杂物的类型

连铸坯中夹杂物的类型是由所浇铸的钢种和脱氧方法所决定的。在连铸坯中较常见的夹杂物有Al_2O_3和以SiO_2为主并含有MnO和CaO的硅酸盐，以及以Al_2O_3为主并含有SiO_2、CaO和CaS等的铝酸盐。此外还有硫化物如MnS、FeS等。

将尺寸小于50μm的夹杂物称为显微夹杂，尺寸大于50μm的夹杂物称为宏观夹杂。显微夹杂多为脱氧产物，而宏观夹杂除来源于耐火材料熔损外，主要是由钢液的二次氧化形成的。

Task Implementation 任务实施

In order to improve the purity of steel, it is necessary to reduce the pollution to the molten steel and to eliminate the inclusions from the molten steel. The following measures should be taken:

(1) Slag-less steel tapping. Converter should keep slag-stopping tapping and electric furnace should adopt eccentric bottom tapping to prevent slag from entering ladle.

(2) Ladle refining. According to the needs of steel grades, the appropriate refining methods should be selected to purify the molten steel and improve the morphology of inclusions.

(3) Oxidation-free casting technology is adopted. After refining, the oxygen content of the molten steel has been reduced to less than 0.002%. Protective casting (shielded casting) is used in ladle, tundish and mold. The molten steel is isolated from air to avoid reoxidation of the molten steel.

(4) Develop the role of tundish metallurgical purifier. The Ar stirring is used to improve the flow of molten steel, increase the capacity and depth of tundish, promote the floating of inclusions and further purify molten steel.

(5) Choose submerged nozzle and add mold powder. It can promote the floating separation of inclusions, and the slag with good performance can absorb inclusions and purify the molten steel.

(6) Electromagnetic stirring technology is used to control the movement of casting flow. It can promote inclusion floating and improve the purity of molten steel.

提高钢的纯净度就应在钢液进入结晶器之前，从各工序着手尽量减少对钢液的污染，并最大限度促使夹杂物从钢液中排除。为此应采取以下措施：

(1) 无渣出钢。转炉应挡渣出钢，电炉采用偏心炉底出钢，阻止钢渣进入钢包。

(2) 钢包精炼。根据钢种的需要选择合适的精炼处理方式，纯净钢液并改善夹杂物的形态。

(3) 采用无氧化浇铸技术。经过精炼处理后的钢液氧含量已降到0.002%以下；在钢包→中间包→结晶器均采用保护浇铸，钢液与空气隔绝，避免钢液的二次氧化。

(4) 充分发挥中间包冶金净化器的作用。采用吹Ar搅拌，改善钢液流动状况，加大中间包容量和深度，促进夹杂物上浮，进一步净化钢液。

(5) 选用浸入式水口并加保护渣。促进夹杂物的上浮分离，良好性能的保护渣能够吸收夹杂净化钢液。

(6) 采用电磁搅拌技术，控制铸流的运动。促进夹杂物上浮，提高钢液的纯净度。

Exercises:

(1) Analyze the type and influence of inclusions in continuous casting billet.

(2) Analyze the methods to improve the purity of billet.

思考与习题：

(1) 分析连铸坯中的夹杂物的种类和影响。

(2) 分析提高铸坯纯净度的方法。

References
参 考 文 献

[1] 冯捷，牛海云. 连续铸钢操作与控制 [M]. 北京：冶金工业出版社，2011.
[2] 冯捷，贾艳. 连续铸钢实训 [M]. 北京：冶金工业出版社，2004.
[3] 时彦林. 连铸工培训教程 [M]. 北京：冶金工业出版社，2013.
[4] 干勇. 现代连续铸钢实用手册 [M]. 北京：冶金工业出版社，2010.
[5] 张岩，刘建斌. 连续铸钢工学习指导 [M]. 北京：冶金工业出版社，2015.
[6] 贺道中. 连续铸钢 [M]. 北京：冶金工业出版社，2009.
[7] 王维. 连续铸钢500问 [M]. 北京：化学工业出版社，2009.
[8] 蔡开科，程士富. 连续铸钢原理与工艺 [M]. 北京：冶金工业出版社，2007.
[9] 田乃媛. 薄板坯连铸连轧 [M]. 北京：冶金工业出版社，2004.
[10] Alan W. Cramb. *The Making, Shaping and Treating of Steel: Casting volume* [M]. The AISE Steel Foundation, 2003.
[11] W. R. Irving. *Continuous Casting of Steel* [M]. Institute of Materials, 1993.